TEUBNER-TEXTE zur Mathematik   Band 137

V. I. Burenkov

Sobolev Spaces on Domains

TEUBNER-TEXTE zur Mathematik

Herausgegeben von
Prof. Dr. Jochen Brüning, Berlin
Prof. Dr. Herbert Gajewski, Berlin
Prof. Dr. Herbert Kurke, Berlin
Prof. Dr. Hans Triebel, Jena

Die Reihe soll ein Forum für Beiträge zu aktuellen Problemstellungen der Mathematik sein. Besonderes Anliegen ist die Veröffentlichung von Darstellungen unterschiedlicher methodischer Ansätze, die das Wechselspiel zwischen Theorie und Anwendungen sowie zwischen Lehre und Forschung reflektieren. Thematische Schwerpunkte sind Analysis, Geometrie und Algebra.

In den Texten sollen sich sowohl Lebendigkeit und Originalität von Spezialvorlesungen und Seminaren als auch Diskussionsergebnisse aus Arbeitsgruppen widerspiegeln.

TEUBNER-TEXTE erscheinen in deutscher oder englischer Sprache.

Sobolev Spaces
on Domains

By Prof. Dr. Victor I. Burenkov

University of Wales, Cardiff

 Springer Fachmedien Wiesbaden GmbH 1998

Prof. Dr. Victor I. Burenkov

Born in Russia (Yaroslavl region) in 1941. Studied mathematics and physics in Moscow Institute of Physics and Technology, gained Ph. D. in Mathematics in 1967 and D. Sc. in Mathematics in 1982 (both in Steklov Institute of Mathematics, Moscow). From 1984 to 1994 Professor of the Department of Differential Equations and Functional Analysis at the Peoples' Friendship University of Russia (Moscow). Since 1994 Professor of the School of Mathematics at the University of Wales, Cardiff. More than 100 publications. Fields of research: real and functional analysis, partial differential equations, especially the theory of function spaces (Sobolev spaces, spaces with fractional order of smoothness) and applications.

Gedruckt auf chlorfrei gebleichtem Papier.

Die Deutsche Bibliothek – CIP-Einheitsaufnahme

Burenkov, Viktor I.:
Sobolev spaces on domains / Viktor I. Burenkov.

(Teubner-Texte zur Mathematik ; Bd. 137)
ISBN 978-3-8154-2068-3   ISBN 978-3-663-11374-4 (eBook)
DOI 10.1007/978-3-663-11374-4

Umschlaggestaltung: E. Kretschmer, Leipzig

Preface

The book is based on the lecture course "Function spaces", which the author gave for more than 10 years in the People's Friendship University of Russia (Moscow). The idea to write this book was proposed by Professors H. Triebel and H.-J. Schmeißer in May-June 1993, when the author gave a short lecture course for post-graduate students in the Friedrich-Schiller University Jena.

The initial plan to write a short book for post-graduate students was transformed to wider aims after the work on the book had started. Finally, the book is intended both for graduate and post-graduate students and for researchers, who are interested in applying the theory of Sobolev spaces. Moreover, the methods used in the book allow us to include, in a natural way, some recent results, which have been published only in journals.

Nowadays there exist numerous variants and generalizations of Sobolev spaces and it is clear that this variety is inevitable since different problems in real analysis and partial differential equations give rise to different spaces of Sobolev type. However, it is more or less clear that an attempt to develop a theory, which includes all these spaces, would not be effective. On the other hand, the basic ideas of the investigation of such spaces have very much in common.

For all these reasons we restrict ourselves to the study of Sobolev spaces themselves. However, we aim to discuss the main ideas in detail, and in such a way that, we hope, it will be clear how to apply them to other types of Sobolev spaces.

We shall discuss the following main topics: approximation by smooth functions, integral representations, embedding and compactness theorems, the problem of traces and extension theorems. The basic tools of investigation will be mollifiers with a variable step and integral representations.

Mollifiers with variable step are used both for approximation by smooth functions and for extension of functions (from open sets in $\mathbb{R}^n$ in Chapter 6 and from manifolds of lower dimensions in Chapter 5). All approximation and extension operators constructed in these chapters are the best possible in the sense that the derivatives of higher orders of approximating and extending functions have the minimal possible growth on approaching the boundary.

Sobolev's integral representation is discussed in detail in Chaper 3. It is used in the proofs of the embedding theorems (Chapter 4) and some essential estimates in Chapter 6. An alternative proof of the embedding theorems, without application of Sobolev's integral representation, is also given.

The direct trace theorems (Chapter 5) are proved on the basis of some elementary identities for the differences of higher orders and the definition of Nikol'skiĭ-Besov spaces in terms of differences only.

The author pays particular attention to all possible "limiting" cases, including the cases $p = \infty$ in approximation theorems, $p = 1$ in embedding theorems and The $1, \infty$ in extension theorems.

There are no references to the literature in the main text (Chapters $1-6$): all relevant references are to be found in Chapter 7, which consists of brief notes and comments on the results presented in the earlier chapters.

The proofs of all statements in the book consist of two parts: the idea of the proof and the proof itself. In some simple or less important cases the proofs are omitted. On the other hand, the proofs of the main results are given in full detail and sometimes alternative proofs are also given or at least discussed. The one-dimensional case is often discussed separately to provide a better understanding of the origin of multi-dimensional statements. Also sharper results for this case are presented.

It is expected that the reader has a sound basic knowledge of functional analysis, the theory of Lebesgue integration and the main properties of the spaces $L_p(\Omega)$. It is desirable, in particular, that he/she is accustomed to applying Hölder's and Minkowski's inequalities for sums and integrals. The book is otherwise self-contained: all necessary references are given in the text or footnotes. Each chapter has its own numeration of theorems, corollaries, lemmas, etc. If you are reading, say, Chapter 4 and Theorem 2 is mentioned, then Theorem 2 of Chapter 4 is meant. If we refer to a theorem in another chapter, we give the number of that chapter, say, Theorem 2 of Chapter 3.

For more than 30 years the author participated in the famous seminar "The theory of differentiable functions of several variables and applications" in the Steklov Institute of Mathematics (Moscow) headed at different times by Professors S.L. Sobolev, V.I. Kondrashov, S.M. Nikol'skiĭ, L.D. Kudryavtsev and O.V. Besov. He was much influenced by ideas discussed during its work and, in particular, by his personal talks with Professors S.M. Nikol'skiĭ and S.L. Sobolev.

It is a pleasure for the author to express his deepest gratitude to the participants of that seminar, to his friends and co-authors. with whom he discussed the general plan and different parts of the book.

I am grateful to my colleagues in the University of Wales Cardiff: Professor W.D. Evans, with whom I have had many discussions, and Mr. D.J. Harris. who has thoroughly read the manuscript of the book.

I would also like to mention Dr. A.V. Kulakov who has actively helped in typing the book in $\TeX$.

Finally, I express my deepest love, respect and gratitude to my wife Dr. T.V. Tararykova who not only typed in $\TeX$ a considerable part of the book but also encouraged me in all possible ways.

Moscow/Cardiff, November 1997 V.I. Burenkov

Contents

Notation and basic inequalities

We shall use the following standard notation for sets:

$\mathbb{N}$ – the set of all natural numbers,

$\mathbb{N}_0$ – the set of all nonnegative integers,

$\mathbb{Z}$ – the set of all integers,

$\mathbb{R}$ – the set of all real numbers,

$\mathbb{C}$ – the set of all complex numbers,

$\mathbb{N}_0^n = \underbrace{\mathbb{N}_0 \times \cdots \times \mathbb{N}_0}_{n}$ – the set of multi-indices (n is the natural number

which will be used exclusively to denote the dimension),

$\mathbb{R}^n = \underbrace{\mathbb{R} \times \cdots \times \mathbb{R}}_{n}$,

$B(x,r)$ – the open ball of radius $r > 0$ centered at the point $x \in \mathbb{R}^n$,

${}^c\Omega \; (\Omega \subset \mathbb{R}^n)$ – the complement of Ω in $\mathbb{R}^n$,

$\overline{\Omega} \; (\Omega \subset \mathbb{R}^n)$ – the closure of Ω,

$\underline{\Omega} \; (\Omega \subset \mathbb{R}^n)$ – the interior of Ω,

$\Omega^\delta \; (\Omega \subset \mathbb{R}^n, \delta > 0)$ – the δ-neighborhood of Ω $(\Omega^\delta = \bigcup_{x \in \Omega} B(x,\delta))$,

$\Omega_\delta = \{x \in \Omega : \operatorname{dist}(x, \partial\Omega) \geq \delta\} \quad (\Omega \subset \mathbb{R}^n, \delta > 0)$ (for each $\Omega \subset \mathbb{R}^n$

$\qquad \Omega_\delta = \{x \in \Omega : \operatorname{dist}(x, \partial\Omega) > \delta\})$.

For $\alpha \in \mathbb{N}_0^n, \alpha \neq 0$, we shall write:

$D^\alpha f \equiv \dfrac{\partial^{\alpha_1 + \cdots + \alpha_n} f}{\partial x_1^{\alpha_1} \cdots \partial x_n^{\alpha_n}}$ – the (ordinary) derivative of the function f of order α

and

$D_w^\alpha f \equiv \left(\dfrac{\partial^{\alpha_1 + \cdots + \alpha_n} f}{\partial x_1^{\alpha_1} \cdots \partial x_n^{\alpha_n}} \right)_w$ – the weak derivative of the function f of order α

(see section 1.2).

For an arbitrary nonempty set $\Omega \subset \mathbb{R}^n$ we shall denote by:

$C(\Omega)$ – the space of functions continuous on Ω,

$C_b(\Omega)$ – the Banach space of functions f continuous and bounded on Ω
with the norm

$$\|f\|_{C(\Omega)} = \sup_{x \in \Omega} |f(x)|,$$

$\overline{C}(\Omega)$ – the Banach space of functions uniformly continuous and bounded on Ω with the same norm.

For a measurable nonempty set $\Omega \subset \mathbb{R}^n$ we shall denote by:

$L_p(\Omega)$ $(1 \leq p < \infty)$ – the Banach space [1] of functions f measurable [2] on Ω such that the norm

$$\|f\|_{L_p(\Omega)} = \left(\int\limits_{\Omega} |f|^p dx \right)^{\frac{1}{p}} < \infty,$$

$L_\infty(\Omega)$ – the Banach space of functions f measurable on Ω such that the norm

$$\|f\|_{L_\infty(\Omega)} = \operatorname{ess\,sup}_{x \in \Omega} |f(x)| = \inf_{\omega:\, \text{meas}\,\omega = 0} \sup_{x \in \Omega \setminus \omega} |f(x)| < \infty$$

(in the case in which $\text{meas}\,\Omega > 0$ [3] ; if $\text{meas}\,\Omega = 0$, then we set $\|f\|_{L_\infty(\Omega)} = 0$). [4]

For an open nonempty set $\Omega \subset \mathbb{R}^n$ we shall denote by:

$L_p^{loc}(\Omega)$ $(1 \leq p \leq \infty)$ – the set of functions defined on Ω such that for each compact $K \subset \Omega$ $f \in L_p(K)$,[5]

$C^l(\Omega)$ $(l \in \mathbb{N})$ – the space of functions f defined on Ω such that $\forall \alpha \in \mathbb{N}_0^n$ where $|\alpha| = \alpha_1 + \cdots + \alpha_n = l$ and $\forall x \in \Omega$ the derivatives $(D^\alpha f)(x)$ exist and $D^\alpha f \in C(\Omega)$,

$C_b^l(\Omega)$ $(l \in \mathbb{N})$ – the Banach space of functions $f \in C_b(\Omega)$ such that $\forall \alpha \in \mathbb{N}_0^n$ where $|\alpha| = l$ and $\forall x \in \Omega$ the derivatives $(D^\alpha f)(x)$ exist and $D^\alpha f \in C_b(\Omega)$, with the norm

$$\|f\|_{C^l(\Omega)} = \|f\|_{C(\Omega)} + \sum_{|\alpha|=l} \|D^\alpha f\|_{C(\Omega)},$$

---

[1] As usual when saying a "Banach space" we ignore here the fact that the condition $\|f\|_{L_p(\Omega)} = 0$ is equivalent to the condition $f \sim 0$ on Ω (i.e., f is equivalent to 0 on $\Omega \iff$ meas $\{x \in \Omega : f(x) \neq 0\} = 0$) and not to the condition $f = 0$ on Ω. To be strict we ought to call it a "semi-Banach space" (and it will be necessary to keep this fact in mind in Section 4.1) or consider classes of equivalent functions instead of functions. The same applies to the spaces $L_\infty(\Omega)$ and $W_p^l(\Omega)$ below.

[2] "Measurable" means "measurable with respect to Lebesgue measure." All the integrals throughout the book are Lebesgue integrals.

[3] We need to do so because otherwise if meas $\Omega = 0$, then by the convention $\sup \varnothing = -\infty$ we have $\operatorname{ess\,sup}_{x \in \Omega} |f(x)| = -\infty$.

[4] If $\Omega \subset \mathbb{R}^n$ is an open set, then for $f \in C(\Omega)$ $\|f\|_{C(\Omega)} = \|f\|_{L_\infty(\Omega)}$.

[5] $f_k \to f$ in $L_p^{loc}(\Omega)$ as $k \to \infty$ means that for each compact $K \subset \Omega$ $f_k \to f$ in $L_p(K)$.

$\overline{C}^l(\Omega)$ ($l \in \mathbb{N}$) – the Banach space of functions $f \in \overline{C}(\Omega)$ such that $\forall \alpha \in \mathbb{N}_0^n$ where $|\alpha| = l$ and $\forall x \in \Omega$ the derivatives $(D^\alpha f)(x)$ exist and $D^\alpha f \in \bar{C}(\Omega)$, with the same norm,

$C^\infty(\Omega) = \bigcap\limits_{l=0}^\infty C^l(\Omega)$ – the space of infinitely continuously differentiable functions on Ω,

$C_0^\infty(\Omega)$ – the space of functions in $C^\infty(\Omega)$ compactly supported in Ω,

$W_p^l(\Omega)$ ($l \in \mathbb{N}, 1 \leq p \leq \infty$) – Sobolev space, which is the Banach space of functions $f \in L_p(\Omega)$ such that $\forall \alpha \in \mathbb{N}_0^n$ where $|\alpha| = l$ the weak derivatives $D_w^\alpha f$ exist on Ω and $D_w^\alpha f \in L_p(\Omega)$, with the norm

$$\|f\|_{W_p^l(\Omega)} = \|f\|_{L_p(\Omega)} + \sum_{|\alpha|=l} \|D_w^\alpha f\|_{L_p(\Omega)}$$

(see Section 1.3),

$w_p^l(\Omega)$ ($l \in \mathbb{N}, 1 \leq p \leq \infty$) – the semi-normed Sobolev space, which is the semi-Banach space of functions $f \in L_1^{loc}(\Omega)$ such that $\forall \alpha \in \mathbb{N}_0^n$ where $|\alpha| = l$ the weak derivatives $D_w^\alpha f$ exist on Ω and $D_w^\alpha f \in L_p(\Omega)$, with the semi-norm

$$\|f\|_{w_p^l(\Omega)} = \sum_{|\alpha|=l} \|D_w^\alpha f\|_{L_p(\Omega)}$$

(see Section 1.3),

$\widetilde{W}_p^l(\Omega)$ ($l \in \mathbb{N}, 1 \leq p \leq \infty$) – the Banach space of functions $f \in L_p(\Omega)$ such that $\forall \alpha \in \mathbb{N}_0^n$ where $|\alpha| \leq l$ the weak derivatives $D_w^\alpha f$ exist on Ω and $D_w^\alpha f \in L_p(\Omega)$, with the norm

$$\|f\|_{\widetilde{W}_p^l(\Omega)} = \sum_{|\alpha|\leq l} \|D_w^\alpha f\|_{L_p(\Omega)}$$

(see Sections 2.3 and 4.4).

$(W_p^l)_0(\Omega)$ ($l \in \mathbb{N}, 1 \leq p \leq \infty$) – the space of functions in $W_p^l(\Omega)$ compactly supported in Ω

and, finally,

$\overset{\circ}{W}_p^l(\Omega)$ ($l \in \mathbb{N}, 1 \leq p \leq \infty$) – the closure of $C_0^\infty(\Omega)$ in $W_p^l(\Omega)$. [6]

Further notation will be introduced in the text.

---

[6] In general, if $Z(\Omega)$ is a space of functions defined on an open set $\Omega \subset \mathbb{R}^n$, then $Z_0(\Omega)$ will denote the space of all functions in $Z(\Omega)$ compactly supported in Ω and $\overset{\circ}{Z}(\Omega)$ – the closure of $C_0^\infty(\Omega)$ in the topology of $Z(\Omega)$ (if $C_0^\infty(\Omega) \subset Z(\Omega)$).

Let $\Omega \subset \mathbb{R}^n$ be a measurable set and $1 \leq p \leq \infty$.

Hölder's inequality. Suppose that $\frac{1}{p'} + \frac{1}{p} = 1$, i.e., $p' = \frac{p}{p-1}$ for $1 < p < \infty$. $p' = \infty$ for $p = 1$ and $p' = 1$ for $p = \infty$. If $f \in L_p(\Omega)$ and $g \in L_{p'}(\Omega)$, then $fg \in L_1(\Omega)$ and

$$\|fg\|_{L_1(\Omega)} \leq \|f\|_{L_p(\Omega)} \, \|g\|_{L_{p'}(\Omega)}.$$

Minkowski's inequality. If $f, g \in L_p(\Omega)$, then $f + g \in L_p(\Omega)$ and

$$\|f + g\|_{L_p(\Omega)} \leq \|f\|_{L_p(\Omega)} + \|g\|_{L_p(\Omega)}.$$

Minkowski's inequality for integrals. In addition, let $A \subset \mathbb{R}^m$ be a measurable set. Suppose that f is measurable on $A \times \Omega$ and $f(\cdot, y) \in L_p(\Omega)$ for almost all $y \in A$. Then

$$\left\| \int_A f(\cdot, y) \, dy \right\|_{L_p(\Omega)} \leq \int_A \|f(\cdot, y) \, dy\|_{L_p(\Omega)} \, dy$$

if the right-hand side is finite.

Similar inequalities hold for finite and infinite sums. Let $a_k, b_k \in \mathbb{C}$. Then

$$\sum_{k=1}^{s} |a_k b_k| \leq \left(\sum_{k=1}^{s} |a_k|^p \right)^{\frac{1}{p}} \left(\sum_{k=1}^{s} |b_k|^{p'} \right)^{\frac{1}{p'}}$$

and

$$\left(\sum_{k=1}^{s} |a_k + b_k|^p \right)^{\frac{1}{p}} \leq \left(\sum_{k=1}^{s} |a_k|^p \right)^{\frac{1}{p}} + \left(\sum_{k=1}^{s} |b_k|^p \right)^{\frac{1}{p}}.$$

Here $s \in \mathbb{N}$ or $s = \infty$. (If $p = \infty$, one should replace $(\sum_k |a_k|^p)^{\frac{1}{p}}$ by $\sup_k |a_k|$.)

Throughout the book we shall often use these basic inequalities (without additional comments).

Chapter 1

Preliminaries

1.1 Mollifiers

Let ω be a *kernel of mollification*, i.e.,

$$\omega \in C_0^\infty(\mathbb{R}^n), \qquad \text{supp } \omega \subset \overline{B(0,1)}, \qquad \int_{\mathbb{R}^n} \omega \, dx = 1. \tag{1.1}$$

For $\delta > 0$ and $\forall x \in \mathbb{R}^n$ we set $\omega_\delta(x) = \frac{1}{\delta^n}\omega(\frac{x}{\delta})$.

Definition 1 *Let $\Omega \subset \mathbb{R}^n$ be a measurable set and $\delta > 0$. For a function f defined on Ω and such that $f \in L_1(\Omega \cap B)$ for each ball B, the operator $A_\delta \equiv A_{\delta,\Omega}$ (a mollifier with step (or radius) δ) is defined by the equality* [1]: $\forall x \in \mathbb{R}^n$

$$(A_\delta f)(x) = (\omega_\delta * f_0)(x) = \frac{1}{\delta^n}\int_\Omega \omega\left(\frac{x-y}{\delta}\right)f(y)dy = \int_{B(0,1)} f_0(x - \delta z)\,\omega(z)dz.$$

$$\tag{1.2}$$

We recall that for each function f under consideration $A_\delta f \in C^\infty(\mathbb{R}^n)$, $\forall \alpha \in \mathbb{N}_0^n$

$$D^\alpha A_\delta f = \delta^{-|\alpha|}(D^\alpha \omega)_\delta * f_0 \tag{1.3}$$

on $\mathbb{R}^n$ and

$$\text{supp } A_\delta f \subset \overline{(\text{supp } f)^\delta}. \tag{1.4}$$

---

[1] Here and in the sequel f_0 denotes the extension of f by zero outside Ω: $f_0(x) = f(x)$ for $x \in \Omega$ and $f_0(x) = 0$ for $x \in {}^c\Omega$.

We note also that on Ω_δ

$$(A_\delta f)(x) = (\omega_\delta * f)(x) = \int\limits_{B(0,1)} f(x - \delta z)\,\omega(z)\,dz,$$

and $\forall \alpha \in \mathbb{N}_0^n$

$$D^\alpha A_\delta f = \delta^{-|\alpha|}(D^\alpha\omega)_\delta * f.$$

If $\Omega \subset \mathbb{R}^n$ is an open set and $f \in L_1^{loc}(\Omega)$, then $A_\delta f \in C^\infty(\Omega_\delta)$ and

$$A_\delta f \to f \quad \text{a.e.}^2 \ \text{on} \ \Omega \tag{1.5}$$

as $\delta \to 0+$ (if $f \in C(\Omega)$, then the convergence holds everywhere on Ω). For $1 \le p \le \infty$ and $\forall f \in L_p(\Omega)$

$$\|A_\delta f\|_{L_p(\mathbb{R}^n)} \le c\,\|f\|_{L_p(\Omega)}. \tag{1.6}$$

Moreover, for each measurable set $G \subset \mathbb{R}^n$

$$\|A_\delta f\|_{L_p(G)} \le c\,\|f\|_{L_p(\Omega \cap G^\delta)}. \tag{1.7}$$

Here $c = \|\omega\|_{L_1(\mathbb{R}^n)}$ ($c = 1$ for a nonnegative kernel ω; if, in addition, the function f is nonnegative, then $\|A_\delta f\|_{L_1(\mathbb{R}^n)} = \|f\|_{L_1(\Omega)}$).

Furthermore, [3]

$$\|A_\delta f - f\|_{L_p(\Omega)} \le c\,\omega(\delta, f)_{L_p(\Omega)}, \tag{1.8}$$

where

$$\omega(\delta, f)_{L_p(\Omega)} = \sup_{|h| \le \delta} \|f_0(x + h) - f(x)\|_{L_p(\Omega)}$$

is the *modulus of continuity* of the function f in $L_p(\Omega)$.

From (1.8) it follows that for $1 \le p < \infty$ and $\forall f \in L_p(\Omega)$

$$A_\delta f \to f \quad \text{in} \ L_p(\Omega) \tag{1.9}$$

as $\delta \to 0+$. For $p = \infty$ for any kernel of mollification this relation in general does not hold.

From (1.9) it follows that for $1 \le p < \infty$

$$\|A_\delta f\|_{L_p(\Omega)} \to \|f\|_{L_p(\Omega)} \tag{1.10}$$

---

[2] a.e.$\equiv$ almost everywhere.
[3] See also Lemma 12 of Chapter 5.

as $\delta \to 0+$. We note that for nonnegative kernels ω this relation holds also for $p = \infty$. (If the kernel ω changes its sign, this relation in general does not hold. For example, if $n = 1, \omega(x) < 0$ on $(-1,0)$ and $\omega(x) > 0$ on (0.1), then $\forall \delta > 0$ we have $\|A_\delta(\operatorname{sgn} x)\|_{L_\infty(\mathbb{R})} = \|\omega\|_{L_1(\mathbb{R})} > 1$.)

If $\Omega \subset \mathbb{R}^n$, then the function [4] $\eta = A_{\frac{\delta}{4}} \chi_{\Omega^{\frac{\delta}{2}}}$ constructed with the help of a nonnegative kernel is a *function of "cap-shaped" type*, i.e.,

$$\eta \in C^\infty(\mathbb{R}^n), \quad 0 \leq \eta \leq 1, \quad \eta = 1 \text{ on } \Omega, \quad \operatorname{supp} \eta \subset \Omega^\delta, \tag{1.11}$$

and

$$|(D^\alpha \eta)(x)| \leq c_\alpha \delta^{-|\alpha|},$$

where c_α depends only on n and α.

If the function f satisfies the *Lipschitz condition* on $\mathbb{R}^n$, i.e., if for some $M \geq 0$ and $\forall x, y \in \mathbb{R}^n$

$$|f(x) - f(y)| \leq M|x - y|, \tag{1.12}$$

then $\forall \delta > 0$ and $\forall x, y \in \mathbb{R}^n$

$$|(A_\delta f)(x) - (A_\delta f)(y)| \leq c\, M|x - y|. \tag{1.13}$$

Thus for nonnegative kernels, in which case $c = 1$, the mollifier A_δ completely preserves the Lipschitz condition. If (1.12) holds for all $x, y \in \Omega$, where $\Omega \subset \mathbb{R}^n$ is an open set, then (1.13) holds on Ω_δ.

The mollifier A_δ^* defined by (1.2) with the kernel of mollification $\omega^*(x) = \omega(-x)$ replacing $\omega(x)$ is the conjugate of the mollifier A_δ in $L_2(\Omega)$. In particular, if the kernel ω is real-valued and even, then the mollifier A_δ is a self-adjoint operator on $L_2(\Omega)$.

Finally, we note that for a measurable set $\Omega \subset \mathbb{R}^n$ and for any function f such that $f \in L_1(\Omega \cap B)$ for each ball B

$$A_\delta A_\gamma f = A_\gamma A_\delta f \quad \text{on} \quad \Omega_{\delta+\gamma}.$$

In particular,

$$A_\delta A_\gamma = A_\gamma A_\delta \quad \text{on} \quad L_1^{loc}(\mathbb{R}^n).$$

---

[4] Here and in the sequel χ_G denotes the characteristic function of a set G.

1.2   Weak derivatives

We shall start with the following observation for the one-dimensional case and for an open interval (a, b), $-\infty \le a < b \le +\infty$. According to well-known theorems in analysis, the differentiation operator

$$\frac{d}{dx} : C^1(a, b) \subset C(a, b) \to C(a, b) \ {}^5$$

is a closed operator in $C(a, b)$, i.e., if $f_k \in C^1(a, b), k \in \mathbb{N}, \ f, g \in C(a, b)$ and

$$f_k \to f, \quad \frac{df_k}{dx} \to g \quad \text{in} \quad C(a, b)$$

as $k \to \infty$, ^{6} then $f \in C^1(a, b)$ and $\frac{df}{dx} = g$ on (a, b).

Suppose now $1 \le p < \infty$. The following simple example shows that the differentiation operator

$$\frac{d}{dx} : C^1(a, b) \subset L_p^{loc}(a, b) \to L_p^{loc}(a, b) \tag{1.14}$$

is not closed in $L_p^{loc}(a, b)$.

Example 1 Let $(a, b) = (-1, 1)$ and $\forall x \in (-1, 1)$ set $f(x) = |x|$, $f_k(x) = (x^2 + \frac{1}{k})^{1/2}$, $k \in \mathbb{N}$. Then $f_k \to |x|$, $f_k' \to \operatorname{sgn} x$ even in $L_p(-1, 1)$, but $|x| \notin C^1(-1, 1)$ (and $|x|'$ does not exist on the whole interval $(-1, 1)$).

Idea of the proof. This follows easily by direct calculation. $\square$

For this reason it is natural to study the closure of the operator (1.14) in $L_p^{loc}(a, b)$. This is one approach leading to a generalization of the notion of differentiation.

On the other hand if $f \in C^1(a, b)$ and $\varphi \in C_0^1(a, b)$, then

$$\int_a^b f\varphi' dx = - \int_a^b f'\varphi dx.$$

This equality can also be naturally used to generalize the notion of differentiation, since for some functions (e.g., $f(x) = |x|$) the ordinary derivative does not

---

^{5} Here and in the sequel we shall write for brevity $C(a, b), \bar{C}(a, b), L_p(a, b), L_p^{loc}(a, b)$ etc instead of $C((a, b)), \bar{C}((a, b)), L_p((a, b)), L_p^{loc}((a, b))$ etc.

^{6} By $f_k \to f$ in $C(a, b)$ we mean that $\|f_k - f\|_{C[\alpha, \beta]} \to 0$ as $k \to \infty$ for each closed interval $[\alpha, \beta] \subset (a, b)$. This definition is similar to that of convergence in $L_p^{loc}(a, b)$ (see footnote 5 on page 12).

exist on (a, b), but a function $g \in L_1^{loc}(a, b)$ exists (in Example 1 $g(x) = \operatorname{sgn} x$) such that $\forall \varphi \in C_0^1(a, b)$

$$\int\limits_a^b f\varphi' dx = -\int\limits_a^b g\varphi dx.$$

These approaches lead to strong, weak respectively, extensions of the differentiation operator.

We give now the corresponding definitions for the multidimensional case and for differentiation of arbitrary order.

Definition 2 *Let $\Omega \subset \mathbb{R}^n$ be an open set, $\alpha \in \mathbb{N}_0^n$, $\alpha \neq 0$ and $f, g \in L_1^{loc}(\Omega)$. The function g is a weak derivative of the function f of order α on Ω (briefly $g = D_w^\alpha f$) if*

$$\forall \varphi \in C_0^\infty(\Omega) \qquad \int\limits_\Omega f D^\alpha \varphi dx = (-1)^{|\alpha|} \int\limits_\Omega g \varphi dx. \tag{1.15}$$

Lemma 1 *Let $\Omega \subset \mathbb{R}^n$ be an open set, $\alpha \in \mathbb{N}_0^n$, $\alpha \neq 0$. Moreover, let f be a function defined on Ω, which $\forall x \in \Omega$ has an (ordinary) derivative $(D^\alpha f)(x)$ and $D^\alpha f \in C(\Omega)$. Then $D^\alpha f = D_w^\alpha f$.*

Idea of the proof. By integrating by parts α_j times with respect to the variables x_j, $j = 1, ..., n$, show that

$$\forall \varphi \in C_0^\infty(\Omega) \qquad \int\limits_\Omega f D^\alpha \varphi dx = (-1)^{|\alpha|} \int\limits_\Omega D^\alpha f \varphi dx. \tag{1.16}$$

(One may assume without loss of generality that Ω is bounded and consider instead of f the extended function f_0 on a cube $(-a, a)^n \supset \Omega$.) $\square$

Remark 1 The assumption about the continuity of $D^\alpha f$ in Lemma 1 is essential. For example, the ordinary derivative of the function $f(x) = x^2 \sin \frac{1}{x^2}$ $(x \neq 0; f(0) = 0)$, which exists everywhere on $\mathbb{R}$, is not a weak derivative of f on $\mathbb{R}$ because it is not locally integrable on $\mathbb{R}$. (See also Example 4.)

From Definition 2 it follows that if $g = D_w^\alpha f$ and the function h is equivalent to g on Ω, then $h = D_w^\alpha f$ also. Thus the weak derivative is not uniquely defined. The following lemma shows that it is the only way in which uniqueness fails.

Lemma 2 *Let $\Omega \subset \mathbb{R}^n$ be an open set, $\alpha \in \mathbb{N}_0^n, \alpha \neq 0$, $f, g, h \in L_1^{loc}(\Omega)$ and $g = D_w^\alpha f$, $h = D_w^\alpha f$ on Ω. Then $g \sim h$ on Ω.*

Idea of the proof. Use the main lemma of the calculus of variations. $\square$

Remark 2 Because of this nonuniqueness, the notation $g = D_w^\alpha f$ in Definition 2 (which is not to be interpreped as equality of the functions g and $D_w^\alpha f$) needs some explanation. To be strict, the binary relation $= D_w^\alpha$ on L_1^{loc} is introduced: "$g = D_w^\alpha f$" means "g is a weak derivative of the function f of order α on Ω". We also use $D_w^\alpha f$ for any weak derivative of the function f of order α on Ω. Thus, for example, the assertion "the function f has a weak derivative $D_w^\alpha f$" means "the function, denoted by $D_w^\alpha f$, is a weak derivative of the function f of order α on Ω". From this point of view the relation $D_w^\alpha f_1 + D_w^\alpha f_2 = D_w^\alpha (f_1 + f_2)$ means the following: if each of $D_w^\alpha f_k$, $k = 1, 2$, is a weak derivative (i.e., any of the weak derivatives) of the function f_k, then the function $D_w^\alpha f_1 + D_w^\alpha f_2$ is a weak derivative of the function $f_1 + f_2$. Finally, we assume that $D_w^\alpha f = g$ means $g = D_w^\alpha f$. This will allow us to rewrite the above relation in the more usual form $D_w^\alpha (f_1 + f_2) = D_w^\alpha f_1 + D_w^\alpha f_2$.

Remark 3 Note that if a function $f \in L_1^{loc}(\Omega)$ has a weak derivative $D_w^\alpha f$ on Ω, then automatically $D_w^\alpha f \in L_1^{loc}(\Omega)$.

Example 2 $(n = 1, \Omega = \mathbb{R})$ $|x|_w' = \mathrm{sgn}\, x$.

Idea of the proof. This was discussed above. $\square$

Example 3 Let $n = 1$ and $f \in L_1^{loc}(\mathbb{R})$, then, as is known from the theory of Lebesgue integral, the function $\int\limits_a^x f(y)dy$ is locally absolutely continuous [7] on $\mathbb{R}$ and $(\int\limits_a^x f(y)dy)' = f(x)$ for almost all $x \in \mathbb{R}$. There can, of course, exist an $x \in \mathbb{R}$, for which either the derivative does not exist or exists but is different from $f(x)$. On the other hand, $\forall f \in L_1^{loc}(\mathbb{R})$ we have $(\int\limits_a^x f(y)dy)_w' = f(x)$ on $\mathbb{R}$.

---

[7] We recall that the function g is absolutely continuous on the closed interval $[\alpha.\beta]$ if $\forall \varepsilon > 0$ there exists $\delta > 0$ such that for each finite collection of disjoint intervals $(\alpha_j, \beta_j) \subset (\alpha, \beta), j = 1, ..., s$, satisfying $\sum\limits_{j=1}^s (\beta_j - \alpha_j) < \delta$ one has $\sum\limits_{j=1}^s |f(\beta_j) - f(\alpha_j)| < \varepsilon$. The function g is locally absolutely continuous on the open set $\Omega \subset \mathbb{R}$ if it is absolutely continuous on each closed interval $[\alpha, \beta] \subset \Omega$.

Idea of the proof. Integrate by parts. This is possible since $\int\limits_a^x f(y)dy$ is locally absolutely continuous on $\mathbb{R}$. $\square$

Example 4 Suppose that $n \geq 2$, $l \in \mathbb{N}$, the function $f \in C^l(\mathbb{R}^n \setminus \{0\})$, $\alpha \in \mathbb{N}_0^n$ and $|\alpha| = l$. Then the weak derivative $D_w^\alpha f$ exists on $\mathbb{R}^n$ if, and only if, the (ordinary) derivative $D^\alpha f$ lies in $L_1(B(0,1)\setminus\{0\})$. If $n = 1$, then this statement holds for $f \in C^l(\mathbb{R}^n \setminus \{0\}) \cap C^{l-1}(\mathbb{R})$.

In particular, for $n \geq 1$, $\mu \in \mathbb{R}$ and $\forall \alpha \in \mathbb{N}_0^n, \alpha \neq 0$, the weak derivative $D_w^\alpha(|x|^\mu)$ exists on $\mathbb{R}^n$ if, and only if, either $\mu > l - n$, or μ is a nonegative even integer $\leq l - n$.

Idea of the proof. For $n \geq 2$ integrate by parts, excluding the origin. For $n = 1$ use Definition 4 below and the properties of absolutely continuous functions. $\square$

Example 5 $(n = 1, \Omega = \mathbb{R})$ The weak derivative $(\mathrm{sgn}\,x)'_w$ does not exist on $\mathbb{R}$.

Idea of the proof. Suppose that $g \in L_1^{loc}(\mathbb{R})$ is a weak derivative. By integrating by parts show that $\forall \varphi \in C_0^\infty(\mathbb{R})$ $\int\limits_\mathbb{R} g\varphi dx = 2\varphi(0)$. Taking $\varphi(x) = x\psi(x)$ with arbitrary $\psi \in C_0^\infty(\mathbb{R})$, prove that $\int\limits_\mathbb{R} xg(x)\psi(x)dx = 0$. Thus $g \sim 0$, which leads to a contradiction. $\square$

Remark 4 For each $f \in L_1^{loc}(\Omega)$ the derivative $D^\alpha f$ exists in the sense of the theory of distributions, i.e., as a functional in $D'(\Omega)$:

$$\forall \varphi \in C_0^\infty(\Omega) \quad (D^\alpha f, \varphi) = (-1)^{|\alpha|}(f, D^\alpha \varphi) = (-1)^{|\alpha|} \int\limits_\Omega f D^\alpha \varphi dx.$$

In Example 5 $(\mathrm{sgn}\,x)' = 2\delta(x)$, where δ is the Dirac δ-function. From the point of view of the theory of distributions the weak derivative $D_w^\alpha f$ of a function $f \in L_1^{loc}(\Omega)$ exists if, and only if, the distributional derivative $D^\alpha f$ is a regular distribution, i.e., a functional represented by a function $g \in L_1^{loc}(\Omega)$:

$$\forall \varphi \in C_0^\infty(\Omega) \quad (D^\alpha f, \varphi) = \int\limits_\Omega g\varphi dx.$$

This function g (defined up to equivalence on Ω) is a weak derivative of the function f of order α on Ω.

Definition 3 *Let $\Omega \subset \mathbb{R}^n$ be an open set, $\alpha \in \mathbb{N}_0^n, \alpha \neq 0$ and $f, g \in L_1^{loc}(\Omega)$. The function g is a weak derivative of the function f of order α on Ω (briefly $g = D_s^\alpha f$) if there exist $\psi_k \in C^\infty(\Omega)$, $k \in \mathbb{N}$, such that*

$$\psi_k \to f, \quad D^\alpha \psi_k \to g \quad \text{in} \ \ L_1^{loc}(\Omega) \tag{1.17}$$

as $k \to \infty$.

Theorem 1 *Definitions 2 and 3 are equivalent.*

Idea of the proof. $2 \Rightarrow 3$. In (1.15) write ψ_k for f and pass to the limit as $k \to \infty$. $3 \Rightarrow 2$. For $k \in \mathbb{N}$ let χ_k be the characteristic function of the set $\{x \in \Omega : |x| < k, \ \text{dist}\,(x, \partial\Omega) > \frac{2}{k}\}$. Functions $\psi_k \in C^\infty(\Omega)$ (and even $\psi_k \in C_0^\infty(\Omega)$) are constructed in the following way: $\psi_k = A_{\frac{1}{k}}(f\chi_k)$. where A_δ is a mollifier as in Section 1.1. $\square$

Definition 4 *Let $\Omega \subset \mathbb{R}$ be an open set, $l \in \mathbb{N}$ and $f, g \in L_1^{loc}(\Omega)$. The function g is a weak derivative of the function f of order l on Ω (briefly $g = D_w^l f \equiv f_w^{(l)}$) if there is a function h equivalent to f on Ω, which has a locally absolutely continuous $(l-1)$-th ordinary derivative $h^{(l-1)}$ and such that its ordinary derivative $h^{(l)}$ is equivalent to g. (Recall that $h^{(l)}$ exists almost everywhere on Ω.)*

Theorem 2 *In the one-dimensional case Definitions 2, 3 and 4 are equivalent.*

Idea of the proof. It is enough to consider the case in which $\Omega = (a, b)$.

$4 \Rightarrow 2$. Since $h^{(l-1)}$ is locally absolutely continuous on (a, b), it is possible $\forall \varphi \in C_0^\infty(\Omega)$ to integrate by parts l times:

$$\int_a^b f\varphi^{(l)} dx = \int_a^b h\varphi^{(l)} dx = (-1)^l \int_a^b h^{(l)}\varphi dx - (-1)^l \int_a^b g\varphi dx \ .$$

$3 \Rightarrow 4$. Let $l = 1$. Since $\psi_k \to f$ in $L_1^{loc}(a, b)$ as $k \to \infty$ there exists a subsequence k_s and a set $G \subset (a, b)$ such that $\text{meas}\,[(a, b) \setminus G] = 0$ and $\psi_{k_s}(x) \to f(x)$ as $s \to \infty$ for each $x \in G$. Choose $z \in G$ and pass to the limit in the equality $\psi_{k_s}(x) = \psi_{k_s}(z) + \int_z^x \psi_{k_s}'(y) dy$. Then $f(x) = f(z) + \int_z^x g(y) dy \equiv h(x)$ for each $x \in G$. By the properties of absolutely continuous functions the function h (which is defined on (a, b) and equivalent to h) is locally absolutely continuous on (a, b) and $g \sim h'$.

If $l > 1$, then apply the averaged Taylor's formula (3.15) with $a < \alpha < x < \beta < b$ to the functions ψ_{k_s}. Write it in the form

$$\psi_{k_s}(x) = \int_\alpha^\beta p(x,y)\psi_{k_s}(y)\,dy \;+\; \frac{1}{(l-1)!}\int_\alpha^x (x-y)^{l-1}\Big(\int_\alpha^y \omega(u)\,du\Big)\psi_{k_s}^{(l)}(y)\,dy$$

$$-\frac{1}{(l-1)!}\int_x^\beta (x-y)^{l-1}\Big(\int_y^\beta \omega(u)\,du\Big)\psi_{k_s}^{(l)}(y)\,dy$$

and argue as above. (Here $p \in C([a,b] \times [a,b])$, $\forall y \in [a,b]$ $p(\cdot,y)$ is a polynomial of order less than or equal to $l-1$ and $\omega \in C_0^\infty(\alpha,\beta)$.) $\square$

The notion of a weak derivative, as the notion of an ordinary derivative, is a local notion in the following sense. If the function $g \in L_1^{loc}(\Omega)$ is a weak derivative of the function $f \in L_1^{loc}(\Omega)$ of order $\alpha \in \mathbb{N}_0^n, \alpha \neq 0$, on Ω locally, i.e., $\forall x \in \Omega$ there exists a neighbourhood U_x of x such that g is a weak derivative of f of order α on U_x, then [8] g is a weak derivative of f of order α on Ω.

For an open set $\Omega \subset \mathbb{R}^n$ and $\alpha \in \mathbb{N}_0^n, \alpha \neq 0$, let us denote by $G_\alpha(\Omega)$ the domain of the operator D_w^α, i.e., the subset of $L_1^{loc}(\Omega)$ consisiting of all functions $f \in L_1^{loc}(\Omega)$, for which the weak derivatives $D_w^\alpha f$ exist on Ω. We note that the weak differentiation operator

$$D_w^\alpha : G_\alpha(\Omega) \to L_1^{loc}(\Omega)$$

is closed, i.e., if the functions $f_k \in G_\alpha(\Omega)$ and the functions $f, g \in L_1^{loc}(\Omega)$ are such that

$$f_k \to f \ \text{ in } \ L_1^{loc}(\Omega), \quad D_w^\alpha f_k \to g \ \text{ in } \ L_1^{loc}(\Omega),$$

---

[8] Indeed, consider for an arbitrary $\varphi \in C_0^\infty(\Omega)$ a finite open covering $\{U_{x_k}\}_{k=1}^s$ of supp φ and the corresponding partition of unity $\{\psi_k\}_{k=1}^s$, i.e., a family of fuctions $\psi_k \in C_0^\infty(U_{x_k})$, which are such that $\sum_{k=1}^s \psi_k = 1$ on supp φ. (See Lemma 3 of Section 2.2.) Then $\varphi = \sum_{k=1}^s \varphi\psi_k$ on Ω and

$$\int_\Omega f D^\alpha \varphi\, dx = \sum_{k=1}^s \int_{U_{x_k}} f D^\alpha(\varphi\psi_k)\, dx = (-1)^{|\alpha|}\sum_{k=1}^s \int_{U_{x_k}} g\varphi\psi_k\, dx = (-1)^{|\alpha|}\int_\Omega g\varphi\, dx.$$

then $f \in G_\alpha(\Omega)$ and $D_w^\alpha f = g$. The operator D_w^α considered as operator

$$D_w^\alpha : G_\alpha(\Omega) \cap L_p(\Omega) \to L_p(\Omega),$$

where $1 \le p \le \infty$, is also closed. In order to prove these statements it is enough to write f_k for f in (1.15) and let $k \to \infty$.

Lemma 3 (Weak differentiation under the integral sign) *Let $\Omega \subset \mathbb{R}^n$ be an open set, $A \subset \mathbb{R}^m$ a measurable set and let $\alpha \in \mathbb{N}_0^n, \alpha \ne 0$. Suppose that the function f is defined on $\Omega \times A$, for almost every $y \in A$ $f(\cdot, y) \in L_1^{loc}(\Omega)$ and there exists a weak derivative $D_w^\alpha f(\cdot, y)$ on Ω. Moreover, suppose that $f, D_w^\alpha f \in L_1(K \times A)$ for each compact $K \subset \Omega$. Then on Ω*

$$D_w^\alpha \left(\int\limits_A f(x,y) dy \right) = \int\limits_A (D_w^\alpha f)(x,y) dy. \tag{1.18}$$

Remark 5 According to Remark 2 formula (1.18) means the following: if for a function denoted by $D_w^\alpha f$ and defined on $\Omega \times A$ for almost every $y \in A$ the function $(D_w^\alpha f)(\cdot, y)$ is a weak derivative of order α of $f(\cdot, y)$ on Ω, then the function $\int\limits_A (D_w^\alpha f)(\cdot, y) dy$ is a weak derivative of order α of $\int\limits_A f(\cdot, y) dy$ on Ω.

Idea of the proof. Use Definition 2 and Fubini's theorem. $\square$

Proof. For all $\varphi \in C_0^\infty(\Omega)$ the functions $f(x,y)(D^\alpha \varphi)(x)$ and $(D_w^\alpha f)(x,y)\varphi(x)$ belong to $L_1(\Omega \times A)$, because, for example,

$$\int\limits_{\Omega \times A} |f(x,y)(D^\alpha \varphi)(x)| dx dy \le M \int\limits_{\mathrm{supp}\,\varphi \times A} |f| dx dy < \infty,$$

where $M = \max\limits_{x \in \Omega} |(D^\alpha \varphi)(x)|$. Therefore, starting from Definition 2, we can use Fubini's theorem twice to change the order of integration and deduce that $\forall \varphi \in C_0^\infty(\Omega)$

$$\int\limits_\Omega \left(\int\limits_A (D_w^\alpha f)(x,y) dy \right) \varphi(x) dx = \int\limits_A \left(\int\limits_\Omega (D_w^\alpha f)(x,y)\varphi(x) dx \right) dy$$

$$= (-1)^{|\alpha|} \int\limits_A \left(\int\limits_\Omega f(x,y)(D^\alpha \varphi)(x) dx \right) dy = (-1)^{|\alpha|} \int\limits_\Omega \left(\int\limits_A f(x,y) dy \right) (D^\alpha \varphi)(x) dx$$

and (1.18) follows. $\square$

Lemma 4 (Commutativity of weak differentiation and the mollifiers) *Let $\Omega \subset \mathbb{R}^n$ be an open set, $\alpha \in \mathbb{N}_0^n, \alpha \neq 0, f \in L_1^{loc}(\Omega)$ and suppose that there exists a weak derivative $D_w^\alpha f$ on Ω. Then $\forall \delta > 0$*

$$D^\alpha(A_\delta f) = A_\delta(D_w^\alpha f) \quad \text{on} \quad \underline{\Omega_\delta}. \tag{1.19}$$

Idea of the proof. Use Lemma 3. $\square$

Proof. Recall that $A_\delta(D_w^\alpha f) \in C^\infty(\underline{\Omega_\delta})$ (see Section 1.1). Moreover, $\forall x \in \Omega_\delta$

$$(A_\delta f)(x) = \int_{B(0,1)} f(x - \delta z)\omega(z)dz.$$

Furthermore, $D_w^\alpha(f(\cdot - \delta z)) = (D_w^\alpha f)(\cdot - \delta z)$, on Ω_δ, which follows from Definition 2.

For $(x, z) \in \underline{\Omega_\delta} \times B(0,1)$, let $F(x, z) = f(x - \delta z)\omega(z)$ and $G(x, z) = (D_w^\alpha f)(x - \delta z)\omega(z)$. Then for each compact $K \subset \underline{\Omega_\delta}$ the functions F, G belong to $L_1(K \times B(0,1))$, because they are measurable on $\underline{\Omega_\delta} \times B(0,1)$ [9] and, for example,

$$\int_K \left(\int_{B(0,1)} |f(x - \delta z)\omega(z)|dz \right) dx \leq M \int_K \left(\int_{B(0,1)} |f(x - \delta z)|dz \right) dx$$

$$= M \int_K \left(\int_{B(x,\delta)} |f(y)|dy \right) dx \leq M \int_K \left(\int_{K^\delta} |f(y)|dy \right) dx$$

$$= M \operatorname{meas} K \int_{K^\delta} |f(y)|dy < \infty.$$

Here $M = \max\limits_{z \in \mathbb{R}^n} |\omega(z)|$ and $\overline{K^\delta} \subset \Omega$ (because $K \subset \underline{\Omega_\delta}$). Now (1.19) follows from Lemmas 1 and 3: $\forall x \in \Omega$

$$D^\alpha((A_\delta f)(x)) = D_w^\alpha \left(\int_{B(0,1)} f(x - \delta z)\omega(z)dz \right) = \int_{B(0,1)} D_w^\alpha(f(x - \delta z))\omega(z)dz$$

$$= \int_{B(0,1)} (D_w^\alpha f)(x - \delta z)\omega(z)dz = (A_\delta(D_w^\alpha f))(x). \;\square$$

---

[9] We use the following fact from the theory of measurable functions: if a function g is measurable on a measurable set $E \subset \mathbb{R}^n$, then the function G, defined by $G(x, y) = g(x - y)$ is measurable on the measurable set $\{(x, y) \in \mathbb{R}^{2n} : x - y \in E\} \subset \mathbb{R}^{2n}$.

Corollary 1 *For $\Omega = \mathbb{R}^n$*

$$D_w^\alpha A_\delta = A_\delta D_w^\alpha. \tag{1.20}$$

Corollary 2 *If $\gamma \in \mathbb{N}_0^n$ and $\gamma \geq \alpha$,[10] then*

$$D^\gamma(A_\delta f) = \delta^{-|\alpha|}(D^{\gamma-\alpha}\omega)_\delta * D_w^\alpha f \quad \text{on} \quad \underline{\Omega_\delta}. \tag{1.21}$$

Idea of the proof. Use Lemma 4. $\square$

Proof. Using the properties of mollifiers (Section 1.1), we can write

$$D^\gamma(A_\delta f) = D^{\gamma-\alpha}(D^\alpha(A_\delta f)) = D^{\gamma-\alpha}(A_\delta(D_w^\alpha f))$$

$$= \delta^{|\alpha|-|\gamma|}(D^{\gamma-\alpha}\omega)_\delta * D_w^\alpha f$$

on $\underline{\Omega_\delta}$ (we note that $D_w^\alpha f \in L_1^{loc}(\Omega)$). $\square$

Example 6 If $\Omega \subset \mathbb{R}^n$ is an open set, $\Omega \neq \mathbb{R}^n$, then (1.20) does not hold on Ω, because, for $f \equiv 1$ on Ω, $A_\delta(D^\alpha f) \equiv 0$ on Ω and $D_w^\alpha(A_\delta f) \not\equiv 0$ on $\Omega \setminus \Omega_\delta$.

In Definition 2 the weak derivative is defined directly (not by induction as the ordinary derivative). Therefore the question arises as to whether a weak derivative $D_w^\beta f$, where $\beta \leq \alpha, \beta \neq \alpha$, exists, when a weak derivative $D_w^\alpha f$ exists. In general the answer is negative as the following example shows.

Example 7 Set $\forall (x_1, x_2) \in \mathbb{R}^2$ $f(x_1, x_2) = \text{sgn } x_1 + \text{sgn } x_2$. Then derivatives $(\frac{\partial f}{\partial x_1})_w$ and $(\frac{\partial f}{\partial x_2})_w$ do not exist (see Example 2, while $(\frac{\partial^2 f}{\partial x_1 \partial x_2})_w = 0$ on $\mathbb{R}^2$.

Idea of the proof. Direct calculation starting with Definition 2. $\square$

Nevertheless, in some important cases we can infer the existence of derivatives of lower order.

Lemma 5 *Let $\Omega \subset \mathbb{R}^n$ be an open set, $l \in \mathbb{N}$, $l \geq 2$, $f \in L_1^{loc}(\Omega)$ and suppose that for some $j = \overline{1, n}$ a weak derivative $(\frac{\partial^l f}{\partial x_j^l})_w$ exists on Ω. Then $\forall m \in \mathbb{N}$ satisfying $m < l$ a weak derivative $(\frac{\partial^m f}{\partial x_j^m})_w$ also exists on Ω.*

---

[10] Here and in the sequel $\gamma \geq \alpha$ means that $\gamma_j \geq \alpha_j$ for $j = \overline{1, n}$. We note also that $j = \overline{1, n}$ means $j \in \{1, ..., n\}$.

Idea of the proof. Apply the inequality

$$\left\|\frac{\partial^m f}{\partial x_j^m}\right\|_{L_1(Q)} \le c_1 \left(\|f\|_{L_1(Q)} + \left\|\frac{\partial^l f}{\partial x_j{}^l}\right\|_{L_1(Q)} \right),$$

where $f \in C^l(Q)$, Q is any open cube with faces parallel to the coordinate planes, which is such that $\overline{Q} \subset \Omega$ and $c_1 > 0$ is independent of f. (See footnote 3 in Section 3.1.) $\square$

Proof. For sufficiently large $k \in \mathbb{N}$ the functions $f_k = A_{\frac{1}{k}} f \in C^\infty(Q)$. By (1.5) and Lemma 4 $f_k \to f$ in $L_1(Q)$ and $\frac{\partial^l f_k}{\partial x_j^l} = A_{\frac{1}{k}}\left(\frac{\partial^l f}{\partial x_j^l}\right) \to \frac{\partial^l f}{\partial x_j^l}$ in $L_1(Q)$. Moreover,

$$\left\|\frac{\partial^m f_k}{\partial x_j^m} - \frac{\partial^m f_s}{\partial x_j^m}\right\|_{L_1(Q)} \le c_1 \left(\|f_k - f_s\|_{L_1(Q)} + \left\|\frac{\partial^l f_k}{\partial x_j^l} - \frac{\partial^l f_s}{\partial x_j^l}\right\|_{L_1(Q)} \right).$$

Consequently,

$$\lim_{k,s\to\infty} \left\|\frac{\partial^m f_k}{\partial x_j^m} - \frac{\partial^m f_s}{\partial x_j^m}\right\|_{L_1(Q)} = 0.$$

Because of the completeness of $L_1(Q)$ there exists a function $g_Q \in L_1(Q)$ such that $\frac{\partial^m f_k}{\partial x_j^m} \to g_Q$ in $L_1(Q)$ as $k \to \infty$. Since $f_k \to f$ in $L_1(Q)$ as well, by Definition 3 it follows that g_Q is a weak derivative of order l with respect to x_j on Q.

We note that if Q_1 and Q_2 are any intersecting admissible cubes then $g_{Q_1} = g_{Q_2}$ almost everywhere on $Q_1 \cap Q_2$, since both g_{Q_1} and g_{Q_2} are weak derivatives of f on $Q_1 \cap Q_2$. Consequently, there exists a function $g \in L_1^{loc}(\Omega)$ such that $g = g_Q$ almost everywhere on each admissible cube Q and g is a weak derivative of f on Q. Hence, by Section 1.2 g is a weak derivative of f of order l with respect to x_j on Ω. $\square$

Lemma 6 *Let $n \ge 2$, $\Omega \subset \mathbb{R}^n$ be an open set, $l \in \mathbb{N}$, $l \ge 2$, $f \in L_1^{loc}(\Omega)$ and suppose that $\forall \alpha \in \mathbb{N}_0^n$ satisfying $|\alpha| = l$ a weak derivative $D_w^\alpha f$ exists on Ω. Then $\forall \beta \in \mathbb{N}_0^n$ satisfying $0 < |\beta| < l$ a weak derivative $D_w^\beta f$ also exists on Ω.*

Idea of the proof. Apply the inequality

$$\|D^\beta f\|_{L_1(Q)} \le c_2 \left(\|f\|_{L_1(Q)} + \sum_{|\alpha|=l} \|D^\alpha f\|_{L_1(Q)} \right),$$

where $f \in C^l(\Omega)$, Q is any cube considered in the case of Lemma 5, $c_2 > 0$ is independent of f, and the proof of Lemma 5. $\square$

Proof. The above inequality, by induction, follows from the inequality considered in the proof of Lemma 5. For, if $Q = (a,b)^n$, then

$$\|D^\beta f\|_{L_1(Q)} = \left\| \cdots \left\| \frac{\partial^{\beta_1}}{\partial x_1^{\beta_1}} \left(\frac{\partial^{\beta_2 + \cdots + \beta_n} f}{\partial x_2^{\beta_2} \cdots \partial x_n^{\beta_n}} \right) \right\|_{L_1(a,b)} \cdots \right\|_{L_1(a,b)}$$

$$\leq c_1 \left(\left\| \frac{\partial^{\beta_2 + \cdots + \beta_n} f}{\partial x_2^{\beta_2} \cdots \partial x_n^{\beta_n}} \right\|_{L_1(Q)} + \left\| \frac{\partial^l f}{\partial x_1^{l - \beta_2 - \cdots - \beta_n} x_2^{\beta_2} \cdots \partial x_n^{\beta_n}} \right\|_{L_1(Q)} \right)$$

$$\leq \cdots \leq c_2 \left(\|f\|_{L_1(Q)} + \sum_{|\alpha|=l} \|D^\alpha f\|_{L_1(Q)} \right).$$

The rest is the same as in the proof of Lemma 5.

By writing f_k for f in this inequality and taking limits we see that it is possible to replace here the ordinary derivatives $D^\beta f, D^\alpha f$ by the weak ones $D_w^\beta f$, $D_w^\alpha f$ respectively. [11] $\square$

Lemma 7 *Let $n \geq 2$, $\Omega \subset \mathbb{R}^n$ be an open set, $l \in \mathbb{N}$, $l \geq 2$, $f \in L_1^{loc}(\Omega)$ and suppose that $\forall j \in \{1, ..., n\}$ a weak derivative $(\frac{\partial^l f}{\partial x_j^l})_w$ exists on Ω. Then $\forall \beta \in \mathbb{N}_0^n$ satisfying $0 < |\beta| < l$ a weak derivative $D_w^\beta f$ also exists on Ω. For $|\beta| = l$ in general a weak derivative $D_w^\beta f$ does not exist, but if, in addition, for some $p > 1$ $\quad (\frac{\partial^l f}{\partial x_j^l})_w \in L_p^{loc}(\Omega)$, then a weak derivative $D_w^\beta f$ does exist for $|\beta| = l$.*

Idea of the proof. This statement is a corollary of Theorem 9 of Chapter 4. $\square$

1.3   Sobolev spaces (basic properties)

Definition 5 *Let $\Omega \subset \mathbb{R}^n$ be an open set, $l \in \mathbb{N}$, $1 \leq p \leq \infty$. The function f belongs to the Sobolev space $W_p^l(\Omega)$ if $f \in L_p(\Omega)$, if it has weak derivatives*

---

[11] Moreover, starting by the appropriate inequality in footnote 3 of Chapter 3, by the same argument it follows that

$$\|D_w^\beta f\|_{L_p(Q)} \leq M \left(\|f\|_{L_p(Q)} + \sum_{|\alpha|=l} \|D_w^\alpha f\|_{L_p(Q)} \right),$$

where $1 \leq p \leq \infty$ and M is independent of f. This inequality holds also for $\Omega = \mathbb{R}^n$. This follows by replacing Q by $Q_0 + k$, where $Q_0 = \{x \in \mathbb{R}^n : 0 < x_j < 1, j = 1, ..., n\}$ and $k \in \mathbb{Z}^n$, raising these inequalities to the power p, applying to the right-hand side Hölder's inequality for sums, adding all of them and raising to the power $\frac{1}{p}$. For more general open sets such inequalities will be proved in Section 4.4.

$D_w^\alpha f$ on Ω for all $\alpha \in \mathbb{N}_0^n$ satisfying $|\alpha| = l$ and

$$\|f\|_{W_p^l(\Omega)} = \|f\|_{L_p(\Omega)} + \sum_{|\alpha|=l} \|D_w^\alpha f\|_{L_p(\Omega)} < \infty. \tag{1.22}$$

Remark 6 In the one-dimensional case this definition is by Definition 4 equivalent to the following. The function f is equivalent to a function h on Ω, for which the (ordinary) derivative $h^{(l-1)}$ is locally absolutely continuous on Ω and

$$\|f\|_{W_p^l(\Omega)} = \|f\|_{L_p(\Omega)} + \|f_w^{(l)}\|_{L_p(\Omega)} = \|h\|_{L_p(\Omega)} + \|h^{(l)}\|_{L_p(\Omega)} < \infty.$$

Moreover, if $\Omega = (a,b)$ is a finite interval, the limits $\lim_{x \to a+} h(x)$ and $\lim_{x \to b-} h(x)$ exist and one may define h on $[a,b]$ by setting $h(a)$ and $h(b)$ to be equal to those limits. Then $h^{(s)}$, $s = 1, ..., l-1$, exist and $h^{(l-1)}$ is absolutely continuous on $[a,b]$. This follows from the Taylor expansion

$$h^{(s)}(x) = \sum_{k=0}^{l-s-1} \frac{h^{(s+k)}(x_0)}{k!}(x - x_0)^k + \frac{1}{(l-s-1)!} \int_{x_0}^{x} (x-u)^{l-s-1} h^{(l)}(u)\, du,$$

where $x, x_0 \in (a,b)$ and $s = 1, ..., l-1$. Since $h^{(l)} \in L_p(a,b)$, hence $h^{(l)} \in L_1(a,b)$, the limits $\lim_{x \to a+} h(x)$ and $\lim_{x \to b-} h(x)$ exist. Consequently, the right derivatives $h^{(s)}(a)$ and the left derivatives $h^{(s)}(b)$ exist and $h^{(s)}(a) = \lim_{x \to a+} h(x)$, $h^{(s)}(b) = \lim_{x \to b-} h(x)$. Finally, since $h^{(l-1)}(x) = h^{(l-1)}(x_0) + \int_{x_0}^{x} h^{(l)}(u)\, du$ for all $x, x_0 \in [a,b]$ and $h^{(l)} \in L_1(a,b)$, it follows that $h^{(l-1)}$ is absolutely continuous on $[a,b]$.

Remark 7 By Lemma 6 $D_w^\alpha f$ exists also for $|\alpha| < l$. Moreover, $D_w^\alpha f \in L_p^{loc}(\Omega)$, but in general $D_w^\alpha f \notin L_p(\Omega)$ (see Section 4.4).

Theorem 3 *Let* $\Omega \subset \mathbb{R}^n$ *be an open set,* $l \in \mathbb{N}$, $1 \le p \le \infty$. *Then* $W_p^l(\Omega)$ *is a Banach space.* [12]

Idea of the proof. Obviously $W_p^l(\Omega)$ is a normed space. To prove completeness, starting with the Cauchy sequence $\{f_k\}_{k \in \mathbb{N}}$ in $W_p^l(\Omega)$, deduce using the completeness of $L_p(\Omega)$ that there exist $f \in L_p(\Omega)$ and $f_\alpha \in L_p(\Omega)$, where $\alpha \in \mathbb{N}_0^n$, $|\alpha| = l$, such that $f_k \to f$ and $D_w^\alpha f_k \to f_\alpha$ in $L_p(\Omega)$. From the closedness of the weak differentiation it follows that $f_\alpha = D_w^\alpha f$. Hence $f_k \to f$ in $W_p^l(\Omega)$. $\square$

---

[12] See footnote 1 on page 12. The same refers to the spaces $L_p^l(\Omega)$ in Remark 9 below.

Remark 8 Norm (1.22) is equivalent to

$$\|f\|_{W_p^l(\Omega)}^{(1)} = \left(\int_\Omega \left(|f|^p + \sum_{|\alpha|=l} |D_w^\alpha f|^p \right) dx \right)^{\frac{1}{p}}$$

for $1 \le p < \infty$ and to

$$\|f\|_{W_\infty^l(\Omega)}^{(1)} = \max\{\|f\|_{L_\infty(\Omega)}, \max_{|\alpha|=l} \|D_w^\alpha f\|_{L_\infty(\Omega)}\}$$

for $p = \infty$, i.e., $\forall f \in W_p^l(\Omega)$

$$c_3 \|f\|_{W_p^l(\Omega)}^{(1)} \le \|f\|_{W_p^l(\Omega)} \le c_4 \|f\|_{W_p^l(\Omega)}^{(1)},$$

where $c_3, c_4 > 0$ are independent of f. This follows, with c_3, c_4 depending only on n, p and l, from Hölder's and Jenssen's inequalities for finite sums. If $p = 2$, then $W_2^l(\Omega)$ is a Hilbert space with the inner product

$$(f, g)_{W_2^l(\Omega)} = \int_\Omega \left(f\bar{g} + \sum_{|\alpha|=l} D_w^\alpha f \overline{D_w^\alpha g} \right) dx$$

and $\|f\|_{W_2^l(\Omega)}^{(1)}$ is a Hilbert norm, i.e., $\|f\|_{W_2^l(\Omega)}^{(1)} = (f, f)_{W_2^l(\Omega)}^{\frac{1}{2}}$.

Let us consider the weak gradient of order l

$$\nabla_w^l f = \left(\left(\frac{\partial^l f}{\partial x_{i_1} \cdots \partial x_{i_l}} \right)_w \right)_{i_1,\ldots,i_l=1}^n.$$

Then

$$|\nabla_w^l f|^2 = \sum_{i_1,\ldots,i_l=1}^n \left| \left(\frac{\partial^l f}{\partial x_{i_1} \cdots \partial x_{i_l}} \right)_w \right|^2 = \sum_{|\alpha|=l} \frac{l!}{\alpha!} |D_w^\alpha f|^2$$

and norm (1.22) is equivalent to

$$\|f\|_{W_p^l(\Omega)}^{(2)} = \left(\int_\Omega \left(|f|^p + |\nabla_w^l f|^p \right) dx \right)^{\frac{1}{p}}.$$

We also note that for even l $\forall f \in C_0^\infty(\Omega)$

$$\int_\Omega |\nabla^l f|^2 \, dx = \int_\Omega |\Delta^{\frac{l}{2}} f|^2 \, dx,$$

where Δ is the Laplacian. Hence, for such f,

$$\|f\|_{W_2^l(\Omega)}^{(2)} = \left(\int_\Omega \left(|f|^2 + |\Delta^{\frac{l}{2}} f|^2 \right) dx \right)^{\frac{1}{2}}.$$

We shall also need the following variant of Sobolev spaces.

Definition 6 *Let $\Omega \subset \mathbb{R}^n$ be an open set, $l \in \mathbb{N}$, $1 \le p \le \infty$. The function f belongs to the* semi-normed Sobolev space $w_p^l(\Omega)$ *if $f \in L_1^{loc}(\Omega)$, if it has weak derivatives $D_w^\alpha f$ on Ω for all $\alpha \in \mathbb{N}_0^n$ satisfying $|\alpha| = l$ and*

$$\|f\|_{w_p^l(\Omega)} = \sum_{|\alpha|=l} \|D_w^\alpha f\|_{L_p(\Omega)} < \infty. \tag{1.23}$$

The space $w_p^l(\Omega)$ is also a complete space (the proof is similar to the proof of Theorem 3). Thus $w_p^l(\Omega)$ is a semi-Banach space, because the condition $\|f\|_{w_p^l(\Omega)} = 0$ is equivalent to the following one: on each connected component of an open set Ω f is equivalent to a polynomial of degree less than or equal to $l - 1$ (in general different polynomials for different components).

Remark 9 Let $\Omega \subset \mathbb{R}^n$ be a bounded domain and B be a ball such that $\overline{B} \subset \Omega$, $l \in \mathbb{N}$, $1 \le p \le \infty$. We denote by $L_p^l(\Omega)$ the Banach space, which is the set $w_p^l(\Omega)$, equipped with the norm

$$\|f\|_{L_p^l(\Omega)} = \|f\|_{L_1(B)} + \|f\|_{w_p^l(\Omega)}.$$

(It is a norm, because if $\|f\|_{L_p^l(\Omega)} = 0$, then from $\|f\|_{w_p^l(\Omega)} = 0$ it follows that f is equivalent to a polynomial of degree less than or equal to $l - 1$, and from $\|f\|_{L_1(B)} = 0$ it follows that $f \sim 0$ on Ω.) For different balls with closure in Ω these norms are equivalent. (This will follow from Section 4.4). One can replace $\|f\|_{L_1(B)}$ by $\|f\|_{L_p(B)}$ and the corresponding norms will again be equivalent. Note that by definition $L_p^l(\Omega) = w_p^l(\Omega)$ [13].

Remark 10 Clearly $W_p^l(\Omega) \subset w_p^l(\Omega)$. In general $W_p^l(\Omega) \ne w_p^l(\Omega)$, but locally they coincide, i.e., for each open set G with compact closure in Ω $W_p^l(\Omega)|_G = w_p^l(\Omega)|_G$. This will follow from the estimates in Section 4.4. In that section the conditions on Ω also will also be given ensuring that $W_p^l(\Omega) = w_p^l(\Omega)$.

Remark 11 The semi-norm $\|\cdot\|_{w_p^l(\mathbb{R}^n)}$ (in contrast to the norm $\|\cdot\|_{W_p^l(\mathbb{R}^n)}$) posesses the following homogenity property: $\forall f \in w_p^l(\mathbb{R}^n)$ and $\forall \varepsilon > 0$

$$\|f(\varepsilon x)\|_{w_p^l(\mathbb{R}^n)} = \varepsilon^{l-n/p} \|f(x)\|_{w_p^l(\mathbb{R}^n)}.$$

---

[13] Here and in the sequel for function spaces $Z_1(\Omega)$, $Z_2(\Omega)$ the notation $Z_1(\Omega) = Z_2(\Omega)$, $Z_1(\Omega) \subset Z_2(\Omega)$ means equality, inclusion respectively, of these spaces considered only as sets of functions (see also Section 4.1).

Moreover, $\forall f \in W_p^l(\mathbb{R}^n)$

$$\|f(\varepsilon x)\|_{W_p^l(\mathbb{R}^n)} \sim \varepsilon^{-n/p}\|f(x)\|_{L_p(\mathbb{R}^n)}$$

as $\varepsilon \to 0+$ and

$$\|f(\varepsilon x)\|_{W_p^l(\mathbb{R}^n)} \sim \varepsilon^{l-n/p}\|f(x)\|_{w_p^l(\mathbb{R}^n)}$$

as $\varepsilon \to +\infty$.

The number $l - n/p$, which is called *the differential dimension* of the spaces $W_p^l(\Omega)$ and $w_p^l(\Omega)$, plays an important role in the formulation of the properties of these spaces (see Chapters 4, 5) [14]. It will also appear in the next statement.

Example 8 Let $n, l \in \mathbb{N}$, $\mu, \nu \in \mathbb{R}$, $1 \le p \le \infty$. Denote by $\mathbb{N}_{0,e}$ the set of all nonnegative even integers. Then $|x|^\mu |\log|x||^\nu \in W_p^l(B(0, 1/2))$ if, and only if, $|x|^\mu |\log|x||^\nu \in w_p^l(B(0, 1/2))$ and if, and only if, the following conditions on the parameters are satisfied. If $1 \le p < \infty$, then in the case $\mu \notin \mathbb{N}_{0,e} : \mu > l - n/p$, $\nu \in \mathbb{R}$ or $\mu = l - n/p$, $\nu < -1/p$ and in the case $\mu \in \mathbb{N}_{0,e} : \nu = 0$ or $\mu > l - n/p$, $\nu \in \mathbb{R}$ or $\mu = l - n/p$, $\nu < 1 - 1/p$. If $p = \infty$, then in the case $\mu \notin \mathbb{N}_{0,e} : \mu > l$, $\nu \in \mathbb{R}$ or $\mu = l$, $\nu \le 0$ and in the case $\mu \in \mathbb{N}_{0,e} : \nu = 0$ or $\mu > l$, $\nu \in \mathbb{R}$ or $\mu = l$, $\nu \le 1$. In particular, for $1 \le p < \infty$

1) $|x|^\mu \in W_p^l(B(0, 1/2))$ if, and only if, either $\mu \notin \mathbb{N}_{0,e}$ and $\mu > l - n/p$, or $\mu \in \mathbb{N}_{0,e}$;

2) $|\log|x||^\nu \in W_p^l(B(0, 1/2))$ where $l = n/p$ if, and only if, $\nu < 1 - 1/p$.

---

[14] Let $Z(\mathbb{R}^n)$ be a semi-normed space of functions defined on $\mathbb{R}^n$. One may define the *differential dimension* of the space $Z(\mathbb{R}^n)$ as a real number μ posessing the following property: $\forall f \in Z_0(\mathbb{R}^n)$ there exist $\varepsilon_0, c_5, c_6 > 0$ such that $\forall \varepsilon \ge \varepsilon_0$

$$c_5 \varepsilon^\mu \|f(x)\|_{Z(\mathbb{R}^n)} \le \|f(\varepsilon x)\|_{Z(\mathbb{R}^n)} \le c_6 \varepsilon^\mu \|f(x)\|_{Z(\mathbb{R}^n)}.$$

If the semi-norm $\|\cdot\|_{Z(\mathbb{R}^n)}$ is homogenuous, i.e., for some $\nu \in \mathbb{R}$ $\forall f \in Z(\mathbb{R}^n)$ and $\forall \varepsilon > 0$ $\|f(\varepsilon x)\|_{Z(\mathbb{R}^n)} = \varepsilon^\nu \|f(x)\|_{Z(\mathbb{R}^n)}$, then the differential dimension of $Z(\mathbb{R}^n)$ is equal to ν. The differential dimension of $L_p(\mathbb{R}^n)$ is equal to $-n/p$, the differential dimensions of both $W_p^l(\mathbb{R}^n)$ and $w_p^l(\mathbb{R}^n)$ are equal to $l - n/p$ (which follows from the above relations).

This notion may be usefull when obtaining the conditions on the parameters necessary for validity of the inequality

$$\|f\|_{Z_1(\mathbb{R}^n)} \le c_7 \|f\|_{Z_2(\mathbb{R}^n)},$$

where $c_7 > 0$ does not depend on f. From this inequality it follows that the differential dimension of $Z_1(\mathbb{R}^n)$ is less than or equal to the differential dimension of $Z_2(\mathbb{R}^n)$. If, in addition, both of the semi-norms $\|\cdot\|_{Z_1(\mathbb{R}^n)}$ and $\|\cdot\|_{Z_2(\mathbb{R}^n)}$ are homogenuous, then their differential dimensions must coincide.

Idea of the proof. Apply Example 4. Let $\mathbb{M} = \mathbb{N}_0$ for $n = 1$ and $\mathbb{M} = \mathbb{N}_{0,e}$ for $n > 1$. Prove by induction that $\forall \alpha \in \mathbb{N}_0^n$, $\alpha \neq 0$, and $\forall x \in \mathbb{R}^n, x \neq 0$,

$$D^\alpha(|x|^\mu |\log|x||^\nu) = |x|^{\mu-|\alpha|} \sum_{k=\sigma}^{|\alpha|} P_{k,\alpha}\left(\frac{x}{|x|}\right) |\log|x||^{\nu-k},$$

where $P_{k,\alpha}$ are polynomials of degree less than or equal to $|\alpha|$, $P_{\sigma,\alpha} \not\equiv 0$ and $\sigma = 0$ for $\mu \notin \mathbb{M}$ or $\mu \in \mathbb{M}$, $|\alpha| \leq \mu$; $\sigma = 1$ for $\mu \in \mathbb{M}$, $|\alpha| > \mu$, $\nu \neq 0$ (the case in which $\mu \in \mathbb{M}$, $|\alpha| > \mu$, $\nu = 0$ is trivial: $D^\alpha(|x|^\mu |\log|x||^\nu) = 0$). Deduce that $\forall x \in \mathbb{R}^n, x \neq 0$,

$$|D^\alpha(|x|^\mu |\log|x||^\nu)| \leq c_8 \, |x|^{\mu-|\alpha|} |\log|x||^{\nu-\sigma},$$

where $c_8 > 0$ does not depend on x. Moreover, if $n \geq 2$, then for some $\xi \in \mathbb{R}^n$, where $|\xi| = 1$, $\varepsilon > 0$ and $\forall x \in K \equiv \{x \in \mathbb{R}^n : x \neq 0, \, |\frac{x}{|x|} - \xi| < \varepsilon\}$

$$|D^\alpha(|x|^\mu |\log|x||^\nu)| \geq c_9 \, |x|^{\mu-|\alpha|} |\log|x||^{\nu-\sigma},$$

where $c_9 > 0$ does not depend on x. Finally, use that for some $c_{10}, c_{11} > 0$

$$\int_{B(0,1/2)} g(|x|)dx = c_{10} \int_0^{1/2} g(\rho)\rho^{n-1}d\rho, \qquad \int_{B(0,1/2)\cap K} g(|x|)dx = c_{11} \int_0^{1/2} g(\rho)\rho^{n-1}d\rho. \quad \square$$

Example 9 Let $1 \leq p < \infty$. Under the suppositions of Example 8 $|x|^\mu(\log|x|)^\nu \in W_p^l({}^cB(0,2))$ if, and only if, $\mu < -n/p$, $\nu \in \mathbb{R}$ or $\mu = -n/p$, $\nu < -1/p$. On the other hand, $|x|^\mu(\log|x|)^\nu \in w_p^l({}^cB(0,2))$ and if, and only if, in the case $\mu \notin \mathbb{N}_{0,e} : \mu < l - n/p$, $\nu \in \mathbb{R}$ or $\mu = l - n/p$, $\nu < -1/p$ and in the case $\mu \in \mathbb{N}_{0,e} : \nu = 0$ or $\mu < l - n/p$, $\nu \in \mathbb{R}$ or $\mu = l - n/p$, $\nu < 1 - 1/p$. For $p = \infty$ the changes are similar to Example 8.

Let Ff denote the *Fourier transform* of the function f : for $f \in L_1(\mathbb{R}^n)$ and $\forall \xi \in \mathbb{R}^n$

$$(Ff)(\xi) = (2\pi)^{-\frac{n}{2}} \int_{\mathbb{R}^n} e^{-ix\cdot\xi} f(x)dx; \tag{1.24}$$

for $f \in L_2(\mathbb{R}^n)$

$$Ff = \lim_{k\to\infty} F(f\chi_k), \tag{1.25}$$

where χ_k is the characteristic function of a ball $B(0,k)$ and the limit is taken in $L_2(\mathbb{R}^n)$. It exists for each $f \in L_2(\mathbb{R}^n)$ and

$$\|Ff\|_{L_2(\mathbb{R}^n)} = \|f\|_{L_2(\mathbb{R}^n)} \tag{1.26}$$

(*Parseval's equality*).

Lemma 8 *For all $l \in \mathbb{N}$ and $f \in W_2^l(\mathbb{R}^n)$*

$$\|\nabla_w^l f\|_{L_2(\mathbb{R}^n)} = \| \,|\xi|^l (Ff)(\xi)\|_{L_2(\mathbb{R}^n)} \tag{1.27}$$

and

$$\|f\|_{W_2^l(\mathbb{R}^n)}^{(2)} = \|(1 + |\xi|^{2l})^{\frac{1}{2}} (Ff)(\xi)\,\|_{L_2(\mathbb{R}^n)}. \tag{1.28}$$

Idea of the proof. For $f \in L_1(\mathbb{R}^n) \cap W_2^l(\mathbb{R}^n)$ starting with Definition 4 prove that $F(D_w^\alpha f)(\xi) = (i\xi)^\alpha (Ff)(\xi)$ on $\mathbb{R}^n$. To obtain (1.27) and (1.28) apply (1.26) and the identity

$$\sum_{|\alpha|=l} \frac{l!}{\alpha!} |\xi^{2\alpha}| = \sum_{|\alpha|=l} \frac{l!}{\alpha_1! \cdots \alpha_n!} (\xi_1^2)^{\alpha_1} \cdots (\xi_n^2)^{\alpha_n} = |\xi|^{2l}. \quad \square$$

Lemma 9 *Let $\Omega \subset \mathbb{R}^n$ be an open set, $M \geq 0$ and suppose that $\forall x, y \in \Omega$*

$$| f(x) - f(y) | \leq M \,| \,x - y \,| \,. \tag{1.29}$$

Then $f \in w_\infty^1(\Omega)$, the gradient $(\nabla f)(x)$ exists for almost every $x \in \Omega$ and

$$| \nabla f(x) | \leq M \quad \text{a.e.} \quad \text{on} \ \ \Omega. \tag{1.30}$$

If, in addition, Ω is a convex set, then the condition (1.29) is equivalent to the following: $f \in C(\Omega) \cap w_\infty^1(\Omega)$ and (1.30) holds.

Idea of the proof. Let $j \in \{1, ..., n\}$, $x = (x^{(j)}, x_j)$, $x^{(j)} = (x_1, ..., x_{j-1}, x_{j+1}, ..., x_n)$, $\Omega^{(j)} = \mathrm{Pr}_{x_j=0}\Omega \subset \mathbb{R}^{n-1}$ and $\forall x^{(j)} \in \Omega^{(j)}$ $\Omega_{(j)}(x^{(j)}) = \mathrm{Pr}_{0x_j}\Omega \cap l_{x^{(j)}} \subset \mathbb{R}$, where $l_{x^{(j)}}$ is a straight line parallel to the axis Ox_j and passing through the point $(x^{(j)}, 0)$. Deduce from (1.29) that for almost every $x_j \in \Omega_{(j)}(x^{(j)})$ there exists $\frac{\partial f}{\partial x_j}(x) = \frac{\partial f}{\partial x_j}(x^{(j)}, x_j)$ and $\left|\frac{\partial f}{\partial x_j}(x)\right| \leq M$. Integrating by parts (which is possible because $\forall x^{(j)} \in \Omega^{(j)}$ the function $f(x^{(j)}, \cdot)$ is locally absolutely continuous on $\Omega_{(j)}(x^{(j)})$) show that the ordinary derivative $\frac{\partial f}{\partial x_j}$ (existing thus almost everywhere on Ω) is a weak derivative $(\frac{\partial f}{\partial x_j})_w$ on Ω.

If Ω is convex, then to obtain the converse result use Lemma 4 and (1.7) to prove that $\forall x, y \in \Omega$ and $0 < \delta < \mathrm{dist}\,([x,y], \partial\Omega)$ the following inequalities for the mollifier A_δ with a nonnegative kernel are satisfied [15]

$$|(A_\delta f)(x) - (A_\delta f)(y)| \leq \| \nabla A_\delta f\|_{C([x,y])}|x - y|$$

---

[15] When writing $\| \nabla g \|_{C(G)}$ we mean that

$$\| \nabla g \|_{C(G)} \equiv \| \,|\nabla g| \,\|_{C(G)} = \left\|\left(\sum_{j=1}^{n} |\frac{\partial g}{\partial x_j}|^2\right)^{1/2}\right\|_{C(G)}$$

$(\| \nabla g \|_{L_\infty(G)}$ is understood in a similar way).

$$= \| A_\delta \bigtriangledown_w f \|_{C([x,y])} |x - y| \leq \| \bigtriangledown_w f \|_{L_\infty([x,y]^\delta)} |x - y|$$

$$\leq \| \bigtriangledown_w f \|_{L_\infty(\Omega)} |x - y| = \| \bigtriangledown f \|_{L_\infty(\Omega)} |x - y| \leq M |x - y|$$

(note also that for $f \in C(\Omega) \cap w^1_\infty(\Omega)$ the gradient $\bigtriangledown f$ exists a.e. on Ω and $\bigtriangledown f = \bigtriangledown_w f$ on Ω). Now it is enough to pass, applying (1.5), to the limit as $\delta \to 0+$. $\square$

Corollary 3 *If $\Omega \subset \mathbb{R}^n$ is a convex open set, then $g \in w^1_\infty(\Omega)$ if, and only if, it is equivalent to a function f satisfying (1.29) with some $M \geq 0$. (Given a function g, the function f is defined uniquely.)*

Moreover, denote by M^* the minimal possible value of M in (1.29). Then $\| \bigtriangledown g \|_{L_\infty(\Omega)} = M^*$ and, hence,

$$M^* \leq \|g\|_{w^1_\infty(\Omega)} \leq n M^*.$$

Idea of the proof. The first statement is just a reformulation of Lemma 9 for the case of convex open sets. The second one follows from the definitions of $\|g\|_{w^1_\infty(\Omega)}$ and $\bigtriangledown_w g$. $\square$

Lemma 10 (Minkowski's inequality for Sobolev spaces) *Let $\Omega \subset \mathbb{R}^n$ be an open set and $A \subset \mathbb{R}^m$ a measurable set, $l \in \mathbb{N}$, $1 \leq p \leq \infty$. Moreover, suppose that f is a function measurable on $\Omega \times A$ and that $f(\cdot, y) \in W^l_p(\Omega)$ for almost every $y \in A$. Then*

$$\left\| \int_A f(x,y) dy \right\|_{W^l_p(\Omega)} \leq \int_A \| f(x,y) \|_{W^l_p(\Omega)} dy \tag{1.31}$$

(the norm $\| f(x,y) \|_{W^l_p(\Omega)}$ is calculated with respect to x).

Idea of the proof. Use Lemma 3 and Minkowski's inequality for $L_p(\Omega)$. $\square$
Proof. Let the right-hand side of (1.31) be finite, then by Hölder's inequality for each compact $K \subset \Omega$

$$\int_A \left(\int_K |f(x,y)| dx \right) dy < \infty \quad \text{and} \quad \int_A \left(\int_K |D^\alpha_w f(x,y)| dx \right) dy < \infty$$

$\forall \alpha \in \mathbb{N}^n_0$ where $|\alpha| = l$. Hence by Fubini's theorem the function f, being measurable on $K \times A$, belongs to $L_1(K \times A)$. Now the inequality (1.31) follows from Lemma 3 and Minkowski's inequality for $L_p(\Omega)$:

$$\left\| \int_A f(x,y) dy \right\|_{W^l_p(\Omega)} = \left\| \int_A f(x,y) dy \right\|_{L_p(\Omega)} + \sum_{|\alpha|=l} \left\| D^\alpha_w \int_A f(x,y) dy \right\|_{L_p(\Omega)}$$

$$\leq \int_A \|f(x,y)\|_{L_p(\Omega)}dy + \sum_{|\alpha|=l} \int_A \|(D_w^\alpha f)(x,y)\|_{L_p(\Omega)}dy = \int_A \|f(x,y)\|_{W_p^l(\Omega)}dy. \quad \square$$

Lemma 11 (Multiplication by C_0^∞-functions) *Let $\Omega \subset \mathbb{R}^n$ be an open set, $l \in \mathbb{N}$, $1 \leq p \leq \infty$. Then $\forall \varphi \in C_0^\infty(\Omega)$ there exists $c_\varphi > 0$ such that $\forall f \in W_p^l(\Omega)$*

$$\|\varphi f\|_{W_p^l(\Omega)} \leq c_\varphi \|f\|_{W_p^l(\Omega)}. \tag{1.32}$$

Idea of the proof. Use Lemma 6, Leibnitz' formula and the L_p-estimates of the derivatives of lower order. $\square$

Proof. Let $\alpha \in \mathbb{N}_0^n$ satisfy $|\alpha| = l$. By Lemma 6 $\forall \beta \in \mathbb{N}_0^n$ where $|\beta| \leq l$ there exist $D_w^\beta f$, therefore on Ω Leibnitz' formula [16] holds:

$$D_w^\alpha(\varphi f) = \sum_{0 \leq \beta \leq \alpha} \binom{\alpha}{\beta} D^{\alpha-\beta}\varphi D_w^\beta f. \tag{1.33}$$

Let $Q_j \subset \Omega. j = 1, ..., s$, be open cubes with faces parallel to the coordinate planes such that supp $\varphi \subset \bigcup_{j=1}^{s} Q_j$. Then, applying twice the inequality in footnote 11, we get

$$\|D_w^\alpha(\varphi f)\|_{L_p(\Omega)} \leq 2^l \max_{|\gamma| \leq l} \|D^\gamma \varphi\|_{C(\text{supp}\,\varphi)} \sum_{|\beta| \leq l} \|D_w^\beta f\|_{L_p(\text{supp}\,\varphi)}$$

$$\leq 2^l \left(\sum_{|\gamma| \leq l} \sum_{j=1}^{s} \|D^\gamma \varphi\|_{C(Q_j)} \right) \left(\sum_{|\beta| \leq l} \sum_{j=1}^{s} \|D_w^\beta \varphi\|_{L_p(Q_j)} \right)$$

$$\leq M \|\varphi\|_{C^l(\Omega)} \|f\|_{W_p^l(\Omega)},$$

where M depends only on l, Ω and supp φ. (See also Lemma 15 of Chapter 4.) $\square$

Lemma 12 *Let $\Omega \subset \mathbb{R}^n$ be an open set, $l \in \mathbb{N}$, $1 \leq p \leq \infty$. Then $\forall \varphi \in C_0^\infty(\Omega)$ and $\forall f \in w_p^l(\Omega)$ $\varphi f \in w_p^l(\Omega)$.*

Idea of the proof. Since locally $w_p^l(\Omega)$ and $W_p^l(\Omega)$ coincide (see Remark 9) and φ is compactly supported in Ω, it is enough to apply Lemma 11. The estimate (1.32) does not hold if $W_p^l(\Omega)$ is replaced by $w_p^l(\Omega)$. (Take any nontrivial polynomial of degree less than or equal to $l - 1$ as f to verify this.) $\square$

---

[16] Here $\binom{\alpha}{\beta} = \frac{\alpha!}{\beta!(\alpha-\beta)!}$, $\alpha! = \alpha_1!...\alpha_n!$; note that $\sum_{0 \leq \beta \leq \alpha} \binom{\alpha}{\beta} = \prod_{j=1}^{n} \sum_{\beta_j=0}^{\alpha_j} \binom{\alpha_j}{\beta_j} = 2^{|\alpha|}$.

Lemma 13 *Let $l \in \mathbb{N}$, $1 \le p < \infty$, $\eta \in C_0^\infty(\mathbb{R}^n)$ be a function of "cap-shaped" type such that $\eta = 1$ on $B(0,1)$ and $\forall s \in \mathbb{N}$, $\forall x \in \mathbb{R}^n$ $\eta_s(x) = \eta\left(\frac{x}{s}\right)$. Then $\forall f \in W_p^l(\mathbb{R}^n)$*

$$\eta_s f \to f \quad \text{in} \quad W_p^l(\mathbb{R}^n) \tag{1.34}$$

as $s \to \infty$.

Idea of the proof. Use the definition of the norm in $W_p^l(\mathbb{R}^n)$ and Leibnitz' formula. $\square$

Proof. First of all $\forall g \in L_p(\mathbb{R}^n)$ where $1 \le p < \infty$

$$\|(\eta_s - 1)g\|_{L_p(\mathbb{R}^n)} \le \|g\|_{L_p({}^c B(0,s))} \to 0$$

as $s \to \infty$. From (1.33) it follows that $\forall \alpha \in \mathbb{N}_0^n$ where $|\alpha| = l$

$$\|D_w^\alpha(\eta_s f - f)\|_{L_p(\mathbb{R}^n)}$$

$$\le \|(\eta_s - 1)D_w^\alpha f\|_{L_p(\mathbb{R}^n)} + \sum_{0 \le \beta \le \alpha, \beta \ne 0} \binom{\alpha}{\beta} \|D^{\alpha-\beta}\eta_s \, D_w^\beta f\|_{L_p(\mathbb{R}^n)}$$

$$\le \|\eta_s D_w^\alpha f - D_w^\alpha f\|_{L_p(\mathbb{R}^n)} + \frac{M}{s} \sum_{0 \le \beta \le \alpha, \beta \ne 0} \|D_w^\beta f\|_{L_p(\mathbb{R}^n)}.$$

where M does not depend on f and s. By footnote 11 $D_w^\beta f \in L_p(\mathbb{R}^n)$, consequently we have (1.34). $\square$

Remark 12 For $p = \infty$ Lemma 13 does not hold, because, for instance, for $f = 1$ on $\mathbb{R}^n$ $\forall s \in \mathbb{N}$ $\|\eta_s f - f\|_{L_\infty(\mathbb{R}^n)} = 1$. However, $\eta_s f \to f$ a.e. in $\mathbb{R}^n$ and $\|\eta_s f\|_{W_\infty^l(\mathbb{R}^n)} \to \|f\|_{W_\infty^l(\mathbb{R}^n)}$ as $s \to \infty$, which sometimes is enough for applications.

It is well-known that if $\Omega \subset \mathbb{R}^n$ is a measurable set and $1 \le p < \infty$, then each function $f \in L_p(\Omega)$ is *continuous with respect to translation* ($\equiv$ *continuous in the mean*), i.e.,

$$\lim_{h \to 0} \| f_0(x + h) - f(x) \|_{L_p(\Omega)} = 0. \tag{1.35}$$

The analogous result is valid for Sobolev spaces. We recall that for an open set $\Omega \subset \mathbb{R}^n$ the space $(W_p^l)_0(\Omega)$ is the set of all functions $f \in W_p^l(\Omega)$ compactly supported in Ω.

Lemma 14 (Continuity with respect to translation for Sobolev spaces) *Let $\Omega \subset \mathbb{R}^n$ be an open set, $l \in \mathbb{N}$, $1 \le p < \infty$. Then $\forall f \in W_p^l(\Omega)$*

$$\lim_{h \to 0} \| f(x+h) - f(x) \|_{W_p^l(\Omega_{\{h\}})} = 0, \tag{1.36}$$

where $h \in \mathbb{R}^n$, $\Omega_{\{h\}} = \{x \in \Omega : x + h \in \Omega\}$, and $\forall f \in (W_p^l)_0(\Omega)$

$$\lim_{h \to 0} \| f_0(x+h) - f(x) \|_{W_p^l(\Omega)} = 0. \tag{1.37}$$

Idea of the proof. Use the definition of the norm in $W_p^l(\mathbb{R}^n)$ and (1.35). $\square$
Proof. (1.36) follows from (1.35) because

$$\| f(x+h) - f(x) \|_{W_p^l(\Omega_{\{h\}})}$$

$$=\| f(x+h) - f(x) \|_{L_p(\Omega_{\{h\}})} + \sum_{|\alpha|=l} \| (D_w^\alpha f)(x+h) - (D_w^\alpha f)(x) \|_{L_p(\Omega_{\{h\}})}$$

$$\le \| f_0(x+h) - f(x) \|_{L_p(\Omega)} + \sum_{|\alpha|=l} \| (D_w^\alpha f)_0(x+h) - (D_w^\alpha f)(x) \|_{L_p(\Omega)} .$$

If $f \in (W_p^l)_0(\Omega)$, then $\forall \alpha \in \mathbb{N}_0^n$ satisfying $|\alpha| = l$ we have $(D_w^\alpha f)_0 = D_w^\alpha(f_0)$ on $\mathbb{R}^n$, which easily follows from Definition 2, and thus $f_0 \in W_p^l(\mathbb{R}^n)$. Therefore

$$\| f_0(x+h) - f(x) \|_{W_p^l(\Omega)} \le \| f_0(x+h) - f_0(x) \|_{W_p^l(\mathbb{R}^n)}$$

and (1.38) follows from (1.37). $\square$

Remark 13 In contrast to the situation in $L_p(\Omega)$-spaces the relation (1.37) is not valid for all functions in $W_p^l(\Omega)$. For example, if $n = 1$, $\Omega = (0,1)$, $f \equiv 1$, then on $(0,1)$ we have $f_0(x+h) - f(x) = -\chi_{(1-h,h)}(x) \notin W_p^l(0,1)$ for every $h \in (0,1)$. Moreover, Lemma 14 does not hold for $p = \infty$. For example, if $n = 1$, $l = 1$, $\Omega = (-1,1)$, $f(x) = |x|$, then

$$\| f(x+h) - f(x) \|_{W_\infty^1(\Omega_{\{h\}})} \ge \| f_w'(x+h) - f_w'(x) \|_{L_\infty(-1,1-h)} = 1$$

for every $h \in (0,1)$.

Chapter 2

Approximation by infinitely differentiable functions

2.1 Approximation by C_0^∞-functions on R^n

Let A_δ be a mollifier with the kernel ω defined in Section 1.1. We start by studying the properties of A_δ in the case of Sobolev spaces.

Lemma 1 *Let $l \in \mathbb{N}$. Then $\forall f \in W_p^l(\mathbb{R}^n)$ for $1 \le p \le \infty$*

$$\| A_\delta f \|_{W_p^l(\mathbb{R}^n)} \le c \| f \|_{W_p^l(\mathbb{R}^n)},$$

where $c = \| \omega \|_{L_1(\mathbb{R}^n)}$.
    Moreover, for $1 \le p < \infty$

$$A_\delta f \ \to \ f \quad \text{in} \quad W_p^l(\mathbb{R}^n) \tag{2.1}$$

as $\delta \to 0+$. For $p = \infty$ (2.1) is valid $\forall f \in \overline{C}^l(\mathbb{R}^n)$. If $f \in W_\infty^l(\mathbb{R}^n)$, then in general $A_\delta f \nrightarrow f$ in $W_\infty^l(\mathbb{R}^n)$, but in the case of nonnegative kernels of mollification

$$A_\delta f \ \to \ f \quad \text{in} \quad W_\infty^{l-1}(\mathbb{R}^n), \quad \|A_\delta f\|_{W_\infty^l(\mathbb{R}^n)} \ \to \ \|f\|_{W_\infty^l(\mathbb{R}^n)} \tag{2.2}$$

as [1] $\delta \to 0+$.

---

[1] By footnote 11 of Chapter 1 it follows that $A_\delta f \to f$ in $W_p^m(\mathbb{R}^n)$, where $m = 0, ..., l$ if $1 \le p < \infty$ and $m = 0, ..., l-1$ if $p = \infty$.

Idea of the proof. Apply (1.6), (1.8), (1.9), (1.10) and (1.20). $\square$

Proof. Using the above properties we find that for $1 \le p \le \infty$

$$\| A_\delta f \|_{W_p^l(\mathbb{R}^n)} = \| A_\delta f \|_{L_p(\mathbb{R}^n)} + \sum_{|\alpha|=l} \| A_\delta D_w^\alpha f \|_{L_p(\mathbb{R}^n)}$$

$$\le c \left(\| f \|_{L_p(\mathbb{R}^n)} + \sum_{|\alpha|=l} \| D_w^\alpha f \|_{L_p(\mathbb{R}^n)} \right) = c \| f \|_{W_p^l(\mathbb{R}^n)} \, .$$

If $1 \le p < \infty$, then

$$\| A_\delta f - f \|_{W_p^l(\mathbb{R}^n)} = \| A_\delta f - f \|_{L_p(\mathbb{R}^n)} + \sum_{|\alpha|=l} \| A_\delta (D_w^\alpha f) - D_w^\alpha f \|_{L_p(\mathbb{R}^n)} \to 0$$

as $\delta \to 0+$.

If $p = \infty$, then the same argument works $\forall f \in \overline{C}^l(\mathbb{R}^n)$. It follows from (1.8), because $\omega(\delta, f)_{L_\infty(\mathbb{R}^n)} \to 0$ as $\delta \to 0+$ for these f. If $f \in W_\infty^l(\mathbb{R}^n)$, then by (1.8)

$$\| A_\delta f - f \|_{W_\infty^{l-1}(\mathbb{R}^n)} = \| A_\delta f - f \|_{L_\infty(\mathbb{R}^n)} + \sum_{|\alpha|=l-1} \| A_\delta D_w^\alpha f - D_w^\alpha f \|_{L_\infty(\mathbb{R}^n)}$$

$$\le c \left(\omega(\delta, f)_{L_\infty(\mathbb{R}^n)} + \sum_{|\alpha|=l-1} \omega(\delta, D_w^\alpha f)_{L_\infty(\mathbb{R}^n)} \right).$$

By Corollary 7 of Section 3.3

$$\omega(\delta, f)_{L_\infty(\mathbb{R}^n)} = \| f(x+h) - f(x) \|_{L_\infty(\mathbb{R}^n)} = \delta \| f \|_{w_\infty^1(\mathbb{R}^n)}.$$

Similarly for $|\alpha| = l - 1$

$$\omega(\delta, D_w^\alpha f)_{L_\infty(\mathbb{R}^n)} \le \delta \| f \|_{w_\infty^l(\mathbb{R}^n)}.$$

Consequently, $\omega(\delta, D_w^\alpha f)_{L_\infty(\mathbb{R}^n)} \to 0$ for $|\alpha| = l - 1$ as $\delta \to 0+$. It also follows that $\omega(\delta, f)_{L_\infty(\mathbb{R}^n)} \to 0$, since by footnote 11 of Chapter 1

$$\| f \|_{w_\infty^1(\mathbb{R}^n)} \le M \| f \|_{W_\infty^l(\mathbb{R}^n)},$$

where M is indepent of f.

The second statement of (2.2) follows from (1.10) with $p = \infty$ and (1.20).

Finally by Remark 2 below it follows that for $f \in W_\infty^l(\mathbb{R}^n)$ in general $A_\delta \not\to f$ in $W_\infty^l(\mathbb{R}^n)$. $\square$

Remark 1 If Ω is a proper open subset of $\mathbb{R}^n$, $1 \le p < \infty$ and $f \in W_p^l(\Omega)$, then we can prove only that $\forall \varepsilon > 0$

$$A_\delta f \to f \quad \text{in} \quad W_p^l(\underline{\Omega_\varepsilon}) \tag{2.3}$$

as $\delta \to 0+$. We next aim to construct more sophisticated mollifiers, which will allow us to prove the analogous assertion for Ω itself.

Lemma 2 *Let $l \in \mathbb{N}$, $1 \le p < \infty$. Then $C_0^\infty(\mathbb{R}^n)$ is dense [2] in $W_p^l(\mathbb{R}^n)$.*

Idea of the proof. Let $f \in W_p^l(\mathbb{R}^n)$ and η_s, $s \in \mathbb{N}$, have the same meaning as in Lemma 9 of Chapter 1. Set $\varphi_s = A_{\frac{1}{s}}(\eta_s f)$. Then $\varphi_s \in C_0^\infty(\mathbb{R}^n)$ and $\varphi_s \to f$ in $W_p^l(\mathbb{R}^n)$ as $s \to \infty$. $\square$

Proof. By (2.1), (2.2) and (1.34)

$$\| A_{\frac{1}{s}}(\eta_s f) - f \|_{W_p^l(\mathbb{R}^n)} \le \| A_{\frac{1}{s}} f - f \|_{W_p^l(\mathbb{R}^n)} + \| A_{\frac{1}{s}}(\eta_s f) - A_{\frac{1}{s}} f \|_{W_p^l(\mathbb{R}^n)}$$

$$\le \| A_{\frac{1}{s}} f - f \|_{W_p^l(\mathbb{R}^n)} + c \| \eta_s f - f \|_{W_p^l(\mathbb{R}^n)} \to 0$$

as $\delta \to 0+$. $\square$

Remark 2 For $p = \infty$ Lemma 2 is not valid. The counter-example is simple: $f = 1$ on $\mathbb{R}^n$. Moreover, $C^\infty(\mathbb{R}^n)$ also is not dense in $W_\infty^l(\mathbb{R}^n)$. In order to prove this fact, for example, for $n = 1$ and $l = 1$, it is enough to consider the function $f(x) = |x|\eta(x)$, where η is the same function as in Lemma 13 of Chapter 1. Then $\forall \varphi \in C^\infty(\mathbb{R})$

$$\| f - \varphi \|_{W_\infty^1(\mathbb{R})} \ge \| f_w' - \varphi' \|_{L_\infty(-1,1)} = \| \operatorname{sgn} x - \varphi' \|_{L_\infty(-1,1)} \ge \tfrac{1}{2}.$$

However, by Lemmas 1–2 it follows that $C_0^\infty(\mathbb{R}^n)$ is dense in $W_\infty^l(\mathbb{R}^n)$ in a weaker sense, namely, $\forall f \in W_\infty^l(\mathbb{R}^n)$ functions $\varphi_s \in C^\infty(\mathbb{R}^n), s \in \mathbb{N}$, exist such that

$$\varphi_s \to f \quad \text{in} \quad W_\infty^{l-1}(\mathbb{R}^n), \quad \|\varphi_s\|_{W_\infty^l(\mathbb{R}^n)} \to \|f\|_{W_\infty^l(\mathbb{R}^n)}$$

as $s \to \infty$.

---

[2] Thus $\mathring{W}_p^l(\mathbb{R}^n) = W_p^l(\mathbb{R}^n)$, where $\mathring{W}_p^l(\mathbb{R}^n)$ is the closure of $C_0^\infty(\mathbb{R}^n)$ in $W_p^l(\mathbb{R}^n)$.

2.2   Nonlinear mollifiers with variable step

We start by presenting four variants of smooth partitions of unity, which will be constructed by mollifying discontinuous ones.

Lemma 3 *Let $K \subset \mathbb{R}^n$ be a compact set, $s \in \mathbb{N}$, $\Omega_k \subset \mathbb{R}^n$, $k = 1, ..., s$, be open sets and*

$$K \subset \bigcup_{k=1}^{s} \Omega_k. \tag{2.4}$$

Then functions $\psi_k \in C_0^\infty(\Omega_k)$, $k = 1, ..., s$, exist such that $0 \le \psi_k \le 1$ and

$$\sum_{k=1}^{s} \psi_k = 1 \quad \text{on} \quad K. \tag{2.5}$$

Idea of the proof. Without loss of generality we may assume that the Ω_k are bounded. There exists $\delta > 0$ such that $K \subset G \equiv \bigcup_{k=1}^{s} (\Omega_k)_\delta$. Set $G_k = (\Omega_k)_\delta \setminus \bigcup_{m=1}^{k-1} (\Omega_m)_\delta$ and consider the discontinuous partition of unity: $\sum_{k=1}^{s} \chi_{G_k} = \chi_G$ on $\mathbb{R}^n$. Mollifying it establishes the equality $\sum_{k=1}^{s} A_{\frac{\delta}{2}} \chi_{G_k} = A_{\frac{\delta}{2}} \chi_G$ on $\mathbb{R}^n$, which implies (2.5), where $\psi_k = A_{\frac{\delta}{2}} \chi_{G_k}$. (Here A_δ is a mollifier with a nonnegative kernel.) $\square$

Lemma 4 *Let $\Omega \subset \mathbb{R}^n$ be an open set and $\Omega_k \subset \mathbb{R}^n, k \in \mathbb{N}$, be bounded open sets such that*

$$\overline{\Omega_k} \subset \Omega_{k+1}, \quad k \in \mathbb{N}, \quad \bigcup_{k=1}^{\infty} \Omega_k = \Omega. \tag{2.6}$$

Then functions $\psi_k \in C_0^\infty(\Omega), k \in \mathbb{N}$, exist such that

$$G_k \subset supp\, \psi_k \subset G_{k-1} \cup G_k \cup G_{k+1}, \tag{2.7}$$

where $G_k = \Omega_k \setminus \Omega_{k-1}$ (for $k = 0$ we set $\Omega_k = \varnothing$), $0 \le \psi_k \le 1$ and

$$\sum_{k=1}^{\infty} \psi_k = 1 \quad \text{on} \quad \Omega. \tag{2.8}$$

Idea of the proof. Starting again with the discontinuous partition of unity $\sum_{k=1}^{\infty} \chi_{G_k} = 1$ on Ω, choose

$$\varrho_k = \tfrac{1}{4} \, \mathrm{dist}\, (G_k, \partial(G_{k-1} \cup G_k \cup G_{k+1}))$$

$$= \tfrac{1}{4} \, \min\{\mathrm{dist}\, (\Omega_{k-1}, \partial\Omega_k), \mathrm{dist}\, (\Omega_k, \partial\Omega_{k+1})\} \tag{2.9}$$

(if $\Omega \neq \mathbb{R}^n$, then $\varrho_k \to 0$ as $k \to \infty$) and set

$$\psi_k = \left\{ \begin{array}{ll} A_{\varrho_{k-1}} \, \chi_{G_k} & \text{on} \quad (\Omega_k)_{\varrho_{k-1}} \, , \\ A_{\varrho_k} \, \chi_{G_k} & \text{on} \quad \Omega \setminus (\Omega_k)^{\varrho_k}, \end{array} \right. \tag{2.10}$$

where A_δ is a mollifier with a nonnegative kernel ω.

So the characteristic function χ_{G_k} is mollified with the step ϱ_{k-1} "in the direction of the set G_{k-1} " and with the step ϱ_k "in the direction of the set G_{k+1}". Let $G_k = G_k' \cup G_k'' \cup G_k'''$, where

$$G_k' = (\Omega_{k-1})^{\varrho_{k-1}} \setminus \Omega_{k-1}, \quad G_k'' = (\Omega_k)_{\varrho_k} \setminus (\Omega_{k-1})^{\varrho_{k-1}}, \quad G_k''' = \Omega_k \setminus (\Omega_k)_{\varrho_k}.$$

Then $\psi_k = 1$ on G_k'', supp $\psi_k \subset G_{k-1}''' \cup G_k \cup G_{k+1}'$, therefore, $\psi_m = 0$ on G_k where $m \neq k-1, k, k+1$. Moreover, on $G_k''' \cup G_{k+1}'$

$$\sum_{m=1}^{\infty} \psi_m = \psi_k + \psi_{k+1} = A_{\varrho_k}(\chi_{G_k} + \chi_{G_{k+1}}) = 1. \quad \square$$

Lemma 5 *Let $\Omega \subset \mathbb{R}^n$ be an open set , $\Omega \neq \mathbb{R}^n$,*

$$G_1 = \left\{x \in \Omega : \mathrm{dist}\, (x, \partial\Omega) > 2^{-2}\right\}$$

and for $k \in \mathbb{N}$, $k > 1$, let

$$G_k = \left\{x \in \Omega : 2^{-k-1} < \mathrm{dist}\, (x, \partial\Omega) \leq 2^{-k}\right\}$$

(for $k \leq 0$ $G_k = \varnothing$) Then functions $\psi_k \in C^\infty(\Omega)$, $k \in \mathbb{Z}$, exist (for $k \leq 0$ we set $\psi_k \equiv 0$) such that $0 \leq \psi_k \leq 1$,

$$G_k \subset supp\,\psi_k \subset \left\{x \in \Omega : \tfrac{7}{8}2^{-k-1} \leq \mathrm{dist}\, (x, \partial\Omega) \leq \tfrac{9}{8}2^{-k}\right\}$$

$$\subset G_{k-1} \cup G_k \cup G_{k+1} \, ,$$

$$\sum_{k=-\infty}^{\infty} \psi_k = \sum_{k=1}^{\infty} \psi_k = 1 \quad \text{on} \quad \Omega \tag{2.11}$$

and $\forall \alpha \in \mathbb{N}_0^n$ there exists $c_\alpha > 0$ such that $\forall x \in \mathbb{R}^n$ and $\forall k \in \mathbb{Z}$

$$|D^\alpha \psi_k(x)| \leq c_\alpha 2^{k|\alpha|}. \tag{2.12}$$

Idea of the proof. The same as in Lemma 4. Now the Ω_k are defined via the G_k: $\Omega_k = \bigcup\limits_{m=-\infty}^{k} G_m$ and $\varrho_k = 2^{-k-3}$. Estimate (2.12) follows from the equality $D^\alpha \psi_k = \varrho_k^{-|\alpha|}(D^\alpha \omega_{\varrho_k}) * \chi_{G_k}$ on $\Omega \setminus (\Omega_{k-1})^{\varrho_k}$ and the analogous equality on $(\Omega_k)_{\varrho_{k-1}}$. $\square$

Remark 3 Sometimes it is more convenient to suppose that the functions ψ_k in Lemmas 4 and 5 are defined on $\mathbb{R}^n$ and supp $\psi_k \subset \Omega$. (We shall use the same notation $\psi_k \in C_0^\infty(\Omega)$ in this case also). Then equality (2.8) can be written in the following form: $\sum\limits_{k=1}^{\infty} \psi_k = \chi_\Omega$ (the same refers to equality (2.11)).

Remark 4 There may exist an integer $k_0 = k_0(\Omega) > 1$ such that $G_k = \varnothing$ for $k < k_0$ (in this case we assume that $\psi_k \equiv 0$) and (2.11) takes the form

$$\sum_{k=-\infty}^{\infty} \psi_k = \sum_{k=k_0}^{\infty} \psi_k = 1 \quad \text{on} \quad \Omega. \tag{2.13}$$

For $\Omega = \mathbb{R}^n$ we shall apply the following analogue of Lemma 5.

Lemma 6 *For nonpositive $k \in \mathbb{Z}$ let*

$$G_0 = \{x \in \mathbb{R}^n : |x| \leq 1\}, \quad G_k = \{x \in \mathbb{R}^n : 2^{-k-1} < |x| \leq 2^{-k}\}, \quad k < 0.$$

Then functions $\psi_k \in C_0^\infty(\mathbb{R}^n)$, $k \in \mathbb{Z}$, exist ($\psi_k \equiv 0$ for $k > 0$) such that the properties (2.7) and $\forall \alpha \in \mathbb{N}_0^n$ (2.12) are satisfied, $0 \leq \psi_k \leq 1$ and

$$\sum_{k=-\infty}^{\infty} \psi_k = \sum_{k=-\infty}^{0} \psi_k = 1 \quad \text{on} \quad \mathbb{R}^n. \tag{2.14}$$

Idea of the proof. The same as in Lemma 5. $\square$

Remark 5 Note that in Lemmas $4 - 6$

$$\overline{(\operatorname{supp} \psi_k)^{\varrho_k}} \subset (G_{k-1} \cup G_k \cup G_{k+1})_{\varrho_k}. \tag{2.15}$$

Moreover, in the case of Lemma 4 for any arbitrarily small $\gamma_k > 0, k \in \mathbb{N}$, one can construct functions $\psi_k, k \in \mathbb{N}$, satisfying the requirements of Lemma 4 such that

$$\operatorname{supp} \psi_k \subset (G_k)^{\gamma_k}, \quad \psi_k = 1 \text{ on } (G_k)_{\gamma_k}. \tag{2.16}$$

To do this it is enough to replace ρ_k defined by (2.9) by

$$\rho_k = \min\{\tfrac{1}{4}\,\mathrm{dist}\,(G_k, \partial(G_{k-1} \cup G_k \cup G_{k+1})), \gamma_k\}.$$

In the case of Lemmas 5 and 6 for any fixed $\gamma > 0$ one can construct functions ψ_k, satisfying the requirements of those lemmas, such that

$$\mathrm{supp}\,\psi_k \subset (G_k)^{\gamma 2^{-k}}, \quad \psi_k = 1 \text{ on } (G_k)_{\gamma 2^{-k}}. \tag{2.17}$$

Remark 6 From (2.15) it follows, in particular, that *the multiplicity of the covering* $\{\mathrm{supp}\,\psi_k\}$ *in Lemmas 4–6 is equal to* 2, i.e., $\forall x \in \Omega$ there are at most 2 sets $\mathrm{supp}\,\psi_k$ containing x and there exists $x \in \Omega$ such that there are exactly 2 sets $\mathrm{supp}\,\psi_k$ containing x. (From (2.7) it follows only that the multiplicity of this covering does not exceed 3.) Of course 2 is the minimal possible value (if $\mathrm{supp}\,\psi_k \supset G_k$ and the multiplicity of covering is equal to 1, then $\psi_k = \chi_{G_k}$). Moreover, from (2.15) it follows that for $\delta \in \left(0, \tfrac{1}{8}\right]$ the multiplicity of the covering $\{(\mathrm{supp}\,\psi_k)^{\delta 2^{-k}}\}$ is also equal to 2.

In Chapter 6 we shall need a variant of Lemma 5 for $\Omega = \{x \in \mathbb{R}^n : x_n > \varphi(x_1, ..., x_{n-1})\}$, where φ is a function of class Lip 1 on $\mathbb{R}^{n-1}$, — that variant will be formulated there.

Let $\Omega \subset \mathbb{R}^n$ be an open set and $\Omega_k \subset \Omega$, $k \in \mathbb{N}$, be bounded open sets, possessing the properties (2.6), $G_k = \Omega_k \setminus \Omega_{k-1}$. Suppose that ϱ_k is defined by (2.9) and $\{\psi_k\}_{k\in\mathbb{N}}$ is the partition of unity in Lemma 4 defined by (2.10).

Definition 1 *Let* $\overline{\delta} = \{\delta_k\}_{k\in\mathbb{N}}$, *where*

$$0 < \delta_k \le \varrho_k \tag{2.18}$$

and $f \in L_1^{loc}(\Omega)$. *Then* $\forall x \in \Omega$

$$(B_{\overline{\delta}}f)(x) = \sum_{k=1}^{\infty}(A_{\delta_k}(\psi_k f))(x) = \sum_{k=1}^{\infty}\int_{B(0,1)} \psi_k(x-\delta_k z)f(x-\delta_k z)\omega(z)\,dz\,, \tag{2.19}$$

where ω *is a kernel of mollification defined by* (1.1).

Remark 7 The functions $\psi_k f \in L_1(\Omega)$, therefore, $A_{\delta_k}(\psi_k f) \in C^\infty(\mathbb{R}^n)$. We note that we assume that $\psi_k(y)f(y) = 0$ for all $y \notin \mathrm{supp}\,\psi_k$ even if $y \notin \Omega$ and $f(y)$ is not defined (for this reason in contrast to (1.2) in (2.19) $\psi_k(x-\delta_k z)f(x-$

$\delta_k z)$ is written instead of $(\psi_k)_0(x - \delta_k z)f_0(x - \delta_k z))$. By (1.4), (2.18) and (2.15) it follows that

$$\text{supp } A_{\delta_k}(\psi_k f) \subset \overline{(\text{supp } \psi_k)^{\delta_k}} \subset (G_{k-1} \cup G_k \cup G_{k+1})_{\delta_k}, \tag{2.20}$$

therefore,

$$A_{\delta_k}(\psi_k f) \in C_0^\infty(\Omega) \tag{2.21}$$

and the sum in (2.19) is finite. For, let $\forall x \in \Omega$ a number $s = s(x)$ be chosen such that $x \in G_s$. Then for $k \neq s-1, s, s+1$ we have $x \,\overline{\in}\, \text{supp } A_{\delta_k}(\psi_k f)$ and [3]

$$(B_{\bar\delta}f)(x) = \sum_{k=s(x)-1}^{s(x)+1} \int_{B(0,1)} \psi_k(x - \delta_k z)f(x - \delta_k z)\omega(z)dz. \tag{2.22}$$

For the same reason $\forall m \in \mathbb{N}$

$$B_{\bar\delta}f = \sum_{k=m-1}^{m+1} A_{\delta_k}(\psi_k f) \quad \text{on} \quad G_m. \tag{2.23}$$

Moreover, by Remark 5, for any given $\gamma_k \in (0, \rho_k], k \in \mathbb{N}$, a partition of unity $\{\psi_k\}_{k \in \mathbb{N}}$ can be chosen in such a way that for all sufficiently small $\delta_k, k \in \mathbb{N}$, we have $\forall f \in L_1^{loc}(\Omega)$ and $\forall m \in \mathbb{N}$

$$B_{\bar\delta}f = A_{\delta_m}f \quad \text{on} \quad (G_m)_{\gamma_m}. \tag{2.24}$$

Lemma 7 *Let $\Omega \subset \mathbb{R}^n$ be an open set and $f \in L_1^{loc}(\Omega)$. Then for each $\bar\delta = \{\delta_k\}_{k \in \mathbb{N}}$ satisfying (2.18) $B_{\bar\delta}f \in C^\infty(\Omega)$ and $\forall \alpha \in \mathbb{N}_0^n$*

$$D^\alpha(B_{\bar\delta}f) = \sum_{k=1}^\infty D^\alpha(A_{\delta_k}(\psi_k f)) \quad \text{on} \quad \Omega. \tag{2.25}$$

Idea of the proof. Apply (2.21) and (2.23).$\square$

Proof. From (2.21) and (2.23) it follows that $\forall m \in \mathbb{N}$ and $\forall \alpha \in \mathbb{N}_0^n$

$$D^\alpha(B_{\bar\delta}f) = \sum_{k=m-1}^{m+1} D^\alpha(A_{\delta_k}(\psi_k f)) = \sum_{k=1}^\infty D^\alpha(A_{\delta_k}(\psi_k f))$$

on G_m. Hence $B_{\bar\delta}f \in C^\infty(\Omega)$ and (2.25) holds on Ω. $\square$

---

[3] Moreover, from (2.20) it follows that $\forall x \in \Omega$ in the sum (2.16) no more than 2 summands are not equal to 0.

When applying the mollifiers $B_{\bar\delta}f$ we shall choose δ_k satisfying (2.18) depending on f. For this reason we call them *nonlinear mollifiers with variable step* (though, of course, $B_{\bar\delta}$ is a linear operator for fixed $\bar\delta$). The variableness of the step follows from (2.19): $\forall x \in \Omega$ the mollification is carried out with the steps δ_{s-1}, δ_s, δ_{s+1} depending on x (and these steps tend to 0 as x approaches the boundary $\partial\Omega$). We can also say that $B_{\bar\delta}$ is, in some sense, a mollifier with piecewise constant step, because by (2.23) the same constant steps δ_{m-1}, δ_m, δ_{m+1} are used for the whole "strip" G_m. Moreover, by (2.24) only one step δ_m is used for the whole "substrip" $(G_m)_{\gamma_m}$.

Remark 8 In a number of cases it is more suitable to apply the mollifiers $C_{\bar\delta}$, which are similar to the mollifiers $B_{\bar\delta}$ and are defined by the equality: $\forall x \in \Omega$

$$(C_{\bar\delta}f)(x) = \sum_{k=1}^\infty \psi_k(x)(A_{\delta_k}f)(x) = \sum_{k=1}^\infty \psi_k(x) \int_{B(0,1)} f(x - \delta_k z)\,\omega(z)\,dz. \quad (2.26)$$

For instance, in contrast to the mollifiers $B_{\bar\delta}$, for $f = 1$ on Ω and arbitrary $\delta_k \in (0, \varrho_k]$ we have $C_{\bar\delta} = 1$ on Ω. On the other hand, for $C_{\bar\delta}f$ the equalities analogous to (2.22), (2.23) and (2.24) are valid and $C_{\bar\delta}f \in C^\infty(\Omega)$.

Furthermore, if the kernel of mollification ω is real-valued and even, then the operator $C_{\bar\delta}$ is the adjoint of $B_{\bar\delta}$ in $L_2(\Omega)$, because $\forall f, g \in L_2(\Omega)$

$$(B_{\bar\delta}f, g) = \sum_{k=1}^\infty (A_{\delta_k}(\psi_k f), g) = \sum_{k=1}^\infty (\psi_k f, A_{\delta_k}g) = \left(f, \sum_{k=1}^\infty \psi_k A_{\delta_k}g\right) = (f, C_{\bar\delta}g)$$

(note that for these kernels ω the operator A_δ is self-adjoint in $L_2(\Omega)$ — see Section 1.2).

2.3 Approximation by C^∞-functions on open sets

In this section our main aim is to prove the following statement.

Theorem 1 *Let $\Omega \subset \mathbb{R}^n$ be an open set, $l \in \mathbb{N}$. Then $C^\infty(\Omega) \cap W_p^l(\Omega)$ is dense in $W_p^l(\Omega)$ where $1 \le p < \infty$ and $C^\infty(\Omega) \cap \overline{C}^l(\Omega)$ is dense in $\overline{C}^l(\Omega)$.*

Moreover, $\forall f \in W_p^l(\Omega)$ if $1 \le p < \infty$ and $\forall f \in \overline{C}^l(\Omega)$ if $p = \infty$ there exist functions $\varphi_k \in C^\infty(\Omega) \cap W_p^l(\Omega)$ such that [4]

$$\varphi_k \to f \quad \text{in} \quad W_p^m(\Omega), \quad m = 0, 1, ..., l.$$

The set $C^\infty(\Omega) \cap W_\infty^l(\Omega)$ is not dense in $W_\infty^l(\Omega)$. However, $\forall f \in W_\infty^l(\Omega)$ there exist functions $\varphi_k \in C^\infty(\Omega) \cap W_\infty^l(\Omega), k \in \mathbb{N}$, such that

$$\varphi_k \to f \quad \text{in} \quad W_\infty^m(\Omega), \quad m = 0, 1, ..., l-1, \quad \|\varphi_k\|_{W_\infty^l(\Omega)} \to \|f\|_{W_\infty^l(\Omega)}$$

as $k \to \infty$.

Later, in Section 2.6, we shall see that $\forall f \in W_p^l(\Omega)$ (or $\overline{C}^l(\Omega)$) functions $\varphi_s \in C^\infty(\Omega) \cap W_p^l(\Omega)$, $\varphi_s \in C^\infty(\Omega) \cap \overline{C}^l(\Omega)$ respectively, exist, which depend linearly on f, do not depend on p and are such that for $1 \le p < \infty$

$$\varphi_s \to f \quad \text{in} \quad W_p^l(\Omega), \tag{2.27}$$

in $\overline{C}^l(\Omega)$ respectively. Moreover, the functions φ_s may possess additional useful properties.

We shall deduce the statement of this theorem, in the case in which $1 \le p < \infty$, from a much more general result, which holds for a wide class of semi-normed linear spaces $Z(\Omega)$ of functions defined on Ω with semi-norms $\|\cdot\|_{Z(\Omega)}$ such that $C_0^\infty(\Omega) \subset Z(\Omega) \subset L_1^{loc}(\Omega)$. Let $Z_0(\Omega)$ denote the subspace of $Z(\Omega)$ that consists of all functions $f \in Z(\Omega)$, which are compactly supported in Ω. Moreover, let $Z^{loc}(\Omega)$ denote the space of all functions $f \in L_1^{loc}(\Omega)$, which are such that $\forall \varphi \in C_0^\infty(\Omega)$ we have $\varphi f \in Z(\Omega)$. From these definitions it follows, in particular, that

$$C_0^\infty(\Omega) \subset Z_0(\Omega) \subset (L_1)_0(\Omega)$$

and

$$C^\infty(\Omega) \subset Z^{loc}(\Omega) \subset L_1^{loc}(\Omega).$$

---

[4] Under additional assumptions on Ω (see Theorem 6 of Chapter 4), $\|f\|_{W_p^m(\Omega)} \le M \|f\|_{W_p^l(\Omega)}$, $m = 1, .., l-1$, where M is independent of f, and this statement follows from the density of $C^\infty(\Omega) \cap W_p^l(\Omega)$ in $W_p^l(\Omega)$. However, for arbitrary opens sets it is not so (see Examples 8–9 of Chapter 4), and this statement needs a separate proof. We also note that it is possible that $f \notin W_p^m(\Omega)$. In that case also $\varphi_k \notin W_p^m(\Omega)$ but $f - \varphi_k \in W_p^m(\Omega)$ and $f - \varphi_k \to 0$ in $W_p^m(\Omega)$ as $k \to \infty$.

Remark 9 For any $l \in \mathbb{N}_0$ we have $(\overline{C}^l)^{loc}(\Omega) = (C_b^l)^{loc}(\Omega) = C^l(\Omega)$. Moreover, for $1 \leq p \leq \infty$ the following equivalent definition of the space $(W_p^l)^{loc}(\Omega)$ can be given: $(W_p^l)^{loc}(\Omega) = \{f \in L_1^{loc}(\Omega)$: for each open set G compactly embedded into Ω $f \in W_p^l(G)\}$.

Theorem 2 *Let $\Omega \subset \mathbb{R}^n$ be an open set and suppose that the semi-Banach space $Z(\Omega)$ satisfies the following conditions:*

1) $C_0^\infty(\Omega) \subset Z(\Omega) \subset L_1^{loc}(\Omega)$,

2) *(Minkowski's inequality) if $A \subset \mathbb{R}^m$ is a measurable set and f is a function measurable on $\Omega \times A$, then*

$$\left\| \int_A f(x,y)dy \right\|_{Z(\Omega)} \leq \int_A \|f(x,y)\|_{Z(\Omega)}dy,$$

3) *if $\varphi \in C_0^\infty(\Omega)$ and $f \in Z(\Omega)$, then $\varphi f \in Z(\Omega)$,*

4) *all functions $f \in Z_0(\Omega)$ are continuous with respect to translation, i.e.,*

$$\lim_{h \to 0} \|f_0(x+h) - f(x)\|_{Z(\Omega)} = 0. \tag{2.28}$$

Then $C^\infty(\Omega)$ is dense in $Z^{loc}(\Omega)$ (and, hence, $C^\infty(\Omega) \cap Z(\Omega)$ is dense in $Z(\Omega)$), i.e., $\forall f \in Z^{loc}(\Omega)$ functions $\varphi_s \in C^\infty(\Omega) \cap Z^{loc}(\Omega)$, $s \in \mathbb{N}$, exist such that

$$\varphi_s \to f \text{ in } Z(\Omega) \tag{2.29}$$

as $s \to \infty$.

Idea of the proof. Apply Minkowski's inequality to the right-hand side of the equality

$$(B_{\bar{\delta}}f)(x) - f(x) = \sum_{k=1}^\infty \int_{B(0,1)} (f_k(x - \delta_k z) - f_k(x))\, \omega(z)\, dz\,, \tag{2.30}$$

where $f_k = \psi_k f$ and the mollifier $B_{\bar{\delta}}$ is constructed with the help of a nonnegative kernel of mollification, and prove the inequality

$$\| B_{\bar{\delta}}f - f \|_{Z(\Omega)} \leq \sum_{k=1}^\infty \omega(\delta_k, f_k)_{Z(\Omega)}. \tag{2.31}$$

Here

$$\omega(\delta, f)_{Z(\Omega)} = \sup_{|h|\leq\delta} \|f_0(x + h) - f(x)\|_{Z(\Omega)}$$

is the *modulus of continuity of the function f in $Z(\Omega)$*. (Compare with (1.8).)
Using condition 4) choose δ_k in such a way that $\omega(\delta_k, f_k) < \varepsilon\, 2^{-k}$. Then
$\|B_{\bar{\delta}}f - f\|_{Z(\Omega)} < \varepsilon$. $\square$

Proof. 1. From 4) it follows that $\forall f \in Z_0(\Omega)$ there exists $\gamma = \gamma(f) > 0$
such that $\forall h \in \mathbb{R}^n$ satisfying $|h| < \gamma$ the function $f_0(\cdot + h) - f(\cdot) \in Z(\Omega)$.
Let us suppose, in addition, that $\gamma < \operatorname{dist}(\operatorname{supp} f, \delta\Omega)$, then $\operatorname{supp} f_0(\cdot + h) \subset$
$(\operatorname{supp} f)^{|h|} \subset \Omega$ and $f_0(\cdot + h) \in Z_0(\Omega)$. For, first of all $f_0(\cdot + h) \in Z^{loc}(\Omega)$.
Consider, furthermore, a function of "cap-shaped" type $\eta \in C_0^\infty(\Omega)$ such that
$\eta = 1$ on $\operatorname{supp} f_0(\cdot + h)$ (see Section 1.1), then by definition of $Z^{loc}(\Omega)$ $f_0(\cdot + h) =$
$\eta f_0(\cdot + h) \in Z_0(\Omega)$.

Let $\lambda(h) = \|f_0(x + h) - f(x)\|_{Z(\Omega)}$ for $h \in B(0, \gamma)$. Condition 4) means that
the function λ is continuous at the point 0. Moreover, $\lambda \in C(B(0, \gamma))$. Indeed,
let $u \in B(0, \gamma)$. Then, by the continuity of the semi-norm, in order to prove
that $\lambda(h) \to \lambda(u)$ as $h \to u$ it is enough to prove that $f_0(x + h) - f(x) \to$
$f_0(x + u) - f(x)$ or $f_0(x + h) - f_0(x + u) = g_0(x + h - u) - g(x) \to 0$ as $h \to u$
where $g(x) = f_0(x + u)$. And this is valid because $g \in Z_0(\Omega)$.

2. Let us consider the mollifiers $B_{\bar{\delta}}$, which are constructed with the help
of any nonnegative kernel. Since $\sum_{k=1}^{\infty} \psi_k = 1$ on Ω and $\int_{B(0,1)} \omega dx = 1$ we have
$\forall x \in \Omega$

$$(B_{\bar{\delta}}f)(x) - f(x) = \sum_{k=1}^{\infty}((A_{\delta_k}(\psi_k f)) - \psi_k(x)f(x))$$

$$= \sum_{k=1}^{\infty} \int_{B(0,1)} (f_k(x - \delta_k z) - f_k(x))\omega(z)dz = \sum_{k=1}^{\infty} F_k(x),$$

where $f_k = \psi_k f$ and $F_k(x) = \int_{B(0,1)} (f_k(x - \delta_k z) - f_k(x))\,\omega(z)dz$.

By 3) and (2.15) we have that $f_k, F_k \in Z_0(\Omega)$ and $\operatorname{supp} f_k$, $\operatorname{supp} F_k \subset$
$G_{k-1} \cup G_k \cup G_{k+1}$. Applying Minkowski's inequality for infinite sums (which
holds besause of the completeness of the space $Z(\Omega)$) we have

$$\|B_{\bar{\delta}}f - f\|_{Z(\Omega)} \leq \sum_{k=1}^{\infty} \|F_k\|_{Z(\Omega)}.$$

Now suppose that, in addition to (2.18),

$$\delta_k < \tfrac{1}{2}\,\gamma(f_k), \tag{2.32}$$

then the function $\|f_k(x - \delta_k z) - f_k(x)\|_{Z(\Omega)}$ is continuous on $\overline{B(0,1)}$ and

$$\int\limits_{B(0,1)} \|f_k(x - \delta_k z) - f_k(x)\|_{Z(\Omega)}\,\omega(z)\,dz < \infty.$$

Moreover, the function $[f_k(x - \delta_k z) - f_k(x)]\omega(z)$ is measurable on $\Omega \times B(0,1)$ (see footnote 9 of Chapter 1). Therefore, we can apply Minkowski's inequality in condition 2) and establish that

$$\|F_k\|_{Z(\Omega)} \leq \int\limits_{B(0,1)} \|f_k(x - \delta_k z) - f_k(x)\|_{Z(\Omega)}\,\omega(z)\,dz$$

$$\leq \sup_{|h| \leq \delta_k} \|f_k(x + h) - f_k(x)\|_{Z(\Omega)} = \omega(\delta_k, f_k)_{Z(\Omega)}.$$

Thus, (2.31) follows.

3. Now $\forall \varepsilon > 0$ by 4) choose δ_k such that, in addition to (2.18) and (2.32),

$$\omega(\delta_k, f_k)_{Z(\Omega)} < \varepsilon\, 2^{-k}. \tag{2.33}$$

(we note that $(f_k)_0 = f_k$). With this choice of $\overline{\delta} = \overline{\delta}(\varepsilon, f)$ (depending on ε and f) we have $B_{\overline{\delta}}f \in C^\infty(\Omega)$ and

$$\|B_{\overline{\delta}}f - f\|_{Z(\Omega)} < \varepsilon. \tag{2.34}$$

Thus, Theorem 2 is proved (in (2.29) one can take $\varphi_s = B_{\overline{\delta}_s}f$, where $\overline{\delta}_s = \overline{\delta}\left(\tfrac{1}{s}, f\right)$). $\square$

Remark 10 The functions φ_s in the given proof are constructed in such a way that they depend on f, in general, nonlinearly. Moreover, they may depend, of course, on the space $Z(\Omega)$. For example, in the case of $Z(\Omega) = W_p^l(\Omega)$ they may depend on n, l, p, Ω and f.

Idea of the proof of Theorem 1. The density of $C^\infty(\Omega) \cap W_p^l(\Omega)$ in $W_p^l(\Omega)$ where $1 \leq p < \infty$ and of $C^\infty(\Omega) \cap \overline{C}^l(\Omega)$ in $\overline{C}^l(\Omega)$ follows directly by applying Theorem 2 to $Z(\Omega) = W_p^l(\Omega), \overline{C}^l(\Omega)$ respectively. In order to prove the second statement of the theorem take $Z(\Omega) = \widetilde{W}_p^l(\Omega)$, where $\widetilde{W}_p^l(\Omega) = \bigcap\limits_{m=0}^{l} W_p^m(\Omega)$,

hence $\|f\|_{\widetilde{W}_p^l(\Omega)} = \sum\limits_{m=0}^{l} \|f\|_{W_p^m(\Omega)}$. Note that by Corollary 14 of Chapter 4

$(\widetilde{W}_p^l)^{loc}(\Omega) \supset W_p^l(\Omega)$. The case of the spaces $\overline{C}^l(\Omega)$ is similar.

In the case of the spaces $W_\infty^l(\Omega)$ take open sets $\Omega_k \subset \Omega$, $k \in \mathbb{N}$, which are such that $\operatorname{meas}\partial\Omega_k = 0$. Consider any $\gamma_{ks} \in (0, \rho_k), k, s \in \mathbb{N}$, such that $\lim\limits_{s\to\infty} \gamma_{ks} = 0$. Choose a partition of unity $\{\psi_k\}_{k\in\mathbb{N}}$ in such a way that for any sufficiently small δ_{ks}

$$B_{\bar{\delta}_s} f = A_{\delta_{ms}} f \quad \text{on} \quad (G_m)_{\gamma_{ms}},$$

where $\delta_s = \{\bar{\delta}_{ks}\}_{k\in\mathbb{N}}$. This is possible by Remark 7. Moreover, assume that the kernel of mollification ω is nonegative. $\square$

Proof of Theorem 1. By Lemmas 10, 11 and 14 of Chapter 1 the conditions of Theorem 2 are satisfied for $Z(\Omega) = \widetilde{W}_p^l(\Omega)$ where $1 \le p < \infty$ and for $\overset{\cdot}{Z}(\Omega) = \overset{\approx}{\overline{C}}^l(\Omega)$. Hence the first two statements of Theorem 1 follow from Theorem 2.

However, if $p = \infty$, then condition 4) of Theorem 2 is not satisfied, and Theorem 2 is not applicable. In this case we need a more sophisticated argument. Let $f \in W_\infty^l(\Omega)$ and $m \le l - 1$. Then for any mollifier $B_{\bar{\delta}}$, which is constructed with the help of a nonnegative kernel, and $m = 0, 1, ..., l - 1$ we have

$$\|B_{\bar{\delta}} f - f\|_{W_\infty^m(\Omega)} \le \sum_{k=1}^{\infty} \omega(\delta_k, f_k)_{W_\infty^m(\Omega)} \le \sum_{k=1}^{\infty} \omega(\delta_k, \psi_k f_0)_{W_\infty^m(\mathbb{R}^n)}.$$

By Lemma 11 of Chapter 1 $\psi_k f_0 \in W_\infty^l(\mathbb{R}^n)$ and as in the proof of Lemma 1, applying, in addition, footnote 11 of Chapter 1, we establish that

$$\omega(\delta_k, \psi_k f_0)_{W_\infty^m(\mathbb{R}^n)} \le M_1 \delta_k \|\psi_k f_0\|_{W_\infty^{m+1}(\mathbb{R}^n)},$$

where M_1 is independent of f and k.

Consequently, $\forall \varepsilon > 0$ there exist $\sigma_k^{(1)} > 0, k \in \mathbb{N}$, such that $\forall \delta_k \in (0, \sigma_k^{(1)})$ we have $\omega(\delta_k, \psi_k f_0)_{W_\infty^m(\mathbb{R}^n)} < \varepsilon\, 2^{-k}$, $m = 0, ..., l - 1$, and hence

$$\|B_{\bar{\delta}} f - f\|_{W_\infty^m(\Omega)} < \varepsilon, \quad m = 0, ..., l - 1.$$

Furthermore, for $\forall \alpha \in \mathbb{N}_0^n$ satisfying $|\alpha| = l$ by (2.25), Lemma 4 of Chapter 1 and Leibnitz' formula we have

$$D^\alpha(B_{\bar{\delta}} f) = \sum_{0 \le \beta \le \alpha} \binom{\alpha}{\beta} \sum_{k=1}^{\infty} A_{\delta_k}(D^{\alpha-\beta}\psi_k\, D_w^\beta f),$$

where $\binom{\alpha}{\beta} = \frac{\alpha!}{\beta!(\alpha-\beta)!} = \prod_{i=1}^{n} \frac{\alpha_i!}{\beta_i!(\alpha_i-\beta_i)!}$. If $\beta \neq \alpha$, then $\sum_{k=1}^{\infty} D^{\alpha-\beta}\psi_k = 0$ on Ω and by (1.8)

$$\left\| \sum_{k=1}^{\infty} A_{\delta_k}(D^{\alpha-\beta}\psi_k\, D_w^\beta f) \right\|_{L_\infty(\Omega)} = \left\| \sum_{k=1}^{\infty} (A_{\delta_k}(D^{\alpha-\beta}\psi_k\, D_w^\beta f) - D^{\alpha-\beta}\psi_k\, D_w^\beta f) \right\|_{L_\infty(\Omega)}$$

$$\leq \sum_{k=1}^{\infty} \|A_{\delta_k}(D^{\alpha-\beta}\psi_k\, D_w^\beta f) - D^{\alpha-\beta}\psi_k\, D_w^\beta f\|_{L_\infty(\Omega)} \leq \sum_{k=1}^{\infty} \omega(\delta_k, D^{\alpha-\beta}\psi_k\, D_w^\beta f)_{L_\infty(\Omega)}$$

$$\leq \sum_{k=1}^{\infty} \omega(\delta_k, D^{\alpha-\beta}\psi_k\, D_w^\beta f_0)_{L_\infty(\mathbb{R}^n)}.$$

Since by Lemma 11 of Chapter 1 $D^{\alpha-\beta}\psi_k\, D_w^\beta f_0 \in W_\infty^{l-|\beta|}(\mathbb{R}^n)$ as in the proof of Lemma 1 we establish that

$$\omega(\delta_k, D^{\alpha-\beta}\psi_k\, D_w^\beta f_0)_{L_\infty(\mathbb{R}^n)} \leq M_2 \delta_k \|D^{\alpha-\beta}\psi_k\, D_w^\beta f_0\|_{W_\infty^{l-|\beta|}(\mathbb{R}^n)},$$

where M_2 is independent of f and k.

Consequently, there exist $\sigma_k^{(2)} \in (0, \sigma_k^{(1)})$, $k \in \mathbb{N}$, such that $\forall \delta_k \in (0, \sigma_k^{(2)})$ we have

$$\omega(\delta_k, D^{\alpha-\beta}\psi_k\, D_w^\beta f_0)_{L_\infty(\mathbb{R}^n)} < \varepsilon\, 2^{-k-n}(1 + \sum_{|\alpha|=l} 1)^{-1}$$

and, hence,

$$\| \sum_{k=1}^{\infty} A_{\delta_k}(D^{\alpha-\beta}\psi_k\, D_w^\beta f) \|_{L_\infty(\Omega)} < \varepsilon\, 2^{-n}(1 + \sum_{|\alpha|=l} 1)^{-1}.$$

If $\beta = \alpha$, then since ψ_k, $k \in \mathbb{N}$, and the kernel of mollification is nonnegative we have

$$\| \sum_{k=1}^{\infty} A_{\delta_k}(\psi_k D_w^\alpha f) \|_{L_\infty(\Omega)} \leq \| \sum_{k=1}^{\infty} |A_{\delta_k}(\psi_k D_w^\alpha f)| \|_{L_\infty(\Omega)}$$

$$\leq \| \sum_{k=1}^{\infty} A_{\delta_k}(\psi_k |D_w^\alpha f|) \|_{L_\infty(\Omega)} \leq \| \sum_{k=1}^{\infty} A_{\delta_k}\psi_k \|_{L_\infty(\Omega)} \|D_w^\alpha f\|_{L_\infty(\Omega)}$$

$$\leq \|1 + \sum_{k=1}^{\infty} (A_{\delta_k}\psi_k - \psi_k) \|_{L_\infty(\Omega)} \|D_w^\alpha f\|_{L_\infty(\Omega)}$$

$$\leq \Big(1 + \sum_{k=1}^{\infty} \|A_{\delta_k}\psi_k - \psi_k\|_{L_\infty(\Omega)}\Big)\|D_w^\alpha f\|_{L_\infty(\Omega)}$$

$$\leq \Big(1 + \sum_{k=1}^{\infty} \omega(\delta_k, \psi_k)_{L_\infty(\Omega)}\Big)\|D_w^\alpha f\|_{L_\infty(\Omega)}.$$

Since $\psi_k \in C_0^\infty(\Omega)$, as in the proof of Lemma 1 we have

$$\omega(\delta_k, \psi_k)_{L_\infty(\Omega)} \leq \omega(\delta_k, \psi_k)_{L_\infty(\mathbb{R}^n)} \leq \delta_k \|\psi_k\|_{W_\infty^1(\mathbb{R}^n)}$$

and it follows as above that there exist $\sigma_k^{(3)} \in (0, \sigma_k^{(2)})$, $k \in \mathbb{N}$, such that $\forall \delta_k \in (0, \sigma_k^{(3)})$

$$\Big\|\sum_{k=1}^{\infty} A_{\delta_k}(\psi_k D_w^\alpha f)\Big\|_{L_\infty(\Omega)} < (1 + \varepsilon)\|D_w^\alpha f\|_{L_\infty(\Omega)}.$$

(This inequality also holds for $\alpha = 0$.)

Thus, if $\delta_k \in (0, \sigma_k^{(3)})$, then

$$\|B_{\bar\delta} f\|_{W_\infty^l(\Omega)} < \varepsilon + (1 + \varepsilon)\|f\|_{W_\infty^l(\Omega)}.$$

(We have applied the equality $\sum\limits_{0 \leq \beta \leq \alpha} \binom{\alpha}{\beta} = 2^n$.)

In particular, if $\delta_{ks} \in (0, \sigma_k^{(3)})$, $k, s \in \mathbb{N}$, then

$$\|B_{\bar\delta_s} f\|_{W_\infty^l(\Omega)} < \varepsilon + (1 + \varepsilon)\|f\|_{W_\infty^l(\Omega)}.$$

On the other hand by construction of the mollifier $\bar\delta_s$ and by Lemma 4 of Chapter 1

$$\|B_{\bar\delta_s} f\|_{W_\infty^l(\Omega)} \geq \|B_{\bar\delta_s} f\|_{W_\infty^l((G_k)_{\gamma_{ks}})} = \|A_{\delta_{ks}} f\|_{W_\infty^l((G_k)_{\gamma_{ks}})}$$

$$= \|A_{\delta_{ks}} f\|_{L_\infty((G_k)_{\gamma_{ks}})} + \sum_{|\alpha|=l} \|A_{\delta_{ks}} D_w^\alpha f\|_{L_\infty((G_k)_{\gamma_{ks}})}.$$

By relation (1.9) for $p = \infty$ there exist $\sigma_k^{(4)} \in (0, \sigma_k^{(3)})$, $k \in \mathbb{N}$, such that for $\delta_{ks} \in (0, \sigma_k^{(4)})$

$$\|B_{\bar\delta_s} f\|_{W_\infty^l(\Omega)} \geq \|f\|_{W_\infty^l((G_k)_{\gamma_{ks}})} - \tfrac{\varepsilon}{2}$$

and, hence,

$$\|B_{\bar\delta_s} f\|_{W_\infty^l(\Omega)} \geq \sup_{k \in \mathbb{N}} \|f\|_{W_\infty^l((G_k)_{\gamma_{ks}})} - \tfrac{\varepsilon}{2} = \|f\|_{W_\infty^l\left(\bigcup\limits_{k=1}^{\infty} (G_k)_{\gamma_{ks}}\right)} - \tfrac{\varepsilon}{2}.$$

Since $\text{meas}\left(\bigcup_{k=1}^{\infty} \partial\Omega_k\right) = 0$ we have

$$\|f\|_{W^l_\infty(\Omega)} = \|f\|_{W^l_\infty(\Omega\setminus\bigcup_{k=1}^{\infty}\partial\Omega_k)} = \lim_{s\to\infty}\|f\|_{W^l_\infty(\bigcup_{k=1}^{\infty}(G_k)_{\gamma_{ks}})}.$$

Consequently, there exists $s \in \mathbb{N}$ such that

$$\|f\|_{W^l_\infty(\bigcup_{k=1}^{\infty}(G_k)_{\gamma_{ks}})} \geq \|f\|_{W^l_\infty(\Omega)} - \tfrac{\varepsilon}{2}.$$

Thus, $\forall \varepsilon > 0$, there exist $s \in \mathbb{N}$ and $\delta_{ks} \in (0, \sigma_k^{(4)})$ such that

$$\|B_{\bar\delta_s}f - f\|_{W^{l-1}_\infty(\Omega)} < \varepsilon$$

and

$$\|f\|_{W^l_\infty(\Omega)} - \varepsilon < \|B_{\bar\delta_s}f\|_{W^l_\infty(\Omega)} < \varepsilon + (1+\varepsilon)\|f\|_{W^l_\infty(\Omega)}$$

and the statement of Theorem 1 in the case $p = \infty$ follows. $\square$

Corollary 1 *Let $\Omega \subset \mathbb{R}^n$ be an open set, $l \in \mathbb{N}_0$. Then $C^\infty(\Omega)$ is dense in $(W^l_p)^{loc}(\Omega)$ where $1 \leq p < \infty$ and in $C^l(\Omega)$.*

Idea of the proof. Apply Theorem 2 to $Z^{loc}(\Omega) = (W^l_p)^{loc}(\Omega)$ and $Z(\Omega) = (\overline{C}^l)^{loc}(\Omega) = C^l(\Omega)$. $\square$

Remark 11 If $p = \infty$, then $C^\infty(\Omega) \cap W^l_\infty(\Omega)$ is not dense in $W^l_\infty(\Omega)$ (see Remark 2).

Remark 12 The crucial condition in Theorem 2 is condition 4). It can be proved that under some additional unrestrictive assumptions on $Z(\Omega)$ the density of $C^\infty(\Omega)$ in $Z^{loc}(\Omega)$ (or the density of $C^\infty(\Omega)\cap Z(\Omega)$ in $Z(\Omega)$) is equivalent to condition 4).

Remark 13 Theorem 2 is applicable to a very wide class of spaces $Z(\Omega)$, which are studied in the theory of function spaces. We give only one example. Consider positive functions $a_0,\ a_\alpha \in C(\Omega)$ $(\alpha \in \mathbb{N}^n_0,\ |\alpha| = l)$ and the weighted Sobolev space $W^l_{p,\{a_\alpha\}}(\Omega)$ characterized by the finiteness of the norm

$$\|a_0 f\|_{L_p(\Omega)} + \sum_{|\alpha|=l} \|a_\alpha D^\alpha_w f\|_{L_p(\Omega)}.$$

By Theorem 2 it follows that $C^\infty(\Omega) \cap W^l_{p,\{a_\alpha\}}(\Omega)$ is dense in this space for $1 \leq p < \infty$ without any additional assumptions on weights a_0 and a_α. Such generality is possible due to the fact that the continuity with respect to translation needs to be proved only for functions in this weighted Sobolev space, which are compactly supported in Ω.

Now we give one more example of an application of Theorem 2, in which the spaces $Z^{loc}(\Omega)$ (and not only $Z(\Omega)$) are used.

Example 1 Let $\Omega \subset \mathbb{R}^n$ be an open set, then $\forall \mu \in C(\Omega)$ and $\forall \varepsilon > 0$ there exist $\mu_\varepsilon \in C^\infty(\Omega)$ such that $\forall x \in \Omega$ we have $\mu(x) < \mu_\varepsilon(x) < \mu(x) + \varepsilon$.

To prove this it is enough to set $\mu_\varepsilon = B_{\overline{\delta}}(\mu + \frac{\varepsilon}{2})$ with $\overline{\delta} = \overline{\delta}(\frac{\varepsilon}{2}, \mu + \frac{\varepsilon}{2})$ in the proof of Theorem 2 for $Z(\Omega) = \overline{C}(\Omega)$ (hence, $Z^{loc}(\Omega) = C(\Omega)$) and apply inequality (2.34).

2.4 Approximation with preservation of boundary values

In Theorem 1 it is proved that for each open set $\Omega \subset \mathbb{R}^n$ and $\forall f \in W_p^l(\Omega)$ $(1 \leq p < \infty)$ functions $\varphi_s \in C^\infty(\Omega) \cap W_p^l(\Omega)$, $s \in \mathbb{N}$, exist such that (2.27) holds. In this section we show that it is possible to choose the approximating functions φ_s in such a way that, in addition, they and their derivatives of order $\alpha \in \mathbb{N}_0^n$ satisfying $|\alpha| \leq l$ have in some sense the same "boundary values" as the approximated function f and its corresponding weak derivatives. The problem of existence and description of boundary values will be discussed in Chapter 5. Here we note only that for a general open set $\Omega \subset \mathbb{R}^n$ it may happen that the boundary values do not exist and even for "good" Ω boundary values of weak derivatives of order α satisfying $|\alpha| = l$, in general, do not exist. For this reason in this section we speak about coincidence of boundary values without studying the problem of their existence — we treat the coincidence as the same behaviour, in some sense, of the functions f and φ_s (and their derivatives) when approaching the boundary of Ω.

Theorem 3 *Let $\Omega \subset \mathbb{R}^n$ be an open set, $l \in \mathbb{N}$, $1 \leq p < \infty$. Then $\forall \mu \in C(\Omega)$ and $\forall f \in W_p^l(\Omega)$ functions $\varphi_s \in C^\infty(\Omega) \cap W_p^l(\Omega)$, $s \in \mathbb{N}$, exist such that, besides (2.27), $\forall \alpha \in \mathbb{N}_0^n$ satisfying $|\alpha| \leq l$*

$$\|(D_w^\alpha f - D^\alpha \varphi_s)\mu\|_{L_p(\Omega)} \to 0 \qquad (2.35)$$

as $s \to \infty$. For $p = \infty$ this assertion is valid $\forall f \in \overline{C}^l(\Omega)$.

Corollary 2 *Let $\Omega \subset \mathbb{R}^n$ be an open set, $\Omega \neq \mathbb{R}^n$ and $l \in \mathbb{N}$. Then $\forall f \in \overline{C}^l(\Omega)$ functions $\varphi_s \in C^\infty(\Omega) \cap \overline{C}^l(\Omega), s \in \mathbb{N}$, exist, which depend linearly on f and* *are such that, besides (2.27) where $p = \infty$,* [5]

$$D^\alpha \varphi_s|_{\partial\Omega} = D^\alpha f|_{\partial\Omega}, \qquad |\alpha| \le l. \tag{2.36}$$

Idea of the proof. Choose any positive $\mu \in C(\Omega)$ such that $\lim\limits_{y \to x, y \in \Omega} \mu(y) = \infty$ for all $x \in \partial\Omega$. $\square$

Proof. One may set, for example, $\mu(x) = \text{dist}\,(x, \partial\Omega)^{-1}$. For a continuous function $\|\cdot\|_{C(\Omega)} = \|\cdot\|_{L_\infty(\Omega)}$, therefore, from (2.35) it follows that for some $M > 0$ $\forall s \in \mathbb{N}$ and $\forall y \in \Omega$

$$|D^\alpha \varphi_s(y) - D^\alpha f(y)| \le M(\mu(y))^{-1}.$$

Passing to the limit as $y \to x \in \partial\Omega$, $y \in \Omega$ we arrive at (2.36). $\square$

Corollary 3 *Let $\Omega \subset \mathbb{R}^n$ be an open set, $l \in \mathbb{N}$, $\partial\Omega \in C^l$ and $1 \le p < \infty$. Then $\forall f \in W_p^l(\Omega)$ functions $\varphi_s \in C^\infty(\Omega) \cap W_p^l(\Omega)$, $s \in \mathbb{N}$, exist such that, besides (2.27),* [6]

$$D^\alpha \varphi_s|_{\partial\Omega} = D^\alpha_w f|_{\partial\Omega}, \qquad |\alpha| \le l - 1. \tag{2.37}$$

Idea of the proof. Take again $\mu(x) = \text{dist}\,(x, \partial\Omega)^{-1}$. By Chapter 5 it is enough to consider the case, in which $\Omega = \mathbb{R}^n_+ = \{x \in \mathbb{R}^n : x_n > 0\}$ and $\mu(x) = x_n^{-1}$. In this case the statement follows by Lemma 13 of Chapter 5. $\square$

Remark 14 The function μ in Theorem 3 can have arbitrarily fast growth when approaching $\partial\Omega$. Let, for instance, $\mu(x) = g(\varrho(x))$, where $\varrho(x) = \text{dist}\,(x, \partial\Omega)$ and $g \in C((0, \infty))$ is any positive, nonincreasing function. Then for $1 \le p \le \infty$

$$\|D^\alpha_w f - D^\alpha \varphi_s\|_{L_p(\Omega\setminus\Omega_\delta)} \le (g(\delta))^{-1}\|(D^\alpha_w f - D^\alpha \varphi_s)g(\varrho)\|_{L_p(\Omega)} \le M(g(\delta))^{-1}$$

with some $M > 0$, which does not depend on s and δ. It implies that for a fixed $f \in W_p^l(\Omega)$ where $1 \le p < \infty$ one can find a sequence of approximating

---

[5] We recall that $\forall f \in \overline{C}^l(\Omega)$, $\forall x \in \partial\Omega$ and $\forall \alpha \in \mathbb{N}_0^n$ satisfying $|\alpha| \le l$ there exists $\lim\limits_{y \to x, y \in \Omega} D^\alpha f(y)$ and, thus, the functions $D^\alpha f$, which are defined on Ω can be extended to $\overline{\Omega}$ as continuous functions. It is assumed that $D^\alpha f|_{\partial\Omega}$ are just restrictions to $\partial\Omega$ of these extensions. The same refers to the functions $\varphi_s \in C^\infty(\Omega)$, because by (2.27) where $p = \infty$ we have $\varphi_s \in \overline{C}^l(\Omega)$. From Theorem 8 below it follows, in particular, that φ_s can be chosen in such a way that they depend linearly on f.

[6] Here by $D^\alpha_w f|_{\partial\Omega}$ and $D^\alpha \varphi_s|_{\partial\Omega}$ the traces of the functions $D^\alpha_w f$ and $D^\alpha \varphi_s$ on $\partial\Omega$ are denoted (in the sense of Chapter 5, they exist if $|\alpha| \le l - 1$). See also Theorem 9 below.

functions φ_s, which is such that, besides (2.27), $\forall \alpha \in \mathbb{N}_0^n$ satisfying $|\alpha| \leq l$ the norm $\|D_w^\alpha f - D^\alpha \varphi_s\|_{L_p(\Omega \setminus \Omega_\delta)}$ tends to 0 arbitrarily fast as $\delta \to +0$.

Thus, condition (2.35) with arbitrary choice of μ means not only coincidence of boundary values, but, moreover, arbitrarily close prescribed behaviour of the functions f and φ_s and their derivatives of order α satisfying $|\alpha| \leq l$ when approaching the boundary $\partial \Omega$.

For unbounded Ω we have the same situation with the behaviour at infinity. Choosing positive $\mu \in C(\Omega)$ growing fast enough at infinity, we can construct the functions $\varphi_s \in C^\infty(\Omega) \cap W_p^l(\Omega)$ such that $\|D_w^\alpha f - D^\alpha \varphi_s\|_{L_p(\Omega \setminus B(0,r))}$ where $|\alpha| \leq l$ tends to 0 arbitrarily fast as $r \to +\infty$, i.e., $D_w^\alpha f$ and $D^\alpha \varphi_s$ have arbitrarily close prescribed behaviour at infinity.

As in Section 2.4 we derive Theorem 3 from a similar result, which holds for general function spaces $Z(\Omega)$.

Theorem 4 *In addition to the assumptions of Theorem 3, let the following condition be satisfied:*

5) $\forall f \in C_0^\infty(\Omega)$ there exists $c_\varphi > 0$ such that $\forall f \in Z_0(\Omega)$

$$\|\varphi f\|_{Z(\Omega)} \leq c_\varphi \|f\|_{Z(\Omega)}.$$

Then $\forall \mu \in C^\infty(\Omega)$ and $\forall f \in Z^{loc}(\Omega)$ functions $\varphi_s \in C^\infty(\Omega) \cap Z^{loc}(\Omega)$, $s \in \mathbb{N}$, exist such that

$$\varphi_s \to f \quad in \ \ Z(\Omega) \tag{2.38}$$

and

$$\|(f - \varphi_s)\mu\|_{Z(\Omega)} \to 0 \tag{2.39}$$

as $s \to \infty$.

Idea of the proof. Starting with the equality that differs from (2.30) by the factor μ show, applying 5), that

$$\|(B_{\bar{\delta}} f - f)\mu\|_{Z(\Omega)} \leq \sum_{k=1}^{\infty} c_k \omega(\delta_k, f_k)_{Z(\Omega)}, \tag{2.40}$$

where the $c_k > 0$ are independent of δ_k. $\square$

Proof. In addition to the proof of Theorem 2, we must estimate the expression

$$(B_{\bar{\delta}} f - f)\mu = \sum_{k=1}^{\infty} \mu F_k.$$

Recall that $F_k \in Z_0(\Omega)$ and supp $F_k \subset \widetilde{G_k} = G_{k-1} \cup G_k \cup G_{k+1}$. Let us denote by $\eta_k \in C_0^\infty(\Omega)$ a function of "cap-shaped" type, which is equal to 1 on $\widetilde{G_k}$ (see Section 1.1), then $\mu F_k = \mu \eta_k F_k$. By condition 5) where $\varphi = \mu \eta_k$ there exists $c_k > 0$ depending only on $\mu \eta_k$ (and, thus, independent of δ_k), such that

$$\|\mu F_k\|_{Z(\Omega)} \leq c_k \|F_k\|_{Z(\Omega)} \leq c_k \omega_{\delta_k}(f_k)_{Z(\Omega)}$$

and (2.40) follows (without loss of generality we can assume that $c_k \geq 1$). Choosing $\forall \varepsilon > 0$ positive numbers δ_k in such a way that in this case $\omega_{\delta_k}(f_k)_{Z(\Omega)} < \varepsilon 2^{-k} c_k^{-1}$ (instead of (2.33)), we establish, besides (2.34), the inequality $\|(B_{\bar\delta} f - f)\mu\|_{Z(\Omega)} < \varepsilon$. $\square$

Remark 15 From the above proof it follows that $\forall \mu_1, ..., \mu_m \in C^\infty(\Omega)$ $(m \in \mathbb{N})$ and $\forall f \in Z^{loc}(\Omega)$ functions $\varphi_s \in C^\infty(\Omega) \cap Z^{loc}(\Omega)$, $s \in \mathbb{N}$, exist such that

$$\|(f - \varphi_s)\mu_i\|_{Z(\Omega)} \to 0, \qquad i = 1, ..., m.$$

as $s \to \infty$. (Theorem 4 corresponds to $m = 2$, $\mu_1 = 1$, $\mu_2 = \mu$.)

Idea of the proof of Theorem 3. Apply Theorem 4 and Remark 15 to the space $Z(\Omega) = \widetilde{W_p^l}(\Omega)$ and to a set of the weight functions $(D^\gamma \mu_1)_{|\gamma| \leq l}$, where $\mu_1 \in C^\infty(\Omega)$ and $|\mu| \leq \mu_1$ on Ω. $\square$

Proof of the Theorem 3. The existence of the function μ_1 follows by Example 1. By Remark 15 $\forall f \in (\widetilde{W_p^l})^{loc}(\Omega)$ [7] functions $\varphi_s \in C^\infty(\Omega)$, $s \in \mathbb{N}$, exist such that $\varphi_s \to f$ in $\widetilde{W_p^l}(\Omega)$ and $\forall \gamma \in \mathbb{N}_0^n$ satisfying $|\gamma| \leq l$.

$$\|(f - \varphi_s)D^\gamma \mu_1\|_{\widetilde{W_p^l}(\Omega)} \to 0$$

as $s \to \infty$. Hence, $\forall \alpha \in \mathbb{N}_0^n$ satisfying $|\alpha| \leq l$

$$\|D_w^\alpha((f - \varphi_s)D^\beta \mu_1)\|_{L_p(\Omega)} \to 0. \tag{2.41}$$

Applying "inverted" Leibnitz' formula [8], we have

$$(D_w^\alpha f - D^\alpha \varphi_s)\mu_1 = \sum_{0 \leq \beta \leq \alpha} (-1)^{|\beta|} \binom{\alpha}{\beta} D_\omega^\beta((f - \varphi_s)D^{\alpha-\beta}\mu_1)$$

---

[7] We recall that this space was also considered in the proof of Theorem 1.

[8] For $n = 1$ and ordinary derivatives it has the form

$$f^{(k)}g = \sum_{m=0}^{k} (-1)^m \binom{m}{k} (fg^{(k-m)})^{(m)}$$

and is easily proved by induction or by Leibnitz' formula.

and from (2.34) it follows that $\forall f \in (\widetilde{W_p^l})^{loc}(\Omega)$ and $\forall \alpha \in \mathbb{N}_0^n$ satisfying $|\alpha| \leq l$

$$\|(D_w^\alpha f - D^\alpha \varphi_s)\mu\|_{L_p(\Omega)} \leq \|(D_w^\alpha f - D^\alpha \varphi_s)\mu_1\|_{L_p(\Omega)} \to 0$$

as $s \to \infty$. Finally, as in the proof of Theorem 1, it is enough to note that $W_p^l(\Omega) \subset (\widetilde{W_p^l})^{loc}(\Omega)$. $\square$

2.5   Linear mollifiers with variable step

We start by studying the mollifiers A_δ having kernels with some vanishing moments. The main property of the mollifier A_δ is that $\forall f \in L_p(\Omega)$ where $1 \leq p < \infty$

$$\|A_\delta f - f\|_{L_p(\Omega)} = o(1) \tag{2.42}$$

as $\delta \to 0+$. As for the rate of convergence of $A_\delta f$ to f, in general, it can be arbitrarily slow. However, under additional assumptions on f one can have more rapid convergence.

Lemma 8 *Let $\Omega \subset \mathbb{R}^n$ be an open set, $1 \leq p \leq \infty$, $\delta > 0$ and $G \subset \Omega$ be a measurable set such that $G^\delta \subset \Omega$. Then $\forall f \in w_p^1(\Omega)$*

$$\|A_\delta f - f\|_{L_p(G)} \leq c_1 \delta \|f\|_{w_p^1(G^\delta)}, \tag{2.43}$$

where

$$c_1 = \max_{i=1,\dots,n} \|z_i \omega(z)\|_{L_1(\mathbb{R}^n)}. \tag{2.44}$$

Idea of the proof. For $f \in C^\infty(\Omega)$ apply Taylor's formula and Minkowski's inequality. For $f \in w_p^1(\Omega)$ approximate f by $A_\gamma f$ and pass to the limit as $\gamma \to 0+$. $\square$

Proof. For $f \in C^\infty(\Omega)$

$$\|A_\delta f - f\|_{L_p(G)} = \left\| \int_{B(0,1)} (f(x - \delta z) - f(x))\, \omega(z)\, dz \right\|_{L_p(G)}$$

$$= \left\| \int_{B(0,1)} \left(\int_0^1 \left(\sum_{i=1}^n \frac{\partial f}{\partial x_i}(x - t\delta z)(-\delta z_i)\, \omega(z) \right) dt \right) dz \right\|_{L_p(G)}$$

$$\leq \delta \int_{B(0,1)} \left(\int_0^1 \left(\sum_{i=1}^n \left\| \frac{\partial f}{\partial x_i}(x - t\delta z) \right\|_{L_p(G)} |z_i \omega(z)| \right) dt \right) dz$$

$$\leq \delta \max_{i=1,\dots,n} \int\limits_{B(0,1)} |\, z_i\, \omega(z)\,|\, dz \sum_{i=1}^{n} \left\| \frac{\partial f}{\partial x_i} \right\|_{L_p(G^\delta)} = c_1 \delta \|f\|_{w_p^1(G^\delta)}.$$

Now let $f \in w_p^1(\Omega)$ and first suppose that $\varrho = \mathrm{dist}\,(G^\delta, \partial\Omega) > 0$. Then for $0 < \gamma < \varrho$ we have that $A_\gamma f \in C^\infty(\Omega)$ on $G_\delta \subset \Omega_\gamma$ and $A_\gamma(A_\delta f) = A_\delta(A_\gamma f)$ on G (see Section 1.1). Consequently,

$$\|A_\gamma(A_\delta f - f)\|_{L_p(G)} = \|A_\delta(A_\gamma f) - A_\gamma f\|_{L_p(G)}$$

$$\leq c_1 \delta \|A_\gamma f\|_{w_p^1(G^\delta)} = c_1 \delta \sum_{j=1}^{n} \left\| \frac{\partial A_\gamma f}{\partial x_j} \right\|_{L_p(G^\delta)} = c_1 \delta \sum_{j=1}^{n} \left\| A_\gamma \left(\frac{\partial f}{\partial x_j} \right)_w \right\|_{L_p(G^\delta)}$$

by Lemma 4. Passing to the limit as $\gamma \to 0+$ (see (1.10)) we obtain (2.43).

If $\mathrm{dist}\,(G^\delta, \partial\Omega) = 0$, we choose measurable sets $G_k, k \in \mathbb{N}$, such that $\overline{G_k^\delta} \subset \Omega$, $G_k \subset G_{k+1}$ and $\bigcup_{k=1}^{\infty} G_k = G$. Inequality (2.43) is already proved for G_k replacing G. Passing to the limit as $k \to \infty$ we obtain (2.43) in this case also. $\square$

Estimate (2.43) is sharp as the following examples show.

Example 2 Let for some $j \in \{1,..,n\}$ $\int\limits_{\mathbb{R}} z_j\omega(z)\, dz \neq 0$ and $0 < \mathrm{meas}\,\Omega < \infty$. Then

$$\|A_\delta x_j - x_j\|_{L_p(G)} = c_2 \delta, \quad c_2 = \left| \int\limits_{\mathbb{R}} z_j \omega(z)\, dz \right| (\mathrm{meas}\, G)^{\frac{1}{p}} > 0.$$

Remark 16 This example shows also that for some kernels of mollification c_1 is the best possible constant in inequality (2.43). Let us choose $j = 1,\dots,n$, such that $\|z_j\omega(z)\|_{L_1(\mathbb{R})} = \max_{i=1,\dots,n} \|z_i\omega(z)\|_{L_1(\mathbb{R})}$. Moreover, let G be a bounded measurable set such that $0 < \mathrm{meas}\, G = \mathrm{meas}\,\overline{G} < \infty$. Then

$$\sup_{\delta>0:\, G^\delta \subset \Omega} \delta^{-1} \sup_{\|f\|_{w_p^1(\Omega)} \neq 0} \|A_\delta f - f\|_{L_p(G)} \|f\|_{w_p^1(G^\delta)}^{-1}$$

$$\geq \lim_{\delta \to 0+} \delta^{-1} \|A_\delta x_j - x_j\|_{L_p(G)} \|x_j\|_{w_p^1(G^\delta)}^{-1}$$

$$= \left| \int\limits_{\mathbb{R}} z_j \omega(z)\, dz \right| \lim_{\delta \to 0+} \left(\frac{\mathrm{meas}\, G}{\mathrm{meas}\, G^\delta} \right)^{\frac{1}{p}} = \left| \int\limits_{\mathbb{R}} z_j \omega(z)\, dz \right|.$$

Thus, if, in addition to (1.1), $\omega(z) \leq 0$ if $z_j \leq 0$ and $\omega(z) \geq 0$ if $z_j \geq 0$, then c_1 is the best possible constant in inequality (2.43).

Example 3 Let $n = 1$, $G = \Omega = \mathbb{R}$, $p = 2$ and $f \in W_2^1(\mathbb{R})$. Then by the properties of the Fourier transform

$$\|A_\delta f - f\|_{L_2(\mathbb{R})} = (2\pi)^{\frac{1}{2}}\delta \left\|\delta^{-1}((F\omega)(\delta\xi) - (F\omega)(0))(Ff)(\xi)\right\|_{L_2(\mathbb{R})}.$$

We have

$$\delta^{-1}((F\omega)(\delta\xi) - (F\omega)(0)) \to (F\omega)'(0)\xi$$

as $\delta \to 0+$ and

$$\sup_{\delta > 0}\delta^{-1}|(F\omega)(\delta\xi) - (F\omega)(0)| \leq \max_{z\in\mathbb{R}}|(F\omega)'(z)|\,|\xi|.$$

Therefore, by the dominated convergence theorem

$$\left\|\delta^{-1}((F\omega)(\delta\xi) - (F\omega)(0))(Ff)(\xi)\right\|_{L_2(\mathbb{R})}$$

$$\to |(F\omega)'(0)|\,\|\xi(Ff)(\xi)\|_{L_2(\mathbb{R})} = \left|\int_{\mathbb{R}} z\omega(z)\,dz\right|\,\|f'\|_{L_2(\mathbb{R})}.$$

Hence, if $\int_{\mathbb{R}} z\omega(z)\,dz \neq 0$ and $f \in W_2^1(\mathbb{R})$ is not equivalent to zero, then for some $c_3 > 0$ (independent of δ) and $\|A_\delta f - f\|_{L_2(\mathbb{R})} \geq c_3\delta$ for sufficiently small δ.

Let us make now a stronger assumption: $f \in W_p^l(\Omega)$ where $l > 1$. In this case, however, in general we cannot get an estimate better than

$$\|A_\delta f - f\|_{L_p(G)} = O(\delta)$$

(which is the same as for $l = 1$), if for some $j \in \{1, ..., n\}$ $\int_{\mathbb{R}} z_j\omega(z)\,dz \neq 0$, as Examples 2 – 3 show. Thus, in order to obtain improvement of the rate of convergence of $A_\delta f$ to f for the functions $f \in W_p^l(\Omega)$ where $l > 1$, some moments of the kernel of mollification need to be equal to zero.

Lemma 9 *Let $\Omega \subset \mathbb{R}^n$ be an open set, $1 \leq p \leq \infty$, $l \in \mathbb{N}$, $\delta > 0$ and $G \subset \Omega$ be a measurable set such that $G^\delta \subset \Omega$. Moreover, assume that the kernel of the mollifier A_δ satisfies, besides (1.1), the following condition:*

$$\int_{B(0,1)} z^\alpha\omega(z)\,dz = 0, \quad \alpha \in \mathbb{N}_0^n, \quad 0 < |\alpha| \leq l-1, \tag{2.45}$$

where $z^\alpha = z_1^{\alpha_1}\cdots z_n^{\alpha_n}$. Then $\forall f \in w_p^l(\Omega)$

$$\|A_\delta f - f\|_{L_p(G)} \leq c_4\delta^l\|f\|_{w_p^l(G^\delta)}, \tag{2.46}$$

where

$$c_4 = \max_{|\alpha|=l} \int_{\mathbb{R}^n} \left| \frac{z^\alpha}{\alpha!} \omega(z) \right| dz \leq \|\omega\|_{L_1(\mathbb{R}^n)}. \tag{2.47}$$

Condition (2.45) is necessary in order that inequality (2.46) be valid for all $f \in W_p^l(G)$ *with some* $c_4 > 0$ *independent of* f *and* δ.

Idea of the proof. By condition (2.45) $\forall f \in C^\infty(\Omega)$

$$(A_\delta f)(x) - f(x) = \int_{B(0,1)} (f(x - \delta z) - f(x))\, \omega(z)\, dz$$

$$= \int_{B(0,1)} \left(f(x) - \sum_{|\alpha|<l} \frac{(D^\alpha f)(z)}{\alpha!}(-\delta z)^\alpha \right) \omega(z)\, dz.$$

Now multidimensional Taylor's formula (see section 3.3), Minkowski's inequality and direct estimates (close to those which were applied in the proof of Lemma 8) imply (2.46). If $f \in w_p^l(\Omega)$, then pass to the limit in the same manner as in the proof of Lemma 8.

As for necessity of condition (2.45) for bounded G it is enough to take in (2.46) successively $f(x) = x_j, j = 1, ..., n$, $f(x) = x_j x_k, j, k = 1, ..., n$, ..., $f(x) = x_{j_1}...x_{j_{l-1}}, j_1, ..., j_{l-1} = 1, ..., n$. If G is unbounded, then one needs to multiply the above functions by a "cap-shaped" function $\eta \in C_0^\infty(\mathbb{R}^n)$, which is equal to 1 on a ball B such that meas $B \cap G > 0$. $\square$

In the sequel we shall apply the following generalization of inequality (2.46).

Lemma 10 *Let* $\Omega \subset \mathbb{R}^n$ *be an open set,* $1 \leq p \leq \infty$, $l \in \mathbb{N}$, $\delta > 0$ *and* $G \subset \Omega$ *be a measurable set such that* $G^\delta \subset \Omega$. *Assume that the kernel of the mollifier* A_δ *satisfies, besides (1.1), condition (2.45). Then* $\forall f \in w_p^l(\Omega)$ *and* $\forall \alpha \in \mathbb{N}_0^n$

$$\|D^\alpha(A_\delta f) - D_w^\alpha f\|_{L_p(G)} \leq c_5 \delta^{l-|\alpha|} \|f\|_{w_p^l(G^\delta)}, \quad |\alpha| < l, \tag{2.48}$$

and

$$\|D^\alpha(A_\delta f)\|_{L_p(G)} \leq c_6 \delta^{l-|\alpha|} \|f\|_{w_p^l(G^\delta)}, \quad |\alpha| \geq l, \tag{2.49}$$

where $c_5, c_6 > 0$ *do not depend on* f, δ, G *and* p. *(For instance, one can set* $c_5 = \|\omega\|_{L_1(\mathbb{R}^n)}$ *and* $c_6 = \max_{|\beta|=|\alpha|-l} \|D^\beta \omega\|_{L_1(\mathbb{R}^n)}.)$

Idea of the proof. Inequality (2.48) follows by Lemma 4 of Chapter 1 and inequality (2.46) applied to $D_w^\alpha f \in w_p^{l-|\alpha|}(\Omega)$. Estimate (2.49) does not use condition (2.45). It is enough to apply Young's inequality to the equality (see (1.21))

$$D^\alpha(A_\delta f) = \delta^{|\alpha|-|\gamma|}(D^{\alpha-\gamma}\omega)_\delta * D_w^\gamma f\,,$$

where $\gamma \in \mathbb{N}_0^n$ is such that $0 \leq \gamma \leq \alpha$ and $|\gamma| = l$. $\square$

Let Ω be an open set and let the "strips" G_k be defined as in Lemma 5 if $\Omega \neq \mathbb{R}^n$ and as in Lemma 6 if $\Omega = \mathbb{R}^n$. Moreover, let $\{\psi_k\}_{k\in Z}$ be partitions of unity constructed in those lemmas.

Definition 2 *Let $0 < \delta \leq \frac{1}{8}$, $l \in \mathbb{N}$ and $f \in L_1^{loc}(\Omega)$. Then $\forall x \in \Omega$*

$$(E_\delta f)(x) \equiv (E_{\delta,l}f)(x) = \sum_{k=-\infty}^{\infty} \psi_k(x)(A_{\delta 2^{-|k|}}f)(x)$$

$$= \sum_{k=-\infty}^{\infty} \psi_k(x) \int_{B(0,1)} f(x - \delta 2^{-|k|}z)\,\omega(z)\,dz\,, \tag{2.50}$$

where ω is a kernel satisfying, besides (1.1), condition (2.45) [9].

Remark 17 For bounded Ω the operator E_δ is a particular case of the operator $C_{\overline{\delta}}$ by Remark 7. As in Section 2.2 in (2.50) in the last term we write f and not f_0, assuming that $\psi_k(x)g(x) = 0$ if $\psi_k(x) = 0$ even if $g(x)$ is not defined. (This can happen if dist $(x, \partial\Omega) < \delta 2^{-|k|}$). Since $0 < \delta \leq \frac{1}{8}$ we have

$$\operatorname{supp}\,\psi_k A_{\delta 2^{-|k|}}f \subset (G_{k-1} \cup G_k \cup G_{k+1})_{\delta 2^{-|k|}} \tag{2.51}$$

and

$$\psi_k A_{\delta 2^{-|k|}}f \in C^\infty(\Omega). \tag{2.52}$$

(If Ω is bounded, then $\psi_k A_{\delta 2^{-|k|}}f \in C_0^\infty(\Omega)$.) As in the case of the operators $B_{\overline{\delta}}$ and $C_{\overline{\delta}}$ the sum in (2.50) is finite. If $\forall x \in \Omega$ the number $s(x)$ is chosen in such a way that $x \in G_s$, then

$$(E_\delta f)(x) = \sum_{k=s(x)-1}^{s(x)+1} \psi_k(x) \int_{B(0,1)} f(x - \delta 2^{-|k|}z)\,\omega(z)\,dz. \tag{2.53}$$

---

[9] If $l = 1$, then there is no additional condition on the kernel ω.

Moreover, $\forall m \in \mathbb{Z}$

$$E_\delta f = \sum_{k=m-1}^{m+1} \psi_k A_{\delta 2^{-|k|}} f \quad \text{on} \quad G_m. \tag{2.54}$$

We call the E_δ a *linear mollifier with variable step*. The quantity $E_\delta(x)$ is an average of ordinary mollifications with the steps $\delta 2^{-|s(x)|-1}$, $\delta 2^{-|s(x)|}$, $\delta 2^{-|s(x)|+1}$, which (in the case $\Omega \neq \mathbb{R}^n$) tend to 0 as x approaches the boundary $\partial\Omega$. Again we can say that the E_δ is a mollifier with a piecewise constant step since the steps of mollification, which are used for the "strip" G_m, namely $\delta 2^{-|m|-1}$, $\delta 2^{-|m|}$, $\delta 2^{-|m|+1}$, do not depend on $x \in G_m$.

Moreover, by Remark 5 for any fixed $\gamma > 0$ we can choose a partition of unity $\{\psi_k\}_{k\in\mathbb{Z}}$ in such a way that, in addition to (2.54), $\forall m \in \mathbb{Z}$

$$E_\delta f = A_{\delta 2^{-|m|}} f \quad \text{on} \quad (G_m)_{\gamma 2^{-m}}.$$

Remark 18 Changing in (2.50) the variables $x - \delta 2^{-|k|} z = y$ we find

$$(E_\delta f)(x) = \int_\Omega K(x, y, \delta) f(y)\, dy,$$

where

$$K(x, y, \delta) = \sum_{k=-\infty}^{\infty} \psi_k(x)(\delta 2^{-|k|})^{-n} \omega\left(\frac{x-y}{\delta 2^{-|k|}}\right).$$

Comparing these formulae with formula (1.2) we see that, similarly to the mollifiers A_δ, the mollifiers E_δ are linear integral operators, however, with a more sophisticated kernel $K(x, y, \delta)$ replacing $\delta^{-n} \omega\left(\frac{x-y}{\delta}\right)$.

The mollifier E_δ inherits the main properties of the mollifier A_δ, but there are some distinctions.

Lemma 11 *Let $\Omega \subset \mathbb{R}^n$ be an open set and $f \in L_1^{loc}(\Omega)$. Then $\forall \delta \in \left(0, \frac{1}{8}\right]$ $E_\delta f \in C^\infty(\Omega)$ and $\forall \alpha \in \mathbb{N}_0^n$*

$$D^\alpha(E_\delta f) = \sum_{k=-\infty}^{\infty} D^\alpha(\psi_k A_{\delta 2^{-|k|}} f) \quad \text{on} \quad \Omega. \tag{2.55}$$

Remark 19 In contrast to the mollifier E_δ we could state existence and infinite differentiability of $A_\delta f$ for $f \in L_1^{loc}(\Omega)$, in general, only on $\underline{\Omega_\delta}$.

Idea of the proof. The same as for Lemma 7. $\square$

Lemma 12 *Let $\Omega \subset \mathbb{R}^n$ be an open set and $f \in L_1^{loc}(\Omega)$. Then*

$$E_\delta f \to f \quad \text{a.e. on} \ \ \Omega \tag{2.56}$$

as $\delta \to 0+$.

Idea of the proof. Apply (2.54) and the corresponding property of the mollifier A_δ. $\square$

Lemma 13 *Let $\Omega \subset \mathbb{R}^n$ be an open set and $1 \leq p \leq \infty$. Then $\forall \delta \in \left(0, \frac{1}{8}\right]$ and $f \in L_p(\Omega)$*

$$\|E_\delta f\|_{L_p(\Omega)} \leq 2c_7 \, \|f\|_{L_p(\Omega)} \, , \tag{2.57}$$

where $c_7 = \|\omega\|_{L_1(\mathbb{R}^n)}$.

In order to prove this lemma we need the following two properties of L_p-spaces where $1 \leq p < \infty$.

1) If $\Omega \subset \mathbb{R}^n$ is a measurable set and $\forall x \in \Omega$ a finite or a denumerable sum $\sum\limits_k a_k(x)$ of functions a_k measurable on Ω contains no more than $\varkappa$ nonzero summands, in other words, if the multiplicity of the covering $\{\operatorname{supp} a_k\}$ does not exceed $\varkappa$, then

$$\Big\|\sum_k a_k\Big\|_{L_p(\Omega)} \leq \varkappa^{1-\frac{1}{p}} \left(\sum_k \|a_k\|_{L_p(\Omega)}^p \right)^{\frac{1}{p}}. \tag{2.58}$$

(This is a corollary of Hölder's inequality.)

2) If $\Omega = \bigcup\limits_k \Omega_k$ is either a finite or a denumerable union of measurable sets Ω_k and the multiplicity of the covering $\{\Omega_k\}$ does not exceed $\varkappa$, then for each function f measurable on Ω

$$\left(\sum_k \|f\|_{L_p(\Omega_k)}^p \right)^{\frac{1}{p}} \leq \varkappa^{\frac{1}{p}} \|f\|_{L_p(\Omega)}. \tag{2.59}$$

In particular, if $p = 1$ and $f \equiv 1$, then we have

$$\sum_k \operatorname{meas} \Omega_k \leq \varkappa \operatorname{meas} \Omega. \tag{2.60}$$

For $p = \infty$ these inequalities take the following form

$$\|\sum_k a_k\|_{L_\infty(\Omega)} \leq \varkappa \sup_k \|a_k\|_{L_\infty(\Omega)}, \quad \sup_k \|f\|_{L_\infty(\Omega_k)} = \|f\|_{L_\infty(\Omega)}.$$

Idea of the proof of Lemma 13. Apply (2.58), (1.7) and (2.59). $\square$
Proof of Lemma 13. By (2.7) and (2.15) $\forall \delta \in \left(0, \frac{1}{8}\right]$

$$\bigcup_{k=-\infty}^{\infty} \operatorname{supp} \psi_k = \bigcup_{k=-\infty}^{\infty} (\operatorname{supp} \psi_k)^{\delta 2^{-|k|}} = \Omega$$

and the multiplicities of the coverings $\{\operatorname{supp} \psi_k\}_{k \in \mathbb{Z}}$ and $\{(\operatorname{supp} \psi_k)^{\delta 2^{-|k|}}\}_{k \in \mathbb{Z}}$ are equal to 2 (see Remark 6). Therefore, by (2.58) and (2.59)

$$\|E_\delta f\|_{L_p(\Omega)} = \| \sum_{k=-\infty}^{\infty} \psi_k A_{\delta 2^{-|k|}} f\|_{L_p(\Omega)}$$

$$\leq 2^{1-\frac{1}{p}} \left(\sum_{k=-\infty}^{\infty} \|\psi_k A_{\delta 2^{-|k|}} f\|_{L_p(\Omega)}^p \right)^{\frac{1}{p}}$$

$$\leq 2^{1-\frac{1}{p}} \left(\sum_{k=-\infty}^{\infty} \|A_{\delta 2^{-|k|}} f\|_{L_p(\operatorname{supp} \psi_k)}^p \right)^{\frac{1}{p}}$$

$$\leq 2^{1-\frac{1}{p}} c_7 \left(\sum_{k=-\infty}^{\infty} \|f\|_{L_p((\operatorname{supp} \psi_k)^{\delta 2^{-|k|}})}^p \right)^{\frac{1}{p}} \leq 2c_7 \|f\|_{L_p(\Omega)}. \quad \square$$

Now for an open set $\Omega \subset \mathbb{R}^n$ and $\forall x \in \Omega$ we set $\varrho(x) = \operatorname{dist}(x, \partial\Omega)$ if $\Omega \neq \mathbb{R}^n$ and [10] $\varrho = (1 + |x|)^{-1}\}$ if $\Omega = \mathbb{R}^n$.

Lemma 14 *Let $\Omega \subset \mathbb{R}^n$ be an open set and $0 < \delta \leq \frac{1}{8}$. Then*
  1) *for $1 \leq p < \infty$ and $\forall f \in L_p(\Omega)$*

$$E_\delta f \to f \quad \text{in} \quad L_p(\Omega) \tag{2.61}$$

as $\delta \to 0+$,
  2) *for $p = \infty$ relation (2.61) holds $\forall f \in \overline{C}(\Omega)$,*

---

[10] It is also possible to consider $\varrho(x) = \min\{\operatorname{dist}(x, \partial\Omega), (1 + |x|)^{-1}\}$. However, in that case one must use a partition of unity constructed on the base of altered ϱ and verify that estimate (2.12) still holds.

3) *for $1 \le p \le \infty$, $l \in \mathbb{N}$ and $\forall f \in w_p^l(\Omega)$*

$$\|E_\delta f - f\|_{L_p(\Omega)} \le c_8 \delta^l \|f\|_{w_p^l(\Omega)}, \tag{2.62}$$

4) *for $1 \le p \le \infty$, $l \in \mathbb{N}$ and $\forall f \in w_p^l(\Omega)$*

$$\|(E_\delta f - f)\varrho^{-l}\|_{L_p(\Omega)} \le c_9 \delta^l \|f\|_{w_p^l(\Omega)}, \tag{2.63}$$

where c_8 and c_9 do not depend on f, δ, Ω and p.

Idea of the proof. To prove the statements 1) and 2) establish, by applying the proof of Lemma 13, that the series (2.50) converges in $L_p(\Omega)$ uniformly with respect to $\delta \in \left(0, \frac{1}{8}\right]$ and use the corresponding properties of the mollifiers A_δ. To prove (2.62) and (2.63) follow the proof of Lemma 13, applying inequality (2.48) with $\alpha = 0$ instead of (1.7). In the case of inequality (2.62) apply, in addition, the fact that there exist $B_1, B_2 > 0$ such that

$$B_1 2^{-k} \le \varrho(x) \le B_2 2^{-k} \quad \text{on supp } \psi_k \tag{2.64}$$

$\forall k \in \mathbb{Z}$ if $\Omega \ne \mathbb{R}^n$ and $\forall k \le 0$ if $\Omega = \mathbb{R}^n$. $\square$

Proof. We start with the proof of inequality (2.62). If $f \in w_p^l(\Omega)$, then using, in addition, Minkowski's inequality for sums we find

$$\|E_\delta f - f\|_{L_p(\Omega)} = \left\| \sum_{k=-\infty}^{\infty} \psi_k (A_{\delta 2^{-|k|}} f - f) \right\|_{L_p(\Omega)}$$

$$\le 2^{1-\frac{1}{p}} \left(\sum_{k=-\infty}^{\infty} \|A_{\delta 2^{-|k|}} f - f\|_{L_p(\text{supp } \psi_k)}^p \right)^{\frac{1}{p}}$$

$$\le 2^{1-\frac{1}{p}} c_5 \delta^l \left(\sum_{k=-\infty}^{\infty} 2^{-|k|lp} \|f\|_{w_p^l((\text{supp } \psi_k)^{\delta 2^{-|k|}})}^p \right)^{\frac{1}{p}}$$

$$\le 2^{1-\frac{1}{p}} c_5 \delta^l \left(\sum_{k=-\infty}^{\infty} \left(\sum_{|\alpha|=l} \|D_w^\alpha f\|_{L_p((\text{supp } \psi_k)^{\delta 2^{-|k|}})} \right)^p \right)^{\frac{1}{p}}$$

$$\le 2^{1-\frac{1}{p}} c_5 \delta^l \sum_{|\alpha|=l} \left(\sum_{k=-\infty}^{\infty} \|D_w^\alpha f\|_{L_p((\text{supp } \psi_k)^{\delta 2^{-|k|}})}^p \right)^{\frac{1}{p}}$$

$$\leq 2c_5\delta^l \sum_{|\alpha|=l} \|D_w^\alpha f\|_{L_p(\Omega)} = c_8\delta^l \|f\|_{w_p^l(\Omega)},$$

because the multiplicity of the covering $\{(\operatorname{supp}\psi_k)^{\delta 2^{-|k|}}\}_{k\in\mathbb{Z}}$ is equal to 2 (see Remark 6).

In the case of inequality (2.63) we find with the help of (2.64) that

$$\|(E_\delta f - f)\varrho^{-l}\|_{L_p(\Omega)} = \|\sum_{k=-\infty}^{\infty} \varrho^{-l}\psi_k(A_{\delta 2^{-|k|}}f - f)\|_{L_p(\Omega)}$$

$$\leq 2^{1-\frac{1}{p}}B_1^{-l}\left(\sum_{k=-\infty}^{\infty} 2^{|k|lp}\|A_{\delta 2^{-|k|}}f - f\|_{L_p(\operatorname{supp}\psi_k)}^p\right)^{\frac{1}{p}}$$

$$\leq 2^{1-\frac{1}{p}}B_1^{-l}c_5\delta^l\left(\sum_{k=-\infty}^{\infty} \|f\|_{w_p^l((\operatorname{supp}\psi_k)^{\delta 2^{-|k|}})}^p\right)^{\frac{1}{p}}.$$

The rest is the same as above ($c_9 = 2B_1^{-l}c_5$). $\square$

Lemma 15 *Let $\Omega \subset \mathbb{R}^n$ be an open set, $l \in \mathbb{N}$ and $0 < \delta \leq \frac{1}{8}$. Then for each polynomial p_{l-1} of degree less than or equal to $l-1$*

$$(E_\delta p_{l-1})(x) = p_{l-1}(x), \quad x \in \Omega.$$

Idea of the proof. Apply multidimensional Taylor's formula (see Section 3.3) to $p_{l-1}(x - \delta 2^{-|k|}z)$ in (2.50) and use (2.45), (1.1) and (2.11) or (2.14). $\square$

Remark 20 In Lemmas $13 - 14$ the property (2.45) of the kernel of mollification was not applied. It was applied in Lemma 15, but this lemma will not be used in the sequel. The main and the only reason for introducing this property is connected with the estimates of norms of commutators $[D_w^\alpha, E_\delta]f$, which will be given in Lemma 20 below. In its turn these estimates are based on Lemmas $9-10$, in which the mollifiers A_δ with kernels satisfying the property (2.45) were studied.

Let us denote the commutator of the weak differentiation of first order and the mollifier E_δ in the following way:

$$\left(\frac{\tilde{\partial}}{\partial x_j}\right)_w (E_\delta) \equiv \left[\left(\frac{\partial}{\partial x_j}\right)_w, E_\delta\right] \equiv \left(\frac{\partial}{\partial x_j}\right)_w E_\delta - E_\delta \left(\frac{\partial}{\partial x_j}\right)_w.$$

This operator is defined on $(W_1^1)^{loc}(\Omega)$.

Furthermore, for $l \in \mathbb{N}$, $l \geq 2$, we define the operators $\widetilde{D}_w^\alpha(E_\delta)$, where $\alpha \in \mathbb{N}_0^n$ and $|\alpha| = l$, with the domain $(W_1^l)^{loc}(\Omega)$:

$$\widetilde{D}_w^\alpha(E_\delta) \equiv \left(\left(\frac{\widetilde{\partial}}{\partial x_1} \right)_w^{\alpha_1} \cdots \left(\frac{\widetilde{\partial}}{\partial x_n} \right)_w^{\alpha_n} \right) (E_\delta).$$

Lemma 16 *Let $\Omega \subset \mathbb{R}^n$ be an open set, $l \in \mathbb{N}$, $0 < \delta \leq \frac{1}{8}$ and $f \in (W_1^l)^{loc}(\Omega)$. Then $\forall \alpha \in \mathbb{N}_0^n$ satisfying $|\alpha| \leq l$*

$$\left(\widetilde{D}_w^\alpha(E_\delta) \right) f = \sum_{k=-\infty}^{\infty} (D^\alpha \psi_k) A_{\delta 2^{-|k|}} f.$$

Idea of the proof. Induction. For $|\alpha| = 1$ by (2.55)

$$\left(\left(\frac{\widetilde{\partial}}{\partial x_j} \right)_w (E_\delta) \right) f = \sum_{k=-\infty}^{\infty} \left(\frac{\partial \psi_k}{\partial x_j} A_{\delta 2^{-|k|}} f + \psi_k \frac{\partial}{\partial x_j} (A_{\delta 2^{-|k|}} f) \right)$$

$$- \sum_{k=-\infty}^{\infty} \psi_k A_{\delta 2^{-|k|}} \left(\frac{\partial f}{\partial x_j} \right)_w = \sum_{k=-\infty}^{\infty} \frac{\partial \psi_k}{\partial x_j} A_{\delta 2^{-|k|}} f$$

on Ω, because by (2.7) and (1.19) $\psi_k \frac{\partial}{\partial x_j} (A_{\delta 2^{-|k|}} f) = \psi_k A_{\delta 2^{-|k|}} \left(\frac{\partial f}{\partial x_j} \right)_w$ on Ω. $\square$

Remark 21 For the mollifiers A_δ we have $(\widetilde{D}_w^\alpha(A_\delta)) f \equiv 0$ but, only on $\underline{\Omega_\delta}$ (see Section 1.5), while for the mollifiers E_δ in general $(\widetilde{D}_w^\alpha(E_\delta)) f \not\equiv 0$ even on $\underline{\Omega_\delta}$, but on the whole of Ω the quantity $(\widetilde{D}_w^\alpha(E_\delta)) f$ is in some sense small (because $\sum_{k=-\infty}^{\infty} D^\alpha \psi_k = 0$ on Ω) and, as we shall see below, it tends to 0 in $L_p(\Omega)$ fast enough under appropriate assumptions on f.

Remark 22 On the base of Lemma 16 we define for $\forall \alpha \in \mathbb{N}_0^n$ satisfying $\alpha \neq 0$ the operator $E_\delta^{(\alpha)}$ with the domain $L_1^{loc}(\Omega)$ directly by the equality

$$E_\delta^{(\alpha)} f \equiv \sum_{k=-\infty}^{\infty} (D^\alpha \psi_k) A_{\delta 2^{-|k|}} f. \tag{2.65}$$

Lemma 17 *Let $\Omega \subset \mathbb{R}^n$ be an open set and $0 < \delta \leq \frac{1}{8}$. Then $\forall \alpha \in \mathbb{N}_0^n$ satisfying $\alpha \neq 0$ and $\forall f \in L_1^{loc}(\Omega)$*

$$E_\delta^{(\alpha)} f \to 0 \quad a.e. \ on \ \Omega. \tag{2.66}$$

Idea of the proof. Since $\forall m \in \mathbb{Z}$ we have $E_\delta^{(\alpha)} f = \sum\limits_{k=m-1}^{m+1} D^\alpha \psi_k A_{\delta 2^{-|k|}} f$ and $\sum\limits_{k=m-1}^{m+1} D^\alpha \psi_k = 0$ on G_m, the relation (2.66) follows from (1.5). $\square$

Lemma 18 *Let $\Omega \subset \mathbb{R}^n$ be an open set, $1 \le p \le \infty$ and $0 < \delta \le \frac{1}{8}$. Then*
1) $\forall f \in w_p^l(\Omega)$ *and* $\forall \alpha \in \mathbb{N}_0^n$ *satisfying* $0 < |\alpha| \le l$

$$\|E_\delta^{(\alpha)} f\|_{L_p(\Omega)} \le c_{10}\, \delta^l \|f\|_{w_p^l(\Omega)}, \tag{2.67}$$

2) $\forall f \in w_p^l(\Omega)$ *and* $\forall \alpha \in \mathbb{N}_0^n$ *satisfying* $|\alpha| > 0$

$$\|(E_\delta^{(\alpha)} f)\varrho^{|\alpha|-l}\|_{L_p(\Omega)} \le c_{11}\delta^l \|f\|_{w_p^l(\Omega)}, \tag{2.68}$$

where $c_{10}, c_{11} > 0$ do not depend on f, δ, Ω and p.

Idea of the proof. Starting for $\alpha \ne 0$ from the equality

$$E_\delta^{(\alpha)} f = \sum_{k=-\infty}^{\infty} D^\alpha \psi_k (A_{\delta 2^{-|k|}} f - f) \tag{2.69}$$

follow the proof of Lemma 13, apply estimate (2.12) of Lemmas 5 and 6 and inequality (2.48), in which $\alpha = 0, G = \operatorname{supp} \psi_k$ and δ is replaced by $\delta 2^{-|k|}$. In the case of inequality (2.68) apply, in addition, (2.63). $\square$

Proof. For $|\alpha| \le l$ from (2.69) it follows that

$$\|E_\delta^{(\alpha)} f\|_{L_p(\Omega)} \le 2^{1-\frac{1}{p}} \left(\sum_{k=-\infty}^{\infty} \|D^\alpha \psi_k (A_{\delta 2^{-|k|}} f - f)\|_{L_p(\Omega)}^p \right)^{\frac{1}{p}}$$

$$\le 2^{1-\frac{1}{p}} c_\alpha \left(\sum_{k=-\infty}^{\infty} 2^{k|\alpha|p} \|A_{\delta 2^{-|k|}} f - f\|_{L_p(\operatorname{supp} \psi_k)}^p \right)^{\frac{1}{p}}$$

$$\le 2^{1-\frac{1}{p}} c_\alpha c_5 \delta^l \left(\sum_{k=-\infty}^{\infty} 2^{|k|(|\alpha|-l)p} \|f\|_{w_p^l((\operatorname{supp} \psi_k)^{\delta 2^{-|k|}})}^p \right)^{\frac{1}{p}}$$

$$\le 2^{1-\frac{1}{p}} c_\alpha c_5 \delta^l \left(\sum_{k=-\infty}^{\infty} \|f\|_{w_p^l((\operatorname{supp} \psi_k)^{\delta 2^{-|k|}})}^p \right)^{\frac{1}{p}}$$

$$\leq 2c_\alpha c_5 \delta^l \|f\|_{w_p^l(\Omega)} = c_{10} \delta^l \|f\|_{w_p^l(\Omega)}.$$

We have taken into account that the multiplicity of the covering $\{(\operatorname{supp}\psi_k)^{\delta 2^{-|k|}}\}_{k\in\mathbb{Z}}$ is equal to 2 (see Remark 6).

In the case of inequality (2.68)

$$\|(E_\delta^{(\alpha)} f)\varrho^{|\alpha|-l}\|_{L_p(\Omega)} \leq 2^{1-\frac{1}{p}} \left(\sum_{k=-\infty}^{\infty} \|\varrho^{|\alpha|-l} D^\alpha \psi_k (A_{\delta 2^{-|k|}} f - f)\|_{L_p(\Omega)}^p \right)^{\frac{1}{p}}$$

$$\leq 2^{1-\frac{1}{p}} \left(\max\left\{ B_1^{-1}, B_2 \right\} \right)^{|\alpha|-l|} c_\alpha \left(\sum_{k=-\infty}^{\infty} 2^{|k|lp} \|A_{\delta 2^{-|k|}} f - f\|_{L_p(\operatorname{supp}\psi_k)}^p \right)^{\frac{1}{p}}.$$

The rest is the same as above $\left(c_{11} = 2c_\alpha c_5 \left(\max\left\{ B_1^{-1}, B_2 \right\} \right)^{|\alpha|-l|} \right)$. $\square$

Lemma 19 *Let $\Omega \subset \mathbb{R}^n$ be an open set, $l \in \mathbb{N}$ and $f \in \left(W_1^l \right)^{loc}(\Omega)$. Then $\forall \alpha \in \mathbb{N}_0^n$ satisfying $0 < |\alpha| \leq l$*

$$D^\alpha(E_\delta f) = \sum_{0 \leq \beta \leq \alpha} \binom{\alpha}{\beta} E_\delta^{(\alpha-\beta)}(D_w^\beta f) \tag{2.70}$$

and

$$[D_w^\alpha, E_\delta] f = \sum_{0 \leq \beta \leq \alpha,\, \beta \neq \alpha} \binom{\alpha}{\beta} E_\delta^{(\alpha-\beta)}(D_w^\beta f). \tag{2.71}$$

Idea of the proof. Apply (2.55), Leibnitz' formula, Lemma 4 of Chapter 1 and the definition of the operator $E_\delta^{(\gamma)}$ (see Remark 22). $\square$

In the sequel we shall estimate $D^\alpha(E_\delta f)$ and $D^\alpha(E_\delta f) - D_w^\alpha f$ with the help of (2.71) and the following obvious identities:

$$D^\alpha(E_\delta f) = [D_w^\alpha, E_\delta] f + E_\delta(D_w^\alpha f) \tag{2.72}$$

and

$$D^\alpha(E_\delta f) - D_w^\alpha f = [D_w^\alpha, E_\delta] f + E_\delta(D_w^\alpha f) - D_w^\alpha f. \tag{2.73}$$

Lemma 20 *Let $\Omega \subset \mathbb{R}^n$ be an open set, $1 \leq p \leq \infty$ and $0 < \delta \leq \frac{1}{8}$. Then $\forall f \in w_p^l(\Omega)$:*
   1) $\forall \alpha \in \mathbb{N}_0^n$ *satisfying* $0 < |\alpha| \leq l$

$$\|[D_w^\alpha, E_\delta] f\|_{L_p(\Omega)} \leq c_{12} \delta^{l-|\alpha|+1} \|f\|_{w_p^l(\Omega)}, \tag{2.74}$$

2) $\forall \alpha \in \mathbb{N}_0^n$ *satisfying* $|\alpha| > 0$

$$\|([D_w^\alpha, E_\delta]f)\varrho^{|\alpha|-l}\|_{L_p(\Omega)} \leq c_{13}\delta^{l-|\alpha|+1}\|f\|_{w_p^l(\Omega)}, \tag{2.75}$$

where $c_{12}, c_{13} > 0$ *do not depend on* f, δ, Ω *and* p.

Idea of the proof. Starting from equality (2.71) apply inequalities (2.67), respectively (2.68), with $l - |\beta|$ replacing l, $\alpha - \beta$ replacing α and $D_w^\beta f$ replacing f. Take into consideration that $\varrho^{|\alpha|-l} = \varrho^{|\alpha-\beta|-(l-|\beta|)}$ and $|\beta| \leq |\alpha| - 1$. $\square$

2.6 The best possible approximation with preservation of boundary values

We start by studying some properties of the mollifiers E_δ in Sobolev spaces.

Theorem 5 *Let* $\Omega \subset \mathbb{R}^n$ *be an open set,* $l \in \mathbb{N}$, $0 < \delta \leq \frac{1}{8}$ *and* $1 \leq p \leq \infty$. *Then* $\forall f \in w_p^l(\Omega)$

$$\|E_\delta f\|_{w_p^l(\Omega)} \leq c_{14}\|f\|_{w_p^l(\Omega)} \tag{2.76}$$

and $\forall f \in W_p^l(\Omega)$

$$\|E_\delta f\|_{W_p^l(\Omega)} \leq c_{14}\|f\|_{W_p^l(\Omega)}, \tag{2.77}$$

where $c_{14} > 0$ *does not depend on* f, δ, Ω *and* p.

Idea of the proof. Apply (2.71) and Lemmas 13 and 20. $\square$
Proof. By (2.72), (2.74) and (2.57)

$$\|E_\delta f\|_{w_p^l(\Omega)} = \sum_{|\alpha|=l} \|D_w^\alpha E_\delta f\|_{L_p(\Omega)}$$

$$\leq \sum_{|\alpha|=l} \|[D_w^\alpha, E_\delta]f\|_{L_p(\Omega)} + \|E_\delta D_w^\alpha f\|_{L_p(\Omega)}$$

$$\leq \sum_{|\alpha|=l} (c_{12}\|f\|_{w_p^l(\Omega)} + 2c_7\|D_w^\alpha f\|_{L_p(\Omega)}) = c_{14}\|f\|_{w_p^l(\Omega)}.$$

Inequality (2.77) follows from (2.57) and (2.76). $\square$

Theorem 6 *Let* $\Omega \subset \mathbb{R}^n$ *be an open set,* $l \in \mathbb{N}$ *and* $0 < \delta \leq \frac{1}{8}$.
1) *If* $1 \leq p \leq \infty$, *then* $\forall f \in W_p^l(\Omega)$ *and* $\forall \alpha \in \mathbb{N}_0^n$ *satisfying* $|\alpha| \leq l$

$$D^\alpha(E_\delta f) \to D_w^\alpha f \quad \text{a.e. on} \quad \Omega \tag{2.78}$$

as $\delta \to 0+$.

2) *If $1 \leq p < \infty$, then $\forall f \in W_p^l(\Omega)$*

$$E_\delta f \to f \quad \text{in} \quad W_p^m(\Omega), \quad m = 0, ..., l, \tag{2.79}$$

as $\delta \to 0+$.

3) *If $p = \infty$, then $\forall f \in W_\infty^l(\Omega)$*

$$E_\delta f \to f \quad \text{in} \quad W_\infty^m(\Omega), \quad m = 0, ..., l-1, \tag{2.80}$$

as $\delta \to 0+$ (if $f \in \overline{C}^l(\Omega)$ then (2.79) holds).

Idea of the proof. Relation (2.78) follows from equalities (2.72), (2.71) and Lemmas 14 and 20; relations (2.79) and (2.80) follow from (2.72) and Lemmas 14 and 20. $\square$

Proof. Let us prove (2.79). From (2.72), (2.74), (2.62) and, in the case $m = l$, (2.61) it follows that

$$\|E_\delta f - f\|_{W_p^m(\Omega)} = \|E_\delta f - f\|_{L_p(\Omega)} + \sum_{|\alpha|=m} \|D^\alpha E_\delta f - D_w^\alpha f\|_{L_p(\Omega)}$$

$$\leq \|E_\delta f - f\|_{L_p(\Omega)} + \sum_{|\alpha|=m} (\|[D_w^\alpha, E_\delta]f\|_{L_p(\Omega)} + \|E_\delta(D_w^\alpha f) - D_w^\alpha f\|_{L_p(\Omega)}) \to 0$$

as $\delta \to 0+$.

The same argument works to prove (2.80). Since in this case $m < l$, it is enough to apply only inequalities (2.74) and (2.62). $\square$

Theorem 7 *Let $\Omega \subset \mathbb{R}^n$ be an open set, $l \in \mathbb{N}$, $1 \leq p \leq \infty$, $0 < \delta \leq \frac{1}{8}$ and $\alpha \in \mathbb{N}_0^n$.*

1) *If $|\alpha| \leq l$, then $\forall f \in w_p^l(\Omega)$*

$$\|(D^\alpha(E_\delta f) - D_w^\alpha f)\varrho^{|\alpha|-l}\|_{L_p(\Omega)} \leq c_{15}\delta^{l-|\alpha|}\|f\|_{w_p^l(\Omega)}, \tag{2.81}$$

where $c_{15} > 0$ does not depend on f, δ, Ω and p.

2) *If $|\alpha| > l$, then $\forall f \in w_p^l(\Omega)$*

$$\|(D^\alpha(E_\delta f))\varrho^{|\alpha|-l}\|_{L_p(\Omega)} \leq c_{16}\delta^{l-|\alpha|}\|f\|_{w_p^l(\Omega)}, \tag{2.82}$$

where $c_{16} > 0$ does not depend on f, δ, Ω and p.

3) *There exists an open set Ω such that for any $\varepsilon > 0$ inequality (2.82) with $\varrho^{|\alpha|-l-\varepsilon}$ replacing $\varrho^{|\alpha|-l}$ does not hold.*

Idea of the proof. Inequality (2.81) follows from equality (2.73) and the inequalities (2.74) and (2.62) with $D_w^\alpha f$ replacing f and $l - |\alpha|$ replacing l. Inequality (2.82) follows from equality (2.72), inequality (2.75) and the inequality

$$\|(E_\delta(D_w^\alpha f))\varrho^{|\alpha|-l}\|_{L_p(\Omega)} \leq c_{17}\delta^{l-|\alpha|}\|f\|_{w_p^l(\Omega)} \tag{2.83}$$

for $|\alpha| > l$, where $c_{17} > 0$ does not depend on f, δ, Ω and p. In order to prove (2.83) apply the proof of inequality (2.63). The third statement will be considered in the proof of the Theorem 8 below. $\square$

Proof. It is enough to prove (2.83). Applying Lemma 4 of Chapter 1 and the inequalities (2.63), (2.58), (2.49) and (2.59) we establish that

$$\|(E_\delta(D_w^\alpha f))\varrho^{|\alpha|-l}\|_{L_p(\Omega)} = \left\| \sum_{k=-\infty}^{\infty} \varrho^{|\alpha|-l}\psi_k D^\alpha A_{\delta 2^{-|k|}} f \right\|_{L_p(\Omega)}$$

$$\leq 2^{1-\frac{1}{p}} B_2^{|\alpha|-l} \left(\sum_{k=-\infty}^{\infty} 2^{|k|(|\alpha|-l)p} \|D^\alpha A_{\delta 2^{-|k|}} f\|_{L_p(\operatorname{supp}\psi_k)}^p \right)^{\frac{1}{p}}$$

$$\leq 2^{1-\frac{1}{p}} B_2^{|\alpha|-l} c_6 \delta^{l-|\alpha|} \left(\sum_{k=-\infty}^{\infty} \|f\|_{w_p^l((\operatorname{supp}\psi_k)^{\delta 2^{-|k|}})}^p \right)^{\frac{1}{p}}$$

$$\leq 2 B_2^{|\alpha|-l} c_6 \delta^{l-|\alpha|} \|f\|_{w_p^l(\Omega)} = c_{17}\delta^{l-|\alpha|}\|f\|_{w_p^l(\Omega)}.$$

(For details see the proof of Lemma 14.) $\square$

Theorem 8 I. *Let $\Omega \subset \mathbb{R}^n$ be an open set, $l \in \mathbb{N}$ and $1 \leq p \leq \infty$. Then $\forall f \in W_p^l(\Omega)$ functions $\varphi_s \in C^\infty(\Omega) \cap W_p^l(\Omega)$, $s \in \mathbb{N}$, exist, which depend linearly on f and satisfy the following properties:*
   1) *for $1 \leq p \leq \infty$*

$$D^\alpha \varphi_s \to D_w^\alpha f \quad \text{a.e. on } \Omega, \quad |\alpha| \leq l,$$

as $s \to \infty$,
   2) *for $1 \leq p < \infty$*

$$\varphi_s \to f \quad \text{in } W_p^m(\Omega), \quad m = 0, ..., l, \tag{2.84}$$

as $s \to \infty$,
   3) *for $p = \infty$*

$$\varphi_s \to f \quad \text{in } W_\infty^m(\Omega), \quad m = 0, ..., l-1, \tag{2.85}$$

as $s \to \infty$ (if $f \in \overline{C}^l(\Omega)$, then relation (2.84) also holds),
   4) for $1 \le p \le \infty$

$$\|(D_w^\alpha f - D^\alpha \varphi_s)\varrho^{|\alpha|-l}\|_{L_p(\Omega)} \to 0, \quad |\alpha| < l, \tag{2.86}$$

as $s \to \infty$,
   5) for $1 \le p \le \infty$

$$\|D^\alpha \varphi_s \varrho^{|\alpha|-l}\|_{L_p(\Omega)} \le c_{\alpha,s}\|f\|_{W_p^l(\Omega)}, \quad |\alpha| > l, \tag{2.87}$$

where $c_{\alpha,s}$ are independent of f, Ω and p.
   II. *There exists an open set $\Omega \subset \mathbb{R}^n$, for which, given $\varepsilon > 0$ and $m > l$, a function $f \in W_p^l(\Omega)$ exists such that, whatever are the functions $\varphi_s \in C^\infty(\Omega) \cap W_p^l(\Omega)$, $s \in \mathbb{N}$, satisfying property 4), for some $\forall \alpha \in \mathbb{N}_0^n$ satisfying $|\alpha| = m$*

$$\|D^\alpha \varphi_s \varrho^{|\alpha|-l-\varepsilon}\|_{L_p(\Omega)} = \infty. \tag{2.88}$$

Idea of the proof. The first part of Theorem 8 is an obvious corollary of Theorems 6 and 7: it is enough to take $\varphi_s = E_{\frac{1}{s}} f$. The second part will be proved in Remark 14 of Chapter 5. $\square$

Remark 23 The second part of Theorem 4 is about the sharpness of condition (2.87). We note that since in (2.87) $\varrho(y)^{|\alpha|-l} \to 0$ as y approaches the boundary $\partial\Omega$, the derivatives $D^\alpha \varphi_s(y)$ where $|\alpha| > l$ *can tend to infinity* as y approaches $\partial\Omega$. By the second part of Theorem 8 for some $\Omega \subset \mathbb{R}^n$ and $f \in W_p^l(\Omega)$ for any appropriate choice of φ_s some of the derivatives $D^\alpha \varphi_s(y)$ where $|\alpha| = m > l$ *do tend to infinity* as y approaches a certain point $x \in \partial\Omega$. Indeed, for bounded Ω from (2.88) it follows that for some $\forall \alpha \in \mathbb{N}_0^n$ satisfying $|\alpha| = m$, for some $x \in \partial\Omega$ and for some $y_k \in \Omega$ such that $y_k \to x$ as $k \to \infty$

$$\lim_{k \to \infty} (D^\alpha \varphi_s)(y_k)\varrho(y_k)^{|\alpha|-l-\varepsilon} = \infty, \tag{2.89}$$

i.e., $(D^\alpha \varphi_s)(y_k)$ tends to infinity faster, than $\varrho^{l-|\alpha|-\varepsilon}(y_k)$. (We note that the higher order of a derivative is the faster is its growth to infinity.)

Remark 24 This reveals validity of the following general fact: if one wants to have "good" approximation by C^∞-functions, in the sense that the boundary values are preserved, then there must be some "penalty" for this higher quality. This "penalty" is the growth of the derivatives of higher order of the approximating functions when approaching the boundary. The "minimal penalty" is given by inequality (2.87).

Remark 25 By Theorems 6 and 7 the functions $\varphi_s = E_{\frac{1}{s}} f$ satisfy the statements of the first part of Theorem 8. Thus, by the statement of the second part of this theorem the mollifier E_δ is the best possible approximation operator, preserving boundary values, in the sense that the derivatives of higher orders of $E_\delta f$ have the minimal possible growth on approaching $\partial\Omega$.

Now we formulate the following corollary of Theorem 8 for open sets with sufficiently smooth boundary, in which the preservation of boundary values takes a more explicit form.

Theorem 9 *Let $l \in \mathbb{N}, 1 \leq p < \infty$ and let $\Omega \subset \mathbb{R}^n$ be an open set with a C^l-boundary (see definition in Section 4.3).*

I. For each $f \in W_p^l(\Omega)$ functions $\varphi_s \in C^\infty(\Omega)$, $s \in \mathbb{N}$, exist, which depend linearly on f and are such that

1) $\varphi_s \to f$ *in* $W_p^l(\Omega)$ *as* $s \to \infty$,

2) $D^\alpha\varphi_s\big|_{\partial\Omega} = D_w^\alpha f\big|_{\partial\Omega}$, $|\alpha| \leq l - 1$,

3) $\|D^\alpha\varphi_s \varrho^{|\alpha|-l}\|_{L_p(\Omega)} < \infty$, $|\alpha| > l$.

II. Given $\varepsilon > 0$ and $m > l$, a function $f \in W_p^l(\Omega)$ exists such that , whatever are the functions $\varphi_s \in C^\infty(\Omega), s \in \mathbb{N}$, satisfying 1) and 2), for some $\forall\alpha \in \mathbb{N}_0^n$ satisfying $|\alpha| = m$

$$\|D^\alpha\varphi_s \varrho^{|\alpha|-l-\varepsilon}\|_{L_p(\Omega)} = \infty. \tag{2.90}$$

Idea of the proof. As in the proof of Corollary 3, by Lemma 13 of Chapter 5, propety 2) follows from relation (2.86). $\square$

The most direct application of Theorem 7, for the case in which $p = \infty$, is a construction of the so-called regularized distance. We note that for an open set $\Omega \subset \mathbb{R}^n$, $\Omega \neq \mathbb{R}^n$, the ordinary distance $\varrho(x) = \text{dist}\,(x, \partial\Omega), x \in \Omega$, satisfies Lipschitz condition with constant equal to 1:

$$|\varrho(x) - \varrho(y)| \leq |x - y|, \quad x, y \in \Omega. \tag{2.91}$$

(This is a consequence of the triangle inequality.) Hence, by Lemma 8 of Chapter 1

$$\varrho \in w_\infty^1(\Omega), \quad |\nabla\varrho| \leq 1 \quad \text{a.e. on} \quad \Omega. \tag{2.92}$$

The simplest examples show (for instance, $\varrho(x) = 1 - |x|$) for $\Omega = (-1, 1) \subset \mathbb{R}$) that in general the function ϱ does not possess any higher degree of smoothness than follows from (2.91) and (2.92).

Theorem 10 *Let $\Omega \subset \mathbb{R}^n$ be an open set, $\Omega \neq \mathbb{R}^n$. Then $\forall \delta \in (0,1)$ a function $\varrho_\delta \in C^\infty(\Omega)$ (a regularized distance) exists, which is such that*

$$(1 - \delta)\varrho(x) \leq \varrho_\delta(x) \leq \varrho(x), \quad x \in \Omega, \tag{2.93}$$

$$|\varrho_\delta(x) - \varrho_\delta(y)| \leq |x - y|, \quad x, y \in \Omega, \tag{2.94}$$

$$|\nabla \varrho_\delta(x)| \leq 1 \quad \text{on} \quad \Omega \tag{2.95}$$

and $\forall \alpha \in \mathbb{N}_0^n$ satisfying $|\alpha| \geq 2$ and $\forall x \in \Omega$

$$|(D^\alpha \varrho_\delta)(x)| \leq c_\alpha \delta^{1-|\alpha|} \varrho(x)^{1-|\alpha|}, \tag{2.96}$$

where c_α depends only on α.

Idea of the proof. In order to construct the regularized distance it is natural to regularize, i.e., to mollify, the ordinary distance. Of course, one needs to apply mollifiers with variable step. Set $\varrho_\delta = a E_{b\delta} \varrho$ and choose appropriate $a, b > 0$. Here $E_{b\delta}$ is a mollifier defined by (2.50) where $l = 1$ and the kernel of mollification ω is nonnegative. $\square$

Proof. First let $\Delta_\delta = E_\delta \varrho$. Since $\varrho \in w_\infty^1(\Omega)$, from (2.81) and footnote 4 on the page 12 it follows that

$$\sup_{x \in \Omega} |\Delta_\delta(x) - \varrho(x)| \varrho(x)^{-1} \leq c_{15}\delta$$

or $\forall x \in \Omega$

$$(1 - c_{15}\delta)\varrho(x) \leq \Delta_\delta(x) \leq (1 + c_{15}\delta)\varrho(x),$$

where $c_{15} > 0$ depends only on n.

Moreover, from (2.82) it follows that $\forall \alpha \in \mathbb{N}_0^n$ satisfying $\alpha \neq 0$

$$\sup_{x \in \Omega} |D^\alpha \Delta_\delta(x)| \varrho(x)^{|\alpha|-1} \leq c_{16}\delta^{1-|\alpha|}$$

or $\forall x \in \Omega$

$$|D^\alpha \Delta_\delta(x)| \leq c_{16}\delta^{1-|\alpha|} \varrho(x)^{1-|\alpha|},$$

where $c_{16} > 0$ depends only on n and α.

Furthermore, by definition of E_δ and by (2.11) or (2.14)

$$\Delta_\delta(x) - \Delta_\delta(y) = \sum_{k=-\infty}^{\infty} \left(\psi_k(x)(A_{\delta 2^{-|k|}}\varrho)(x) - \psi_k(y)(A_{\delta 2^{-|k|}}\varrho)(y) \right)$$

$$= \sum_{k=-\infty}^{\infty} \psi_k(x)((A_{\delta 2^{-|k|}}\varrho)(x) - (A_{\delta 2^{-|k|}}\varrho)(y))$$

$$+ \sum_{k=-\infty}^{\infty} (\psi_k(x) - \psi_k(y))((A_{\delta 2^{-|k|}}\varrho)(y) - \varrho(y)).$$

Hence,

$$|\Delta_\delta(x) - \Delta_\delta(y)| \le \sum_{k=-\infty}^{\infty} \psi_k(x)|(A_{\delta 2^{-|k|}}\varrho)(x) - (A_{\delta 2^{-|k|}}\varrho)(y)|$$

$$+ \sum_{k\in S(x,y)} |\psi_k(x) - \psi_k(y)| \int_{B(0,1)} |\varrho(y - \delta 2^{-|k|}z) - \varrho(y)|\,\omega(z)\,dz.$$

Here by (2.53) $S(x,y) = \{s(x) - 1, s(x), s(x) + 1, s(y) - 1, s(y), s(y) + 1\}$. From (2.12) it follows that

$$|\psi_k(x) - \psi_k(y)| = \left| \sum_{j=1}^{n} (x_j - y_j) \int_0^1 \frac{\partial \psi_k}{\partial x_j}(x + t(y - x))\,dt \right| \le c_{18} 2^k |x - y|,$$

where $c_{18} = \left(\sum_{|\alpha|=1} c_\alpha^2 \right)^{1/2}$ with c_α from (2.12) depends only on n. Now, applying (1.13), (2.11) or (2.14), and (2.91) we have

$$|\Delta_\delta(x) - \Delta_\delta(y)| \le |x - y| \left(1 + c_{18} \sum_{k\in S(x,y)} 2^k(\delta 2^{-|k|}) \int_{B(0,1)} |z|\,\omega(z)\,dz \right)$$

$$\le (1 + 6c_{18}\delta)|x - y|.$$

Finally, it is enough to set $\varrho_\delta = aE_{b\delta}\varrho$, where, for instance, $a = \left(1 + \frac{\delta}{2}\right)^{-1}$ and $b = \frac{1}{2}\min\{c_{15}^{-1}, (6c_{18})^{-1}\}$. $\square$

Remark 26 The regularized distance can be applied to the construction of linear mollifiers with variable step. It is quite natural to replace the constant step δ in the definition of the mollifiers A_δ by the variable step $\delta\varrho(x)$, i.e., to consider the mollifiers

$$(H_\delta f)(x) = (A_{\delta\varrho(x)}f)(x) = \int_{B(0,1)} f(x - \delta\varrho(x)z)\,\omega(z)\,dz$$

for $0 < \delta < 1$. (In this case $B(x, \delta\varrho(x)) \subset \Omega$ for each $x \in \Omega$ and, therefore, the function f is defined at the point $x - \delta\varrho(x)$.) If $\varrho \in C^\infty(\Omega)$, it can be proved that $H_\delta f \in C^\infty(\Omega)$ for $f \in L_1^{loc}(\Omega)$ and that $H_\delta f \to f$ a.e. on Ω. This is so, for instance, for $\Omega = \mathbb{R}^n \setminus \mathbb{R}^m$, $1 \leq m < n$, in which case $\varrho(x) = (x_{m+1}^2 + ... + x_n^2)^{1/2}$. However, as it was pointed out above "usually" $\varrho \bar{\in} C^\infty(\Omega)$. This drawback can be removed by replacing the ordinary distance ϱ by the regularized distance $\widetilde{\varrho} = \varrho_{\delta_0}$ with some fixed $0 < \delta_0 < 1$ (say, $\delta_0 = \frac{1}{2}$). We set

$$(\widetilde{H}_\delta f)(x) = (A_{\delta\widetilde{\varrho}(x)} f)(x) = \int\limits_{B(0,1)} f(x - \delta\widetilde{\varrho}(x)z)\,\omega(z)\,dz.$$

Then $\forall f \in L_1^{loc}(\Omega)$ we have $\widetilde{H}_\delta f \in C^\infty(\Omega)$ and $\widetilde{H}_\delta f \to f$ a.e. on Ω. As for results related to the properties of the derivatives $D^\alpha \widetilde{H}_\delta f$, in this case estimate (2.96) is essential. Some statements of Theorems 8–9 can be proved for the operator $\widetilde{H}_\delta$ as well. The main difficulty, which arises on this way is the necessity to work with the superposition $f(x - \delta\varrho_{\delta_0}(x)z)$. For this reason the mollifiers E_δ with piecewise constant step are more convenient, because in their construction superpositions are replaced by locally finite sums of products. Another advantage of the mollifiers with piecewise constant step is that it is possible to choose steps depending on f. This is sometimes is convenient inspite of the fact that the mollifiers become nonlinear. (See the proofs of Theorems 1–4 of this chapter and Theorems 5–7 of Chapter 5.)

Example 4 For each open set $\Omega \subset \mathbb{R}^n$ a function $f \in C^\infty(\mathbb{R}^n)$ exists such that it is positive on Ω and equal to 0 on $\mathbb{R}^n \setminus \Omega$. The function f can be constructed in the following way: $f(x) = \exp(-\frac{1}{\varrho_\delta(x)})$ with some fixed $\delta \in (0, 1)$. The property $(D^\alpha f)(x) = \lim\limits_{y \to x, y \in \Omega} (D^\alpha f)(y) = 0$ for $x \in \partial\Omega$ follows from (2.96). This function f possesses, in addition, the following property, which sometimes is of importance: $\forall \gamma > 1$ and $\forall \alpha \in \mathbb{N}_0^n$ there exists $c_{19} = c_{19}(\gamma, \alpha) > 0$ such that $\forall x \in \mathbb{R}^n$

$$|(D^\alpha f)(x)|^\gamma \leq c_{19} f(x).$$

This also follows from (2.96).

Another application of a regularized distance for extensions will be given in Remark 17 of Chapter 6.

Chapter 3

Sobolev's integral representation

3.1 The one-dimensional case

Let $-\infty < a < b < \infty$,

$$\omega \in L_1(a,b), \qquad \int_a^b \omega\, dx = 1 \tag{3.1}$$

and suppose that the function f is absolutely continuous on $[a,b]$. Then the derivative f' exits almost everywhere on $[a,b]$, $f' \in L_1(a,b)$ and $\forall x, y \in [a,b]$ we have $f(x) = f(y) + \int_y^x f'(u)du$. Multiplying this equality by $\omega(y)$ and integrating with respect to y from a to b we get

$$f(x) = \int_a^b f(y)\omega(y)\, dy + \int_a^b \left(\int_y^x f'(u)\, du \right)\omega(y)\, dy.$$

Interchanging the order of integration we obtain

$$\int_a^b \left(\int_y^x f'(u)\, du \right)\omega(y)\, dy = \int_a^x \left(\int_y^x f'(u)\, du \right)\omega(y)\, dy - \int_x^b \left(\int_x^y f'(u)\, du \right)\omega(y)\, dy$$

$$= \int_a^x \left(\int_a^u \omega(y)\, dy \right) f'(u)\, du - \int_x^b \left(\int_u^b \omega(y)\, dy \right) f'(u)\, du = \int_a^b \Lambda(x,y) f'(y)\, dy,$$

where

$$\Lambda(x,y) = \begin{cases} \displaystyle\int_a^y \omega(u)\,du, & a \leq y \leq x \leq b, \\[4mm] \displaystyle-\int_y^b \omega(u)\,du, & a \leq x < y \leq b. \end{cases} \qquad (3.2)$$

Hence $\forall x \in (a,b)$

$$f(x) = \int_a^b f(y)\,\omega(y)\,dy + \int_a^b \Lambda(x,y)f'(y)\,dy. \qquad (3.3)$$

This formula may be regarded as the simplest case of Sobolev's integral representation.

We note that Λ is bounded:

$$\forall x,y \in [a,b] \quad |\Lambda(x,y)| \leq \|\omega\|_{L_1(a,b)} \qquad (3.4)$$

and if, in addition to (3.1) $\omega \geq 0$, then [1]

$$\forall x,y \in [a,b] \quad |\Lambda(x,y)| \leq \Lambda(b,b) = 1. \qquad (3.5)$$

Let us consider two limiting cases of (3.3). The first one corresponds to $\omega = \text{const}$, hence, $\forall x \in (a,b)$ we have $\omega(x) = (b-a)^{-1}$. Then $\forall x \in [a,b]$

$$f(x) = \frac{1}{b-a}\int_a^b f(y)\,dy + \int_a^x \frac{y-a}{b-a}f'(y)\,dy - \int_x^b \frac{b-y}{b-a}f'(y)\,dy. \qquad (3.6)$$

To obtain another limiting case we take $\omega = \frac{1}{2m}(\chi_{(a,a+\frac{1}{m})} + \chi_{(b-\frac{1}{m},b)})$, where $\chi_{(\alpha,\beta)}$ denotes the characteristic function of an interval (α,β), $m \in \mathbb{N}$ and $m \geq 2(b-a)^{-1}$. Letting $m \to \infty$ we find: $\forall x \in [a,b]$

$$f(x) = \frac{f(a)+f(b)}{2} + \frac{1}{2}\int_a^b \text{sgn}(x-y)f'(y)\,dy. \qquad (3.7)$$

Of course both of formulas (3.6) and (3.7) can be deduced directly by integration by parts or the Newton-Leibnitz formula.

---

[1] If ω is symmetric with respect to the point $\frac{a+b}{2}$, then $\forall y \in [a,b]$ we have $|\Lambda(\frac{a+b}{2},y)| \leq \frac{1}{2}$.

Obviously, from (3.6) it follows that

$$|f(x)| \le \frac{1}{b-a} \int_a^b |f|\, dy + \int_a^b |f'|\, dy \tag{3.8}$$

for all $x \in [a, b]$. [2]

If $f \in (W_1^1)^{loc}(a, b)$, then f is equivalent to a function, which is locally absolutely continuous on (a, b) (its ordinary derivative, which exists almost everywhere on (a, b), is a weak derivative f'_w of f – see Section 1.2). Consequently, (3.3), (3.6) and (3.8) hold for almost every $x \in (a, b)$ if f' is replaced by f'_w.

Let now $-\infty \le a < b \le \infty$, $x_0 \in (a, b)$, $l \in \mathbb{N}$ and suppose that the derivative $f^{(l-1)}$ exists and is locally absolutely continuous on (a, b). Then the derivative $f^{(l)}$ exists almost everywhere on (a, b), $f^{(l)} \in L_1^{loc}(a, b)$ and by Taylor's formula with the remainder written in an integral form $\forall x, x_0 \in (a, b)$

$$f(x) = \sum_{k=0}^{l-1} \frac{f^{(k)}(x_0)}{k!}(x - x_0)^k + \frac{1}{(l-1)!} \int_{x_0}^x (x - u)^{l-1} f^{(l)}(u)\, du$$

$$= \sum_{k=0}^{l-1} \frac{f^{(k)}(x_0)}{k!}(x - x_0)^k + \frac{(x - x_0)^l}{(l-1)!} \int_0^1 (1 - t)^{l-1} f^{(l)}(x_0 + t(x - x_0))\, dt. \tag{3.10}$$

Theorem 1 *Let $l \in \mathbb{N}$, $-\infty \le a < \alpha < \beta < b \le \infty$ and*

$$\omega \in L_1(\mathbb{R}), \quad \operatorname{supp}\omega \subset [\alpha, \beta], \quad \int_{\mathbb{R}} \omega\, dx = 1. \tag{3.11}$$

Moreover, suppose that the derivative $f^{(l-1)}$ exists and is locally absolutely continuous on (a, b).

---

[2] By the limiting procedure inequality (3.8) can be extended to functions f, which are of bounded variation on $[a, b]$: $\forall x \in [a, b]$

$$|f(x)| \le \frac{1}{b-a} \int_a^b |f|\, dy + \operatorname*{Var}_{[a,b]} f. \tag{3.9}$$

One can easily prove it directly: it is enough to integrate the inequality $|f(x)| \le |f(y)| + |f(x) - f(y)| \le |f(y)| + \operatorname*{Var}_{[a,b]} f$ with respect to y from a to b.

Then $\forall x \in (a, b)$

$$f(x) = \sum_{k=0}^{l-1} \frac{1}{k!} \int_a^b f^{(k)}(y)(x-y)^k \omega(y)\, dy + \frac{1}{(l-1)!} \int_a^b (x-y)^{l-1} \Lambda(x,y) f^{(l)}(y)\, dy$$

$$(3.12)$$

$$= \sum_{k=0}^{l-1} \frac{1}{k!} \int_\alpha^\beta f^{(k)}(y)(x-y)^k \omega(y)\, dy + \frac{1}{(l-1)!} \int_{a_x}^{b_x} (x-y)^{l-1} \Lambda(x,y) f^{(l)}(y)\, dy,$$

$$(3.13)$$

where $a_x = x, b_x = \beta$ for $x \in (a, \alpha]$; $\;a_x = \alpha, b_x = \beta$ for $x \in (\alpha, \beta)$; $\;a_x = \alpha$, $b_x = x$ for $x \in [\beta, b)$.

Idea of the proof. Multiply (3.10) with $x_0 = y$ by $\omega(y)$, integrate with respect to y from a to b and interchange the order of integration (as above). $\square$

Proof. The integrated remainder in (3.10) takes the form in (3.12) after interchanging the order of integration:

$$\int_a^b \left(\int_y^x (x-u)^{l-1} f^{(l)}(u)\, du \right) \omega(y)\, dy = \int_a^x \omega(y) \left(\int_y^x (x-u)^{l-1} f^{(l)}(u)\, du \right) dy$$

$$- \int_x^b \omega(y) \left(\int_y^x (x-u)^{l-1} f^{(l)}(u)\, du \right) dy = \int_a^x (x-u)^{l-1} \left(\int_a^u \omega(y)\, dy \right) f^{(l)}(u)\, du$$

$$- \int_x^b (x-u)^{l-1} \left(\int_u^b \omega(y)\, dy \right) f^{(l)}(u)\, du = \int_a^b (x-y)^{l-1} \Lambda(x,y) f^{(l)}(y)\, dy.$$

Finally, since $\operatorname{supp} \omega \subset [\alpha, \beta]$, it follows that $\Lambda(x,y) = 0$ if $y \in (a, a_x) \cup (b_x, b)$ and, hence, (3.13) holds. $\square$

Remark 1 If in Theorem 1 $a > -\infty$ and $f^{(l-1)}$ exists on $[a, b)$ and is absolutely continuous on $[a, b_1)$ for each $b_1 \in (a, b)$, then equality $(3.12) - (3.13)$ holds for $x = a$ and $\alpha = a$ as well. To verify this one needs to pass to the limit as $x \to a+$ and $\alpha \to a+$, noticing that in this case $f^{(l)} \in L_1(a, b_1)$ for each $b_1 \in (a, b)$. The analogous statement holds for the right endpoint of the interval (a, b).

If, in particular, $-\infty < a < b < \infty$, $f^{(l-1)}$ exists and is absolutely continuous on $[a, b]$, then equality $(3.12) - (3.13)$ holds $\forall x \in [a, b]$ and for any interval $(\alpha, \beta) \subset (a, b)$.

Remark 2 Suppose that $-\infty < a < b < \infty$, $f^{(l-1)}$ exists on $[a,b]$ and is absolutely continuous on $[a,b]$. Then the right-hand side of (3.12) is actually defined for any $x \in \mathbb{R}$, if to assume that $\Lambda(x,y)$ is defined by (3.2) for any $x \in \mathbb{R}$ and for any $y \in [a,b]$. Since in this case for any $x \le a$ and for any $y \in [a,b]$ we have $\Lambda(x,y) = -\int_a^b \omega(u)\,du$, the right-hand side of (3.12) for $x \le a$ is the polynomial p_a of order less than or equal to $l-1$ such that $p_a^{(k)}(a) = f^{(k)}(a), k = 0,1,...,l-1$. Respectively, for $x \ge b$ the right-hand side is the polynomial p_b of order less than or equal to $l-1$, which is such that $p_b^{(k)}(b) = f^{(k)}(b), k = 0,1,...,l-1$. Thus the function

$$F(x) = \sum_{k=0}^{l-1} \frac{1}{k!} \int_a^b f^{(k)}(y)(x-y)^k \omega(y)\,dy + \frac{1}{(l-1)!} \int_a^b (x-y)^{l-1}\Lambda(x,y)f^{(l)}(y)\,dy$$

is an extension of the function f with preservation of differential properties, since $F^{(l-1)}$ is locally absolutely continuous on $\mathbb{R}$. See also Section 6.1, where the one-dimensional extensions are studied in more detail.

Corollary 1 *Suppose that $l > 1$, condition (3.11) is replaced by*

$$\omega \in C^{(l-2)}(\mathbb{R}), \quad \operatorname{supp}\omega \subset [\alpha,\beta], \quad \int_\mathbb{R} \omega\,dx = 1 \tag{3.14}$$

and the derivative $\omega^{(l-2)}$ is absolutely continuous on $[a,b]$.
Then for the same f as in Theorem 1 $\forall x \in (a,b)$

$$f(x) = \int_\alpha^\beta \left(\sum_{k=0}^{l-1} \frac{(-1)^k}{k!}[(x-y)^k\omega(y)]_y^{(k)} \right) f(y)\,dy$$

$$+\frac{1}{(l-1)!} \int_{a_x}^{b_x} (x-y)^{l-1}\Lambda(x,y)f^{(l)}(y)\,dy. \tag{3.15}$$

Idea of the proof. Integrate by parts. $\square$
From (3.14) it follows, in particular, that

$$\omega(\alpha) = \ldots = \omega^{(l-2)}(\alpha) = \omega(\beta) = \ldots = \omega^{(l-2)}(\beta) = 0. \tag{3.16}$$

Corollary 2 *Suppose that $l, m \in \mathbb{N},\ m < l$. Then for the same f and ω as in Corollary 1 $\forall x \in (a, b)$*

$$f^{(m)}(x) = \int\limits_{\alpha}^{\beta} \Big(\sum_{k=0}^{l-m-1} \frac{(-1)^{k+m}}{k!} [(x-y)^k \omega(y)]_y^{(k+m)} \Big) f(y)\, dy$$

$$+ \frac{1}{(l-m-1)!} \int\limits_{a_x}^{b_x} (x-y)^{l-m-1} \Lambda(x, y) f^{(l)}(y)\, dy. \tag{3.17}$$

Idea of the proof. Apply (3.15), with $l - m$ replacing l, to $f^{(m)}$ and integrate by parts in the first summand taking into account (3.16). $\square$

Remark 3 The first summand in (3.15) may be written in the following form:

$$\int\limits_{\alpha}^{\beta} \Big(\sum_{s=0}^{l-1} \sigma_s (x-y)^s \omega^{(s)}(y) \Big) f(y)\, dy, \quad \sigma_s = \frac{(-1)^s}{s!} \sum_{k=s}^{l-s-1} \binom{s+k}{k}. \tag{3.18}$$

It is enough to apply Leibnitz' formula and change the order of summation in order to see this.

By the similar argument the first summand of (3.17) may be written in the following form:

$$\int\limits_{\alpha}^{\beta} \Big(\sum_{s=m}^{l-1} \sigma_{s,m} (x-y)^{s-m} \omega^{(s)}(y) \Big) f(y)\, dy, \tag{3.19}$$

where

$$\sigma_{s,m} = \frac{(-1)^s}{(s-m)!} \sum_{k=s}^{l-s-1} \binom{s+k}{k}. \tag{3.20}$$

From (3.18) and (3.19) it is clearly seen that the first summand in (3.17) is a derivative of order m of the first summand of (3.15) and thus (3.17) can be directly obtained from (3.15) by differentiation. (In order to differentiate the second summand one needs to split the integral into two parts – see the proof of Theorem 1.)

Corollary 3 *Let $-\infty < a < b < \infty, l \in \mathbb{N}, m \in \mathbb{N}_0, m < l$. Moreover, suppose that the derivative $f^{(l-1)}$ is absolutely continuous on $[a, b]$. Then $\forall x \in [a, b]$*

$$|f^{(m)}(x)| \le c_1 \Big((b-a)^{l-m-1} (\beta - \alpha)^{-l} \int\limits_{\alpha}^{\beta} |f|\, dy + \int\limits_{a}^{b} |x-y|^{l-m-1} |f^{(l)}(y)|\, dy \Big)$$

$$\leq c_1(b-a)^{l-m-1}\left((\beta-\alpha)^{-l}\int_\alpha^\beta |f|\,dy + \int_a^b |f^{(l)}|\,dy\right) \qquad (3.21)$$

and, consequently,

$$|f^{(m)}(x)| \leq c_1\left((b-a)^{-m-1}\int_a^b |f|\,dy + \int_a^b |x-y|^{l-m-1}|f^{(l)}(y)|\,dy\right)$$

$$\leq c_1\left((b-a)^{-m-1}\int_a^b |f|\,dy + (b-a)^{l-m-1}\int_a^b |f^{(l)}|\,dy\right) \qquad (3.22)$$

and [3]

$$|f^{(m)}(x)| \leq c_2\left(\int_\alpha^\beta |f|\,dy + \int_a^b |x-y|^{l-m-1}|f^{(l)}(y)|\,dy\right)$$

$$\leq c_3\left(\int_\alpha^\beta |f|\,dy + \int_a^b |f^{(l)}|\,dy\right), \qquad (3.23)$$

where $c_1 > 0$ depends only on l, while $c_2, c_3 > 0$ depend on l and, in addition, depend on $\beta - \alpha$ and $b - a$.

Idea of the proof. In (3.17) take $\omega(x) = \frac{1}{r}\mu(\frac{x-x_0}{r})$, where $x_0 = \frac{\alpha+\beta}{2}, r = \frac{\beta-\alpha}{2}$ and $\mu \in C_0^\infty(\mathbb{R})$ is a fixed nonnegative function, for which $\operatorname{supp}\mu \subset [-1, 1]$ and $\int_\mathbb{R} \mu\,dx = 1$. In order to estimate the first summand in (3.17) apply (3.19) and the estimate $|\omega^{(s)}(x)| \leq M\,r^{-s-1}$ for $m \leq s \leq l-1$, where M depends only on l. To estimate the second summand in (3.17) apply (3.5). $\square$

---

[3] From (3.23) it follows, by Hölder's inequality, that for $1 \leq p \leq \infty$

$$\|f^{(m)}\|_{L_p(a,b)} \leq M_1\left(\|f\|_{L_p(a,b)} + \|f^{(l)}\|_{L_p(a,b)}\right),$$

where $M_1 = c_3(b-a)$, and, after additional integration, that

$$\left\|\frac{\partial^m f}{\partial x_j^m}\right\|_{L_p(Q)} \leq M_2\left(\|f\|_{L_p(Q)} + \left\|\frac{\partial^l f}{\partial x_j^l}\right\|_{L_p(Q)}\right),$$

where $Q \subset \mathbb{R}^n$ is any cube, whose faces are parallel to the coordinate planes, $f \in \overline{C}^l(Q)$ and $M_2 > 0$ is independent of f. These inequalities were used in the proof of Lemmas 5–6 of Chapter 1.

Remark 4 We note a simple particular case of the integral representation (3.17): if ω is absolutely continuous on $[a, b]$, $\omega(a) = \omega(b) = 0$ and $\int_a^b \omega\, dx = 1$, then for each f such that f' is absolutely continuous on $[a, b]$, for all $x \in [a, b]$

$$f'(x) = -\int_a^b \omega'(y)f(y)\, dy + \int_a^b \Lambda(x, y)f''(y)\, dy. \tag{3.24}$$

It follows that

$$|f'(x)| \leq \|\omega'\|_{L_\infty(a,b)} \int_a^b |f|\, dy + \|\Lambda(x, \cdot)\|_{L_\infty(a,b)} \int_a^b |f''|\, dy.$$

Choosing ω in such a way that $\|\omega'\|_{L_\infty(a,b)}$ is minimal we find

$$\omega(x) = \frac{4}{(b-a)^2}\left(\frac{b-a}{2} - \left|x - \frac{a+b}{2}\right|\right)$$

and, hence,

$$|f'(x)| \leq \frac{4}{(b-a)^2} \int_a^b |f|\, dy + \left(1 - 2\left(\frac{\min\{x - a, b - x\}}{b - a}\right)^2\right) \int_a^b |f''|\, dy. \tag{3.25}$$

In particular

$$|f'(a)|, |f'(b)| \leq \frac{4}{(b-a)^2} \int_a^b |f|\, dy + \int_a^b |f''|\, dy$$

and

$$\left|f'\left(\frac{a+b}{2}\right)\right| \leq \frac{4}{(b-a)^2} \int_a^b |f|\, dy + \frac{1}{2}\int_a^b |f''|\, dy.$$

From (3.25) it follows that $\forall x \in [a, b]$

$$|f'(x)| \leq 4\left(\frac{1}{(b-a)^2} \int_a^b |f|\, dy + \int_a^b |f''|\, dy\right). \tag{3.26}$$

This is a particular case of (3.22) with the minimal possible constant $c_1 = 4$. The latter follows from setting $f(y) = y - \frac{a+b}{2}$. The same test-function shows that the constant multiplying $\int_a^b |f|\,dy$ in (3.25), (3.26) also cannot be diminished even if the constant multiplying $\int_a^b |f''|\,dy$ is enlarged.

We note that the constant muliplying $\int_a^b |f''|\,dy$ in (3.25) also cannot be diminished. [4] This can be proved in the following way. For $a \leq x \leq b$ and $\delta > 0$ consider the function [5] $g_{\delta,x}(y) = (x - y + \delta)_+$, $y \in [a,b]$, if $a \leq x \leq \frac{a+b}{2}$ and $g_{\delta,x}(y) = (y - x + \delta)_+$, $y \in [a,b]$, if $\frac{a+b}{2} < x \leq b$. In (3.24) take $f = A_{\frac{\delta}{2}}g_{\delta,x}$, where A_δ is a mollifier, and pass to the limit as $\delta \to 0+$.

Finally, as in the case of the integral representation (3.3), we consider a limiting case of (3.24). We write ω_m for ω, where $m \in \mathbb{N}$, $m \geq \frac{2}{b-a}$, $\omega_m(x) = m(x-a)(b-a-\frac{1}{m})^{-1}$ for $a < x \leq a + \frac{1}{m}$, $\omega_m(x) = (b - a - \frac{1}{m})^{-1}$ for $a + \frac{1}{m} \leq x \leq b - \frac{1}{m}$ and $\omega_m(x) = m(b-x)(b-a-\frac{1}{m})^{-1}$ for $b - \frac{1}{m} \leq x < b$. Taking limits we get the equality

$$f'(x) = \frac{f(b) - f(a)}{b - a} + \int_a^x \frac{y - a}{b - a} f''(y)\,dy - \int_x^b \frac{b - y}{b - a} f''(y)\,dy. \qquad (3.27)$$

Here $x \in [a,b]$ and f is such that f' exists and is absolutely continuous on $[a,b]$. Again, as in the case of representations (3.6) and (3.7), (3.27) can be deduced directly.

Corollary 4 *Let* $l \in \mathbb{N}, m \in \mathbb{N}_0, l \geq 2$ *and* $m < l - 1$.

1. If $-\infty < a < b < \infty$ *and the derivative* $f^{(l-1)}$ *is absolutely continuous on* $[a,b]$, *then* $\forall x \in [a,b]$ *and* $\forall \varepsilon \in (0, c_1(b-a)^{l-m-1}]$,

$$|f^{(m)}(x)| \leq c_4 K(\varepsilon) \int_a^b |f|\,dy + \varepsilon \int_a^b |f^{(l)}|\,dy, \qquad (3.28)$$

where $c_4 > 0$ *depends only on* l *and*

$$K(\varepsilon) = \varepsilon^{-\frac{m+1}{l-m-1}}. \qquad (3.29)$$

---

[4] In contrast to the constant multiplying $\int_a^b |f|\,dy$ it can be diminished if to enlarge appropriately the constant multiplying $\int_a^b |f|\,dy$ – see Corollary 4.

[5] Here and in the sequel $a_+ = a$ for $a \geq 0$ and $a_+ = 0$ for $a < 0$.

2. *If $I = [a, \infty)$ where $-\infty < a < \infty$, $I = (-\infty, b]$ where $-\infty < b < \infty$ or $I = (-\infty, \infty)$ and the derivative $f^{(l-1)}$ is absolutely continuous on each closed interval in I, then $\forall x \in I$ and $\forall \varepsilon \in (0, \infty)$*

$$|f^{(m)}(x)| \leq c_5 \, K(\varepsilon) \int_I |f| \, dy + \varepsilon \int_I |f^{(l)}| \, dy, \tag{3.30}$$

where $c_5 > 0$ depends on l only.

Idea of the proof. In the first case for $x \in [a, b]$ apply (3.22) replacing $[a, b]$ by any closed interval $[a_1, b_1] \subset [a, b]$ containing x, whose length is equal to δ, where $0 < \delta \leq b - a$, and set $c_1 \delta^{l-m-1} = \varepsilon$. The second case follows from the first one.

There is an alternative way of proving (3.28). Given a function f, it is enough to apply (3.22) to the functions $f_{\delta,x}$, where $0 < \delta \leq b - a$ and $x \in [a, b]$, which are defined for $y \in [a, b]$ by $f_{\delta,x}(y) = f(x + \delta(\frac{y-x}{b-a}))$, change the variables putting $x + \delta(\frac{y-x}{b-a}) = z$ and set $c_1 \delta^{l-m-1} = \varepsilon$. $\square$

Corollary 5 *If $-\infty < a < b < \infty$, $l \in \mathbb{N}$ and $f^{(l)}$ is absolutely continuous on $[a, b]$, then there exists a polynomial $p_{l-1}(x, f)$ of degree less than or equal to $l - 1$ such that for each $m \in \mathbb{N}_0$, $m < l$ and $\forall x \in [a, b]$*

$$|f^{(m)}(x) - p_{l-1}^{(m)}(x, f)| \leq \frac{(b-a)^{l-m-1}}{(l-m-1)!} \int_a^b |f^{(l)}| \, dx. \tag{3.31}$$

Idea of the proof. It is enough to take Taylor's polynomial $T_{l-1}(x, f)$ as $p_{l-1}(x, f)$:

$$p_{l-1}(x, f) = T_{l-1}(x, f) \equiv \sum_{k=0}^{l-1} \frac{f^{(k)}(x_0)}{k!} (x - x_0)^k$$

with an arbitrary $x_0 \in [a, b]$, apply to $f^{(m)}$ Taylor's formula with $l - m$ replacing l (with the same x_0) and take into account that

$$T_{l-m-1}(x, f^{(m)}) = T_{l-1}^{(m)}(x, f). \tag{3.32}$$

It is also possible to take

$$p_{l-1}(x, f) = S_{l-1}(x, f) \equiv \sum_{k=0}^{l-1} \frac{1}{k!} \int_a^b f^{(k)}(y)(x - y)^k \omega(y) \, dy,$$

where ω satisfies (3.11) and is nonnegative. (Notice that this is the first summand in (3.12).) One must take into account that in this case also $S_{l-m-1}(x, f^{(m)}) = S_{l-1}^{(m)}(x, f)$. Both of the choices lead to (3.31) (in the second case according to (3.5)). $\square$

Theorem 2 *Let $l \in \mathbb{N}$, $-\infty \le a < \alpha < \beta \le b \le \infty$, ω satisfy condition (3.11) and $f \in (W_1^l)^{loc}(a, b)$. Then for almost every $x \in (a, b)$*

$$f(x) = \sum_{k=0}^{l-1} \frac{1}{k!} \int_{\alpha}^{\beta} f_w^{(k)}(y)(x - y)^k \omega(y) \, dy + \frac{1}{(l-1)!} \int_{a_x}^{b_x} (x - y)^{l-1} \Lambda(x, y) f_w^{(l)}(y) \, dy,$$

$$(3.33)$$

where a_x and b_x are defined in Theorem 1.

Idea of the proof. Set $a(\delta) = \max\{a+\delta, -\frac{1}{\delta}\}$, $b(\delta) = \min\{b-\delta, \frac{1}{\delta}\}$ for sufficiently small $\delta > 0$, write (3.13) for $A_\gamma f \in C^\infty(a(\delta), b(\delta))$, where $0 < \gamma < \delta$, and pass to the limit as $\gamma \to 0+$. $\square$

Proof. Since for sufficiently small $\delta > 0$ $[\alpha, \beta] \subset (a(\delta), b(\delta))$ and $(a(\delta))_x = a_x$, $(b(\delta))_x = b_x$ for each $x \in (a(\delta), b(\delta))$, we have $\forall x \in (a(\delta), b(\delta))$

$$(A_\gamma f)(x) = \sum_{k=0}^{l-1} \frac{1}{k!} \int_{\alpha}^{\beta} (A_\gamma f)^{(k)}(y)(x - y)^k \omega(y) \, dy$$

$$= \frac{1}{(l-1)!} \int_{a_x}^{b_x} (x - y)^{l-1} \Lambda(x, y)(A_\gamma f)^{(l)}(y) \, dy.$$

By Lemma 5 of Chapter 1 $f_w^{(k)}$ exists on (a, b) where $k = 1, ..., l - 1$, and by Lemma 4 of Chapter 1 $(A_\gamma f)^{(k)} = A_\gamma(f_w^{(k)})$ on $(a(\delta), b(\delta))$ where $k = 1, ..., l$. Consequently, $\forall x \in (a(\delta), b(\delta))$

$$\left| \int_{\alpha}^{\beta} (A_\gamma f)^{(k)}(y)(x - y)^k \omega(y) \, dy - \int_{\alpha}^{\beta} f_w^{(k)}(y)(x - y)^k \omega(y) \, dy \right|$$

$$\le \int_{\alpha}^{\beta} |A_\gamma(f_w^{(k)})(y) - f_w^{(k)}(y)| \, |(x - y)^k \omega(y)| \, dy \le M_1 \int_{\alpha}^{\beta} |A_\gamma(f_w^{(k)}) - f_w^{(k)}| \, dy \to 0$$

as $\gamma \to 0+$, where $k = 1, ..., l - 1$ and M_1 is independent of γ and x_0.

Analogously, in view of (3.4), $\forall x \in (a(\delta), b(\delta))$

$$\left| \int_{a_x}^{b_x} (x-y)^{l-1} \Lambda(x,y)(A_\gamma f)^{(l)}(y)\, dy - \int_{a_x}^{b_x} (x-y)^{l-1} \Lambda(x,y) f_w^{(l)}(y)\, dy \right|$$

$$\leq M_2 \int_{a(\delta)}^{b(\delta)} |A_\gamma(f_w^{(l)}) - f_w^{(l)}|\, dy \to 0$$

as $\gamma \to 0+$, where M_2 is independent of γ and x.

Finally, by (1.5) $A_\gamma f \to f$ almost everywhere on $(a(\delta), b(\delta))$. Thus (3.33) is valid almost everywhere on $(a(\delta), b(\delta))$ and, hence, on (a,b) since $\bigcup_{\delta>0}(a(\delta), b(\delta)) = (a,b)$. $\square$

Remark 5 By Theorem 2 it follows that if in Corollaries $1-2$ $f \in (W_1^l)^{loc}(a,b)$ and in Corollaries $3-5$ $f \in W_1^l(a,b)$, then equalities (3.15), (3.17) and inequalities $(3.21)-(3.23)$, (3.28), (3.30) and (3.31) hold almost everywhere on (a,b), if to replace the ordinary derivatives $f^{(l)}$ and $f^{(m)}$ by the weak derivatives $f_w^{(l)}$ and $f_w^{(m)}$.

3.2   Star-shaped sets and sets satisfying the cone condition

A domain $\Omega \subset \mathbb{R}^n$ is called *star-shaped with respect to the point* $y \in \Omega$ if $\forall x \in \Omega$ the closed interval $[x,y] \subset \Omega$. A domain $\Omega \subset \mathbb{R}^n$ is called *star-shaped with respect to a point* if for some $y \in \Omega$ it is star-shaped with respect to the point y. A domain $\Omega \subset \mathbb{R}^n$ is called *star-shaped with respect to the ball* [6] $B \subset \Omega$ if $\forall y \in B$ and $\forall x \in \Omega$ we have $[x,y] \subset \Omega$. A domain $\Omega \subset \mathbb{R}^n$ is called *star-shaped with respect to a ball* if for some ball $B \subset \Omega$ it is star-shaped with respect to the ball B. If $0 < d \leq \operatorname{diam} B \leq \operatorname{diam}\Omega \leq D$, we say that Ω is star-shaped with respect to a ball with the parameters d, D.

We call the set

$$V_x \equiv V_{x,B} = \bigcup_{y \in B}(x,y)$$

---

[6] Recall that by "ball" we always mean "open ball".

a *conic body* with the vertex x constructed on the ball B (if $x \in B$, then $V_x = B$). A domain Ω star-shaped with respect to a ball B can be equivalently defined in the following way: $\forall x \in \Omega$ the conic body $V_x \subset \Omega$.

Let us consider now the cone

$$K \equiv K(r, h) = \left\{ x \in \mathbb{R}^n : 0 < \left(\sum_{i=1}^{n-1} x_i^2 \right)^{\frac{1}{2}} < \frac{r x_n}{h} < r \right\}. \qquad (3.34)$$

We say also that an open set $\Omega \subset \mathbb{R}^n$ satisfies *the cone condition with the parameters $r > 0$ and $h > 0$* if $\forall x \in \Omega$ there exists [7] a cone $K_x \subset \Omega$ with the point x as vertex congruent to the cone K. Moreover, an open set $\Omega \subset \mathbb{R}^n$ satisfies *the cone condition* if for some $r > 0$ and $h > 0$ it satisfies the cone condition with the parameters r and h.

Example 1 The one-dimensional case is trivial. Each domain $\Omega = (a, b) \subset \mathbb{R}$ is star-shaped with respect to a ball ($\equiv$ interval). An open set $\Omega = \bigcup_{k=1}^{s} (a_k, b_k)$, where $s \in \mathbb{N}$ or $s = \infty$ and $(a_k, b_k) \cap (a_m, b_m) = \varnothing$ for $k \neq m$, satisfies the cone condition if, and only if, $\inf_{k} (b_k - a_k) > 0$.

Example 2 A star (with arbitrary number of end-points) in $\mathbb{R}^2$ is star-shaped with respect to its center and with respect to sufficiently small balls ($\equiv$ circles) centered at its center. It also satisfies the cone condition.

Example 3 A convex domain $\Omega \subset \mathbb{R}^n$ is star-shaped with respect to each point $y \in \Omega$ and each ball $B \subset \Omega$. A domain Ω is convex if, and only if, it is star-shaped with respect to each point $y \in \Omega$.

Example 4 The domain $\Omega \subset \mathbb{R}^2$ inside the curve described by the equation $|x_1|^\gamma + |x_2|^\gamma = 1$ where $0 < \gamma < 1$ (the astroid for $\gamma = 2/3$) is star-shaped with respect to the origin, but it is not star-shaped with respect to any ball $B \subset \Omega$. It does not also satisfy the cone condition.

Example 5 The union of domains, which are star-shaped with respect to a given ball, is star-shaped with respect to that ball. The union (even of a finite number) of domains star-shaped with respect to different balls in general is not star-shaped with respect to a ball. In contrast to it the union of a finite number of open sets satisfying the cone condition satisfies the cone condition. Moreover, the union of an arbitrary number of open sets satisfying the cone condition with the same parameters r and h satisfies the cone condition.

---

[7] "$\forall x \in \Omega$" can be replaced by "$\forall x \in \overline{\Omega}$" or by "$\forall x \in \partial\Omega$" and this does not affect the definition.

Example 6 The domain $\Omega = \{x \in \mathbb{R}^n : |\bar{x}|^\gamma < x_n < 1, \ |\bar{x}| < 1\}$, where $\bar{x} = (x_1, ..., x_{n-1})$, for $\gamma \geq 1$ is star-shaped with respect to a ball and satisfies the cone condition. For $0 < \gamma < 1$ it is not star-shaped with respect to a ball. Furthermore, it cannot be represented as a union of a finite number of domains, which are star-shaped with respect to a ball, and does not satisfy the cone condition.

Example 7 The domain $\Omega = \{x \in \mathbb{R}^n : -1 < x_n < |\bar{x}|^\gamma, \ |\bar{x}| < 1\}$ satisfies the cone condition for each $\gamma > 0$. It is not star-shaped with respect to a ball, but can be represented as a union of a finite number of domains, which are star-shaped with respect to a ball.

Example 8 The domain $\Omega = \{(x_1, x_2) \in \mathbb{R}^2 : \text{either} \ -2 < x_1 < 1 \ \text{and} \ -2 < x_2 < 2, \ \text{or} \ 1 \leq x_1 < 2 \ \text{and} \ -2 < x_2 < 1\}$ is star-shaped with respect to the ball $B(0, 1)$. For $0 < \delta < \sqrt{2} - 1$ the domain $\underline{\Omega_\delta} \supset B(0, 1)$, but it is not star-shaped with respect to the ball $B(0, 1)$. (It is star-shaped with respect to some smaller ball.)

Lemma 1 *An open set $\Omega \subset \mathbb{R}^n$ satisfies the cone condition if, and only if, there exist $s \in \mathbb{N}$, cones $K_k, k = 1, ..., s$, with the origin as vertex, which are mutually congruent and open sets $\Omega_k, k = 1, ..., s$, such that*

1) $\Omega = \bigcup\limits_{k=1}^{s} \Omega_k,$

2) $\forall x \in \Omega_k$ *the cone* [8] $x + K_k \subset \Omega.$

Idea of the proof. Sufficiency is clear. To prove necessity choose a finite number of congruent cones $K_k, k = 1, ..., s$, with the origin as a vertex, whose openings are sufficiently small and which cover a neighbourhood of the origin, and consider the sets of all $x \in \Omega$ for which $x + K_k \subset \Omega$. $\square$

Proof. Necessity. Let Ω satisfy the cone condition with the parameters $r, h > 0$. We consider the cone $K(r_1, h_1)$ defined by (3.34), where $h_1 < h$ and $r_1 < r$ is such that the opening of the cone $K(r_1, h_1)$ is half that of the cone $K(r, h)$. Furthermore, we choose the cones $K_k, k = 1, ..., s$, with the origin as a vertex, which are congruent to $K(r_1, h_1)$ and are such that $B(0, h_1) \subset \bigcup\limits_{k=1}^{s} K_k$. Hence, $\forall x \in \Omega$ the cone K_x of the cone condition contains $x + K_k$ for some k. Denote by G_k the set of all $x \in \Omega$, for which K_x contains $x + K_k$. Finally, there exists $\delta_x > 0$ such that $\forall y \in B(x, \delta_x)$ we have $y + K_k \subset \Omega$. Consequently, the open

---

[8] Here the sign $+$ denotes a vector sum. The cone $x + K_k$ is a translation of the cone K_k and its vertex is x.

sets $\Omega_k = \bigcup\limits_{x \in G_k} B(x, \delta_x), k = 1, ..., s$, satisfy conditions 1) and 2). $\square$

Let a domain $\Omega \subset \mathbb{R}^n$ be star-shaped with respect to the point x_0. For $\xi \in S^{n-1}$, where S^{n-1} is the unit sphere in $\mathbb{R}^n$, set $\varphi(\xi) = \sup\{\varrho \geq 0 : x_0 + \varrho\xi \in \Omega\}$. Then

$$\Omega = \{x \in \mathbb{R}^n : x = x_0 + \varrho\xi \text{ where } \xi \in S^{n-1}, \ 0 \leq \varrho < \varphi(\xi)\}.$$

Moreover, set $R_1 = \inf\limits_{\xi \in S^{n-1}} \varphi(\xi)$, $R_2 = \sup\limits_{\xi \in S^{n-1}} \varphi(\xi)$ and for $\xi, \eta \in S^{n-1}$ denote by $d(\xi, \eta)$ the distance between ξ and η along the sphere S^{n-1}, which is equal to the angle γ between the vectors $\overrightarrow{O\xi}$ and $\overrightarrow{O\eta}$, where O is the origin.

Lemma 2 *Let a bounded domain $\Omega \subset \mathbb{R}^n$ be star-shaped with respect to the point $x_0 \in \Omega$. Then it is star-shaped with respect to a ball centered at x_0 if, and only if, the function φ satisfies the Lipschitz condition on S^{n-1}, i.e., for some $M \geq 0$ and* [9] *$\forall \xi, \eta \in S^{n-1}$*

$$|\varphi(\xi) - \varphi(\eta)| \leq M \, d(\xi, \eta).$$

Idea of the proof. Sufficiency. Consider the conic surface $C(\xi)$ with the point $f = x_0 + \varphi(\xi)\xi$ as vertex, which is tangent to the ball $B(x_0, r)$. Suppose that $0 < \gamma < \beta = \arcsin \frac{r}{\varphi(\xi)}$. Then the ray $R(\eta) = \{x \in \mathbb{R}^n : x = x_0 + \varrho\eta, \ 0 \leq \varrho < \infty\}$ intersects $C(\xi)$ at two points a and e. Denote $d = x_0 + \varphi(\eta)\eta$. Since $f, d \in \partial\Omega$ it follows that $f \notin V_d$ and $d \notin V_f$. Therefore, $d \in [a, e]$.
Necessity. For fixed $\xi \in S^{n-1}$ consider two closed rotational surfaces L_+ and L_- defined by the equations $\varrho = F_\pm(\eta)$, where $F_\pm(\eta) = \varphi(\xi) \pm M \, d(\xi, \eta)$. Then the boundary $\partial\Omega$ lies between L_+ and L_-. Let the $(n-2)$-dimensional sphere E be an intersection of L_- and the surface of the ball $B(x_0, R_1)$. Consider two conic surfaces, which both pass through E and whose verticies are x_0, f respectively. Let δ denote the angle at the vertex of the conic surface D_{x_0}, then $\delta = \frac{\varphi(\xi) - R_1}{M}$. (We assume that $M > 0$ and $\varphi(\xi) > R_1$, since the cases, in which $M = 0$ or $\varphi(\xi) = R_1$, are trivial.) If $\delta \geq \delta_0 = \arccos \frac{R_1}{\varphi(\xi)}$, set $r(\xi) = R_1$.

---

[9] Since

$$|\xi - \eta| \leq d(\xi, \eta) = \frac{\gamma}{2\sin\frac{\gamma}{2}} \, |\xi - \eta| \leq \frac{\pi}{2} |\xi - \eta|$$

this condition is equivalent to: for some $M_1 \geq 0$ and $\forall \xi, \eta \in S^{n-1}$

$$|\varphi(\xi) - \varphi(\eta)| \leq M_1 |\xi - \eta|.$$

Otherwise, let $r(\xi)$ be such that the ball $B(x_0, r(\xi))$ is tangent to D_f. Then the conic body with the point f as vertex constucted on the ball $B(x_0, r(\xi))$ lies in Ω. $\square$

Proof. Sufficiency. Denote $c = x_0 + \frac{\varphi(\xi)}{r}\eta$. Since $d \in [a, c]$ or $d \in [c, e]$ we have

$$|\varphi(\xi) - \varphi(\eta)| \leq \max\{|\overrightarrow{ac}|, |\overrightarrow{ce}|\}.$$

Since $|\overrightarrow{ac}| < |\overrightarrow{ce}|$ [10] we establish that

$$|\varphi(\xi) - \varphi(\eta)| \leq |\overrightarrow{ce}| = \frac{\varphi(\xi)\sin\beta}{\sin(\beta-\gamma)} - \varphi(\xi)$$

$$= \frac{2\varphi(\xi)\sin\frac{\gamma}{2}\cos(\beta-\frac{\gamma}{2})}{\sin(\beta-\gamma)} \leq \frac{\varphi(\xi)\gamma\cos(\beta-\frac{\gamma}{2})}{\sin(\beta-\gamma)} = \varphi(\xi)\frac{\cos(\beta-\frac{\gamma}{2})}{\sin(\beta-\gamma)}d(\xi,\eta).$$

Consequently, $\forall \xi, \eta \in S^{n-1}$ such that $\gamma < \beta$

$$|\varphi(\xi) - \varphi(\eta)| = R_2\frac{\cos(\beta-\frac{\gamma}{2})}{\sin(\beta-\gamma)}d(\xi,\eta).$$

Hence, given $\varepsilon > 0$, there exists $\delta(\varepsilon) > 0$ such that $\forall \xi, \eta \in S^{n-1}$: $\gamma < \delta(\varepsilon)$ we have

$$|\varphi(\xi) - \varphi(\eta)| \leq \left(R_2\cot\beta + \varepsilon\right)d(\xi,\eta) = \left(\frac{R_2}{r}\sqrt{R_2^2 - r^2} + \varepsilon\right)d(\xi,\eta).$$

Now let ξ and η be arbitrary points in S^{n-1}, $\xi \neq \eta$. We choose on the circle, centered at x_0 and passing through ξ and η, the points $\xi_0 = \xi \prec \xi_1 \prec \ldots \prec \xi_{m-1} \prec \xi_m = \eta$ such that all the angles between the vectors $\overrightarrow{O\xi_{i-1}}$ and $\overrightarrow{O\xi_i}$, $i = \overline{1, m}$, are less than $\delta(\varepsilon)$. Then

$$|\varphi(\xi) - \varphi(\eta)| \leq \sum_{i=1}^{m} |\varphi(\xi_{i-1}) - \varphi(\xi_i)|$$

---

[10] One can see that $|\overrightarrow{ac}| = |\overrightarrow{bf}|(\cot(\beta+\gamma) + \tan\frac{\gamma}{2})$ while $|\overrightarrow{ce}| = |\overrightarrow{bf}|(\cot(\beta-\gamma) - \tan\frac{\gamma}{2})$, where $\overrightarrow{bf} \perp \overrightarrow{ae}$. The inequality $|\overrightarrow{ac}| < |\overrightarrow{ce}|$ follows from the inequality

$$\cot(\beta+\gamma) + \tan\frac{\gamma}{2} < \cot(\beta-\gamma) - \tan\frac{\gamma}{2},$$

which is valid for all β and γ satisfying $0 < \gamma < \beta < \frac{\pi}{2}$. This inequality is equvalent to

$$2\tan\frac{\gamma}{2} < \frac{\sin 2\gamma}{\sin(\beta-\gamma)\sin(\beta+\gamma)} = \frac{2\sin 2\gamma}{\cos 2\gamma - \cos 2\beta},$$

to $\cos 2\gamma - \cos 2\beta < 4\cos^2\frac{\gamma}{2}\cos\gamma$ and to $-\cos 2\beta < 2\cos\gamma + 1$, which is obvious since $0 < \gamma < \frac{\pi}{2}$.

$$\leq \left(\frac{R_2}{r}\sqrt{R_2^2 - r^2} + \varepsilon\right) \sum_{i=1}^{m} d(\xi_{i-1}, \xi_i) = \left(\frac{R_2}{r}\sqrt{R_2^2 - r^2} + \varepsilon\right) d(\xi, \eta).$$

Passing to the limit as $\varepsilon \to 0+$ we find that the Lipshitz condition is satisfied with

$$M = \frac{R_2}{r}\sqrt{R_2^2 - r^2}. \tag{3.35}$$

Necessity. If $0 < \delta < \delta_0$, then

$$r(\xi) = \frac{\varphi(\xi)R_1 \sin\delta}{\sqrt{(\varphi(\xi) - R_1\cos\delta)^2 + (R_1\sin\delta)^2}} = \frac{\varphi(\xi)R_1 \sin\delta}{\sqrt{(\varphi(\xi) - R_1)^2 + 4\varphi(\xi)R_1 \sin^2\frac{\delta}{2}}}$$

$$= \frac{\varphi(\xi)R_1 \sin\delta}{\sqrt{\delta^2 M^2 + 4\varphi(\xi)R_1 \sin^2\frac{\delta}{2}}} \geq \frac{2}{\pi}\frac{\varphi(\xi)R_1}{\sqrt{M^2 + \varphi(\xi)R_1}} \geq \frac{2}{\pi}\frac{R_1^2}{\sqrt{M^2 + R_1^2}} \equiv r_0.$$

One can verify that for any point $g \in L_- \setminus B(x_0, R_1)$, $g \neq f$, the interval (g, f) lies [11] in Ω. Therefore, the conic body V_f with the point f as vertex constructed on the ball $B(x_0, r_0)$ lies in Ω. Hence, Ω is star-shaped with respect to the ball $B(x_0, r_0)$. $\square$

Remark 6 The constant M given by (3.35) is the minimal possible, because, for example, for any conic body V_x defined by (3.34) we have

$$\sup_{\forall \xi, \eta \in S^{n-1}, \xi \neq \eta} \frac{|\varphi(\xi) - \varphi(\eta)|}{d(\xi, \eta)} = \frac{R_2}{r}\sqrt{R_2^2 - r^2}.$$

If a domain $\Omega \subset \mathbb{R}^n$, which is star-shaped with respect to the ball $B(x_0, r)$, is unbounded, then set $S' = \{\xi \in S^{n-1} : \varphi(\xi) < \infty\}$.

---

[11] Consider the curve l_- obtained by intersecting $L_- \setminus B(x_0, R_1)$ by the two-dimensional plane passing through g and the ray going from x_0 through f. Let this ray be the axis Ox of a Cartesian system of coordinates in this plane. Suppose that $y = \psi(x)$ is a Cartesian equation of the curve l_-. We recall that its polar equation is $\varrho = \varphi(\xi) - M|\gamma|$ and note that $|\gamma| \leq \delta$. The part of the curve l_-, for which $0 \leq \gamma \leq \delta$, is convex and the part of l_-, for which $-\delta \leq \gamma \leq 0$, is concave since, for example, for $0 \leq \gamma \leq \delta$

$$\psi''_{xx} = -\frac{2M^2 + (\varphi(\xi) - M\gamma)^2}{((\varphi(\xi) - M\gamma)\sin\varphi + M\cos\varphi)^3} < 0.$$

Hence, for any $g \in l_-, g \neq f$, the interval (g, f) lies in Ω.

Corollary 6 *Let an unbounded domain $\Omega \subset \mathbb{R}^n$ be star-shaped with respect to the ball $B(x_0, r)$. Then S' is an open set (in S^{n-1}) and the function φ satisfies the Lipschitz condition locally* [12] *on S'.*

Idea of the proof. Note that if $\varphi(\xi) = \infty$ for $\xi \in S^{n-1}$, then the whole semi-infinite cylinder, whose axis is the ray $R(\xi)$ and whose bottom is the hyperball $\{x \in B(x_0, r) : \overrightarrow{x_0 x} \perp \overrightarrow{O \xi}\}$, is contained in Ω. Deduce from this that S' is open and apply Lemma 1. $\square$

Example 9 For the domain $\Omega \subset \mathbb{R}^2$, that is obtained from the unit circle $B(0, 1)$ by throwing out the segment $\{x_1 = 0, \ \frac{1}{2} \le x_2 < 1\}$ and which is star-shaped with respect to origin, but is not star-shaped with respect to a ball, the function φ is not even continuous.

Example 10 For the domain $\Omega = \{x_1, x_2) \in \mathbb{R}^2 : |x_1 x_2| < 1\}$, which is star-shaped with respect to the origin, but is not star-shaped with respect to a ball, the function φ is locally Lipschitz on the set $S' = S^1 \setminus \{(0, \pm 1), (\pm 1, 0)\}$.

Lemma 3 *If a bounded domain $\Omega \subset \mathbb{R}^n$ is star-shaped with respect to a ball, then it satisfies the cone condition.*

Idea of the proof. Let Ω be star-shaped with respect to the ball $B(x_0, r)$. Then Ω satisfies the cone condition with the parameters $\frac{r^2}{R_2}$ and r. (It follows because the cone K_x with the point x as vertex and with axis that of the conic body V_x, which is congruent to the cone $K\left(\frac{r^2}{R_2}, r\right)$, is contained in Ω). $\square$

Now we give characterization of the open sets, which satisfy the cone condition with the help of bounded domains star-shaped with respect to a ball.

Lemma 4 1. *A bounded open set $\Omega \subset \mathbb{R}^n$ satisfies the cone condition if, and only if, there exist $s \in \mathbb{N}$ and bounded domains Ω_k, which are star-shaped with respect to the balls $B_k \subset \overline{B_k} \subset \Omega_k, k = 1, ..., s$, such that $\Omega = \bigcup\limits_{k=1}^{s} \Omega_k$.*

2. *An unbounded open set $\Omega \subset \mathbb{R}^n$ satisfies the cone condition if, and only if, there exist bounded domains $\Omega_k, k \in \mathbb{N}$, which are star-shaped with respect to the balls $B_k \subset \overline{B_k} \subset \Omega_k$, $k \in \mathbb{N}$, and are such that*

1) $\Omega = \bigcup\limits_{k=1}^{\infty} \Omega_k,$

---

[12] I.e., $\forall \xi \in S'$ there exist $M(\xi) \ge 0$ and $\nu(\xi) > 0$ such that $\forall \eta \in S'$, for which $|\xi - \eta| \le \nu(\xi)$ we have $|\varphi(\xi) - \varphi(\eta)| \le M(\xi) \, d(\xi, \eta)$.

2) $0 < \inf\limits_{k \in \mathbb{N}} \operatorname{diam} B_k \le \sup\limits_{k \in \mathbb{N}} \operatorname{diam} \Omega_k < \infty$

and

3) *the multiplicity of the covering* $\varkappa\left(\{\Omega_k\}_{k=1}^{\infty}\right)$ *is finite.*

Idea of the proof. Sufficiency. By Lemma 3 Ω satisfies the cone condition with the parameters $c_6^2 c_7^{-1}$ and c_6, where [13]

$$c_6 = \inf\limits_{k=\overline{1,s}} \operatorname{diam} B_k, \qquad c_7 = \sup\limits_{k=\overline{1,s}} \operatorname{diam} \Omega_k,$$

$s \in \mathbb{N}$ for bounded Ω and $s = \infty$ for unbounded Ω.

Necessity. Consider for $x \in \Omega$, in addition to the cone K_x, the conic body $\widetilde{K_x}$ with the point x as a vertex, which is constructed on the ball $B(y(x), r_1)$ inscribed into the cone K_x (here $r_1 = rh/(r + \sqrt{r^2 + h^2})$) and the conic body K_x^* with the point $z(x) = x + \varepsilon_x \frac{x - y(x)}{|x - y(x)|}$ as a vertex, where $\varepsilon_x = \frac{1}{2} \min\{r_1, \operatorname{dist}(x, \partial\Omega)\}$, which is constructed on the same ball $B(y(x), r_1)$. Then $\Omega = \bigcup\limits_{x \in \Omega} K_x^*$. Choose $x_k \in \mathbb{R}^n, k \in \mathbb{N}$, such that $\mathbb{R}^n = \bigcup\limits_{k \in \mathbb{N}} B(x_k, \frac{r_1}{2})$ and the multiplicity of the covering [14] $\varkappa(\{B(x_k, \frac{r_1}{2})\}_{k \in \mathbb{N}}) \le 2^n$. Set

$$\omega_k = \Omega \bigcap B\left(x_k, \frac{r_1}{2}\right), \qquad G_k = \bigcup\limits_{x \in \Omega:\, y(x) \in \omega_k} K_x^*.$$

Then $\Omega = \bigcup\limits_{k=1}^{\infty} G_k$. Renumber those of G_k which are nonempty and denote them by $\Omega_1, \Omega_2, \ldots$. $\square$

Proof. Necessity. Suppose that $G_k \ne \varnothing$ and $\xi \in G_k$, then there exists $x \in \Omega$ such that $y(x) \in \omega_k$ and $\xi \in K_x^*$. Let us consider the conic body $\widetilde{K_\xi}$ with the point ξ as a vertex, which is constructed on the ball $B\left(x_k, \frac{r_1}{2}\right)$. Since $y(x) \in B\left(x_k, \frac{r_1}{2}\right)$ we have $B(y(x), r_1) \supset B\left(x_k, \frac{r_1}{2}\right)$ and $\widetilde{K_\xi} \subset K_x^* \subset \Omega$. Hence, the set G_k is star-shaped with respect to the ball $B\left(x_k, \frac{r_1}{2}\right)$. Furthermore,

$$|\xi - x_k| \le |z(x) - x_k| = |z(x) - x| + |x - y(x)| + |y(x) - x_k| \le h$$

because $|x - y(x)| = h - r_1$, therefore $G_k \subset B(x_k, h)$ and $\operatorname{diam} G_k \le 2h$.

---

[13] Here and in the sequel $k = \overline{1,s}$ where $s \in \mathbb{N}$ means $k \in \{1, \ldots, s\}$ and $k = \overline{1,\infty}$ means $k \in \mathbb{N}$.

[14] This is possible because the minimal multiplicity of the covering of $\mathbb{R}^n$ by balls of the same radius does not exceed 2^n.

Let us consider those of the sets G_k which are nonempty. If Ω is bounded, then there is a finite number of these sets – denote this number by s. If Ω is unbounded, then there is a countable number of these sets ($s = \infty$). Renumbering them and denoting by $\Omega_1, \Omega_2, ...$, we have $\Omega = \bigcup_{k=1}^{\infty} G_k = \bigcup_{k=1}^{s} \Omega_k$. Thus, for Ω_k, $k = \overline{1, s}$, the properties 1) and 2) are satisfied. Finally, [15]

$$\varkappa\left(\{\Omega_k\}_{k=1}^{s}\right) \leq \varkappa\left(\{B(x_k, h)\}_{k=1}^{\infty}\right) \leq 2^n\left(1 + \frac{2h}{r_1}\right)^n. \quad \square$$

Remark 7 In the above proof $c_6 = r_1$ and $c_7 \leq 2h$. Furthermore, $\dfrac{\operatorname{diam} \Omega_k}{\operatorname{diam} B_k} \leq 4\left(1 + \frac{h}{r}\right), k = \overline{1, s}$. It is also not difficult to verify that $\varkappa\left(\{\Omega_k\}_{k=1}^{s}\right) \leq c_8 \equiv 6^n\left(1 + \frac{h}{r}\right)^n$.

3.3   Multidimensional Taylor's formula

Theorem 3 *Let $\Omega \subset \mathbb{R}^n$ be a domain star-shaped with respect to the point $x_0 \in \Omega$, $l \in \mathbb{N}$ and $f \in C^l(\Omega)$. Then $\forall x \in \Omega$*

$$f(x) = \sum_{|\alpha| < l} \frac{(D^\alpha f)(x_0)}{\alpha!}(x - x_0)^\alpha$$

$$+l \sum_{|\alpha| = l} \frac{(x - x_0)^\alpha}{\alpha!} \int_0^1 (1 - t)^{l-1}(D^\alpha f)(x_0 + t(x - x_0))\, dt \qquad (3.36)$$

(here in addition to multi-notation used earlier we mean that $x_0 + t(x - x_0) = (x_{01} + t(x_1 - x_{01}), ..., x_{0n} + t(x_n - x_{0n})))$.

---

[15] We use the inequality

$$\varkappa\left(\{B(z_k, \varrho_2)\}_{k=1}^{s}\right) \leq \left(1 + \frac{\varrho_2}{\varrho_1}\right)^n \varkappa\left(\{B(z_k, \varrho_1)\}_{k=1}^{s}\right),$$

where $0 < \varrho_1 < \varrho_2 < \infty$. Since for $x \in \mathbb{R}^n$ the number $\varkappa(x)$ of the balls $B(z_k, \varrho_2) \ni x$ is equal to the number of the points $z_k \in B(x, \varrho_2)$, by inequality (2.60) we have

$$\varkappa(x)\operatorname{meas} B(0, \varrho_1) = \sum_{k:\, B(z_k, \varrho_2) \ni x} \operatorname{meas} B(z_k, \varrho_1)$$

$$\leq \varkappa\left(\{B(z_k, \varrho_1)\}_{k=1}^{s}\right)\operatorname{meas} \bigcup_{k:\, B(z_k, \varrho_2) \ni x} B(z_k, \varrho_1) \leq \varkappa\left(\{B(z_k, \varrho_1)\}_{k=1}^{s}\right)\operatorname{meas} B(x, \varrho_1 + \varrho_2),$$

and the desired inequality follows.

Idea of the proof. Consider for fixed x and x_0 the function φ of one variable defined for $0 \le t \le 1$ by $\varphi(t) = f(x_0 + t(x - x_0))$ and apply the one-dimensional Taylor's formula (3.10) with the remainder in integral form:

$$\varphi(1) = \sum_{k=0}^{l-1} \frac{\varphi^{(k)}(0)}{k!} + \frac{1}{(l-1)!} \int_0^1 (1-t)^{l-1} \varphi^{(l)}(t)\, dt. \quad \square$$

If $l = 1$, then (3.36) takes the form: $\forall f \in C^1(\Omega)$ and $\forall x \in \Omega$

$$f(x) = f(x_0) + \sum_{j=1}^n (x_j - x_{0j}) \int_0^1 \left(\frac{\partial f}{\partial x_j}\right)(x_0 + t(x - x_0))\, dt.$$

The analogue of this formula for the functions f, which have all the weak derivatives $\left(\frac{\partial f}{\partial x_j}\right)_w$ of the first order on Ω cannot have the same form, even for almost every $x \in \Omega$ because it contains the value $f(x_0)$ at a fixed point x_0. (For, suppose that this formula is valid for some such function f. Then it will not be valid for any function g, which coincides with f on Ω excluding the point x_0.)

For this reason we write the above formula in a different way. Suppose, that $\Omega \subset \mathbb{R}^n$ is an arbitrary open set and $h \in \mathbb{R}^n$, then it follows that $\forall f \in C^1(\Omega)$ and $\forall x \in \Omega_{|h|}$

$$f(x + h) = f(x) + \sum_{j=1}^n h_j \int_0^1 \left(\frac{\partial f}{\partial x_j}\right)(x + th)\, dt.$$

Lemma 5 *Let $\Omega \subset \mathbb{R}^n$ be an open set and $h \in \mathbb{R}^n$. If $f \in L_1^{loc}(\Omega)$ and for each $j = 1, ..., n$ the weak derivative $\left(\frac{\partial f}{\partial x_j}\right)_w$ exists on Ω, then for almost every $x \in \Omega_{|h|}$*

$$f(x + h) = f(x) + \sum_{j=1}^n h_j \int_0^1 \left(\frac{\partial f}{\partial x_j}\right)_w (x + th)\, dt$$

$$= f(x) + |h| \int_0^{|h|} \left(\frac{\partial f}{\partial \xi}\right)_w (x + \xi\tau)\, \tau = f(x) + \int_0^1 \left((\nabla_w f)(x + th) \cdot h\right) dt,$$

where $\xi = \frac{h}{|h|}$ and $\left(\frac{\partial f}{\partial \xi}\right)_w$ and $\nabla_w f$ are the weak derivative in the direction of ξ, the weak gradient respectively.

Idea of the proof. Apply mollification and pass to the limit. $\square$

Proof. Since $A_\delta f \in C^\infty(\Omega_\gamma)$ for each $0 < \delta \leq \gamma$, by Lemma 4 of Section 1.2 we have: $\forall x \in \Omega_{|h|+\gamma}$

$$(A_\delta f)(x + h) = (A_\delta f)(x) + \sum_{j=1}^{n} h_j \int_0^1 \left(A_\delta\left(\frac{\partial f}{\partial x_j}\right)_w\right)(x + th)\, dt.$$

We claim that

$$\int_0^1 \left(A_\delta\left(\frac{\partial f}{\partial x_j}\right)_w\right)(x + th)\, dt \to \int_0^1 \left(\frac{\partial f}{\partial x_j}\right)_w(x + th)\, dt$$

in $L_1^{loc}(\Omega_\gamma)$ as $\delta \to 0 +$. Indeed, for each compact $K \subset \Omega_\gamma$ by Minkowski's inequality

$$\left\|\int_0^1 \left(A_\delta\left(\frac{\partial f}{\partial x_j}\right)_w\right)(x + th)\, dt - \int_0^1 \left(\frac{\partial f}{\partial x_j}\right)_w(x + th)\, dt\right\|_{L_1(K)}$$

$$\leq \int_0^1 \left\|\left(A_\delta\left(\frac{\partial f}{\partial x_j}\right)_w\right)(x + th) - \left(\frac{\partial f}{\partial x_j}\right)_w(x + th)\right\|_{L_1(K)} dt$$

$$\leq \left\|A_\delta\left(\frac{\partial f}{\partial x_j}\right)_w - \left(\frac{\partial f}{\partial x_j}\right)_w\right\|_{L_1(K^{|h|})} \to 0$$

by (1.9) as $\delta \to 0 +$.

Consequently, there exists a sequence $\delta_k > 0$ such that $\delta_k \to 0+$ as $k \to \infty$ and

$$\int_0^1 \left(A_{\delta_k}\left(\frac{\partial f}{\partial x_j}\right)_w\right)(x + th)\, dt \to \int_0^1 \left(\frac{\partial f}{\partial x_j}\right)_w(x + th)\, dt$$

almost everywhere on Ω_γ. Moreover, by (1.5) $(A_{\delta_k} f)(x + h) \to f(x + h)$ and $(A_{\delta_k} f)(x) \to f(x)$ almost everywhere on Ω_γ. Thus, passing to the limit as $k \to \infty$, we obtain the desired equality for almost every $x \in \Omega_\gamma$ and, since $\gamma > 0$ was arbitrary, for almost every $x \in \Omega$. $\square$

We note that one can prove similarly that if $f \in L_1^{loc}(\Omega)$ and $\forall \alpha \in \mathbb{N}_0^n$ satisfying $|\alpha| = l$ there exist weak derivatives $D_w^\alpha f$ on Ω, then for almost every $x \in \Omega_{|h|}$

$$f(x + h) = \sum_{|\alpha|<l} \frac{(D_w^\alpha f)(x)}{\alpha!} h^\alpha + l \sum_{|\alpha|=l} \frac{h^\alpha}{\alpha!} \int_0^1 (1 - t)^{l-1} (D_w^\alpha f)(x + th)\, dt.$$

Corollary 7 *Let $\Omega \subset \mathbb{R}^n$ be an open set, $1 \le p \le \infty$, $h \in \mathbb{R}^n$ and $f \in w_p^l(\Omega)$. Then*

$$\|f(x+h) - f(x)\|_{L_p(\Omega_{|h|})} \le \| |\nabla_w f| \|_{L_p(\Omega)} |h| \le \|f\|_{w_p^l(\Omega)} |h|.$$

Idea of the proof. Apply Lemma 5 and Minkowsli's inequality. $\square$
Proof. By Lemma 5

$$\|f(x+h) - f(x)\|_{L_p(\Omega_{|h|})} = \left\| \int_0^1 ((\nabla_w f)(x+th) \cdot h)\, dt \right\|_{L_p(\Omega_{|h|})}$$

$$\le \int_0^1 \|(\nabla_w f)(x+th) \cdot h\|_{L_p(\Omega_{|h|})}\, dt \le |h| \int_0^1 \| |\nabla_w f|(x+th)\|_{\Omega_{|h|}}\, dt$$

$$= |h| \int_0^1 \| |\nabla_w f| \|_{\Omega_{|h|}+th)}\, dt \le \| |\nabla_w f| \|_{L_p(\Omega)} |h| \le \|f\|_{w_p^l(\Omega)} |h|$$

since $\Omega_{|h|} + th \subset \Omega$ and $|\nabla_w f| \le \sum_{j=1}^n \left| \left(\frac{\partial f}{\partial x_l}\right)_w \right|$. $\square$

Next consider for $l \in \mathbb{N}$ and $h \in \mathbb{R}^n$ the difference of order l of the function f with step h:

$$(\Delta_h^l f)(x) = \sum_{k=0}^l (-1)^{l-k} \binom{l}{k} f(x+kh).$$

Corollary 8 *Let $\Omega \subset \mathbb{R}^n$ be an open set, $l \in \mathbb{N}$, $1 \le p \le \infty$, $h \in \mathbb{R}^n$ and $f \in W_p^l(\Omega)$. Then*

$$\|\Delta_h^l f\|_{L_p(\Omega_{l|h|})} \le 2^l \|f\|_{L_p(\Omega)}, \quad \|\Delta_h^l f\|_{L_p(\Omega_{l|h|})} \le n^{l-1} |h|^l \|f\|_{w_p^l(\Omega)}.$$

Idea of the proof. The first inequality follows by Minkowski's inequality. To prove the second one apply Corollary 7 and take into account that $\frac{l!}{\alpha!} \le n^{l-1}$ if $\alpha \in \mathbb{N}_0^n$ satisfies $|\alpha| = l$. $\square$
Proof. By induction we get

$$\|\Delta_h^l f\|_{L_p(\Omega_{l|h|})} = \|\Delta_h(\Delta_h^{l-1} f)\|_{L_p((\Omega_{(l-1)|h|})_{|h|})}$$

$$\le |h| \sum_{j_1=1}^n \left\| \left(\frac{\partial(\Delta_h^{l-1} f)}{\partial x_{j_1}}\right)_w \right\|_{L_p(\Omega_{(l-1)|h|})} \le |h|^l \sum_{j_1=1}^n \cdots \sum_{j_l=1}^n \left\| \left(\frac{\partial^l f}{\partial x_{j_1} \cdots \partial x_{j_l}}\right)_w \right\|_{L_p(\Omega)}$$

$$= |h|^l \sum_{|\alpha|=l} \frac{l!}{\alpha!} \|D_w^\alpha f\|_{L_p(\Omega)} \le n^{l-1} |h|^l \|f\|_{w_p^l(\Omega)}. \square$$

3.4   Sobolev's integral representation

Theorem 4 *Let $\Omega \subset \mathbb{R}^n$ be a domain star-shaped with respect to the ball $B = B(x_0, r)$ such that $\bar{B} \subset \Omega$,*

$$\omega \in L_1(\mathbb{R}^n), \quad \operatorname{supp}\omega \subset \bar{B}, \quad \int_{\mathbb{R}^n} \omega \, dx = 1, \tag{3.37}$$

$l \in \mathbb{N}$ and $f \in C^l(\Omega)$. Then for every $x \in \Omega$

$$f(x) = \sum_{|\alpha| < l} \frac{1}{\alpha!} \int_B (D^\alpha f)(y)(x-y)^\alpha \omega(y) \, dy + \sum_{|\alpha| = l} \int_{V_x} \frac{(D^\alpha f)(y)}{|x-y|^{n-l}} w_\alpha(x,y) \, dy,$$

$$\tag{3.38}$$

where for $x, y \in \mathbb{R}^n$, $x \neq y$,

$$w_\alpha(x,y) = \frac{|\alpha|}{\alpha!} \frac{(x-y)^\alpha}{|x-y|^{|\alpha|}} w(x,y) \tag{3.39}$$

and

$$w(x,y) = \int_{|x-y|}^{\infty} \omega\left(x + \varrho \frac{y-x}{|y-x|}\right) \varrho^{n-1} \, d\varrho \tag{3.40}$$

(for $x = y \in \Omega$ we define $w_\alpha(x,x) = w(x,x) = 0$).

Remark 8 The first summand in right-hand side of (3.38) is a polynomial of degree less than or equal to $l - 1$ while the second one (the remainder) has the form of an integral of potential type.

Both summands in right-hand side of (3.38) consist of integrals containing the function f and its derivatives and does not involve the values of the function f and its derivative at particular points – thus, in contrast to Taylor's formula, this is an integral representation of the function f via the function f and its derivatives up to the order l.

Let $\xi = \frac{x-y}{|x-y|}$ and $k \in \mathbb{N}$. Consider the derivative of the function f in the direction of ξ of order k: $(\frac{\partial^k f}{\partial \xi^k})(y) = \sum_{|\alpha|=k} \binom{k}{\alpha}(D^\alpha f)(y)\xi^\alpha$. Then one may write (3.38) in the following way

$$f(x) = \int_B \left(\sum_{k=0}^{l-1} \frac{|x-y|^k}{k!} \left(\frac{\partial^k f}{\partial \xi^k}\right)(y) \right) \omega(y) \, dy$$

$$+\frac{1}{(l-1)!}\int\limits_{V_x}\left(\frac{\partial^l f}{\partial \xi^l}\right)(y)\,\frac{w(x,y)}{|x-y|^{n-l}}\,dy.$$

In particular,

$$f(x)=\int\limits_{B} f(y)\,\omega(y)\,dy+\int\limits_{V_x}\big((\nabla f)(y)\cdot(x-y)\big)\frac{w(x,y)}{|x-y|^n}\,dy.$$

Remark 9 Let us denote for $x\neq y$ by $M_{x,y}$ the ray, which goes from the point x through the point y, and by $L_{x,y}$ the "subray" of $M_{x,y}$, which goes from the point y. As the variable ϱ in (3.40) changes from $|x-y|$ to infinity, the argument $z=x+\varrho\frac{y-x}{|y-x|}$ of the function ω runs along the ray $L_{x,y}$. We note that $\varrho=|z-x|$ and that (3.40) may be written with the help of a line integral:

$$w(x,y)=\int\limits_{L_{x,y}}\omega(z)|z-x|^{n-1}\,dL.$$

The ray $L_{x,y}$ intersects the ball B if, and only if, $y\in K_x$. For this reason $\forall x\in\mathbb{R}^n$

$$\operatorname{supp}_y w_\alpha(x,y)=\operatorname{supp}_y w(x,y)\subset\overline{K_x}\tag{3.41}$$

(if ω is positive on B, then there is equality in (3.41)). Furthermore, $\forall x\in\mathbb{R}^n$ and $\forall y\in K_x$

$$\operatorname{supp}_\varrho \omega\left(x+\varrho\,\frac{y-x}{|y-x|}\right)\subset\overline{B\cap L_{x,y}}=[d_1,d_2].$$

Here $d_2=d_2(x,y)$ is the length of the segment of the ray $M_{x,y}$ contained in $\overline{K_x}$ while $d_1=d_1(x,y)=\max\{|x-y|,\widetilde{d_1}\}$, where $\widetilde{d_1}=\widetilde{d_1}(x,y)$ is the length of the segment of the same ray contained in $\overline{K_x\setminus B}$. [16]

Therefore, actually, the integral in (3.40) is equal to 0 if $y\notin K_x$ and is an integral over the finite segment $[d_1,d_2]$ if $y\in K_x$. We note that

$$d_2\le D,\quad d_2-d_1\le d,\tag{3.42}$$

where $D=\operatorname{diam}\Omega$ and $d=\operatorname{diam}B$.

---

[16] If $|x-x_0|=h$ and φ is the angle between the vectors $\overrightarrow{xx_0}$ and $\overrightarrow{xy}$, then $0\le\tan\varphi<\frac{r}{h}$ and $\widetilde{d_1}$ and d_2 are the minimal root, the maximal respectively, of the quadratic equation $d^2-2dh\cos\varphi+h^2-r^2=0$.

Remark 10 If Ω is bounded, then

$$\|w(x,y)\|_{C(\mathbb{R}^n \times \mathbb{R}^n)} \leq \|\omega\|_{L_\infty(\mathbb{R}^n)} \frac{d_2^n - d_1^n}{n} \leq \|\omega\|_{L_\infty(\mathbb{R}^n)} D^{n-1} d.$$

Moreover, $\forall \alpha \in \mathbb{N}_0^n$ satisfying $|\alpha| = l$

$$\|w_\alpha(x,y)\|_{C(\mathbb{R}^n \times \mathbb{R}^n)} \leq \|\omega\|_{L_\infty(\mathbb{R}^n)} n D^{n-1} d. \tag{3.43}$$

We have taken into account that

$$\left| \frac{(x-y)^\alpha}{|x-y|^l} \right| = \left(\frac{|x_1 - y_1|}{|x-y|} \right)^{\alpha_1} \cdots \left(\frac{|x_n - y_n|}{|x-y|} \right)^{\alpha_n} \leq 1$$

and that $\alpha! \geq \frac{l}{n}$ for $|\alpha| = l$.

Hence, if ω is bounded, then for bounded Ω the functions w and w_α are bounded on $\mathbb{R}^n \times \mathbb{R}^n$. If Ω is unbounded, then these functions are bounded on $K \times \mathbb{R}^n$ for each compact K.

If $\omega \in C^\infty(\mathbb{R}^n)$, then the functions $w(x,y)$ and $w_\alpha(x,y)$ have continuous derivatives of all orders $\forall x, y \in \mathbb{R}^n$ such that $x \neq y$. [17]

Remark 11 In the one-dimensional case for $\Omega = (a,b)$ and $B = (x_0 - r, x_0 + r) \equiv (\alpha, \beta)$, where $-\infty \leq a < \alpha < \beta \leq b \leq \infty$, we have $V_x = (a_x, b_x)$, where a_x and b_x are defined in Theorem 1. Moreover, $L_{x,y} = (-\infty, y)$ if $x < y$ and $L_{x,y} = (y, \infty)$ if $x > y$. Furthermore,

$$w(x,y) = \int_{L_{x,y}} \omega(z)\, dL = \begin{cases} \int_a^y \omega(u)\, du, & a \leq y \leq x, \\[2ex] \int_y^b \omega(u)\, du, & x < y \leq b, \end{cases}$$

---

[17] At the points (x,x), where $x \notin \bar{B}$ they are discontinuous. For $n > 1$ it follows from the fact that for each $y \in K_x \setminus B$ lying in some ray going from the point x (for all these y the vectors $\frac{y-x}{|y-x|}$ have the same value, say, $\nu = (\nu_1, \ldots, \nu_n)$) the function $w(x,y)$ has the same value $\gamma(x,\nu) = \int_0^\infty \omega(x + \varrho\nu)\varrho^{n-1}\, d\varrho$. Hence, the limit of $w(x,y)$ as y tends to x along this ray is also equal to $\gamma(x,\nu)$. Respectively for the function $w_\alpha(x,y)$ this limit is equal to $\frac{|\alpha|}{\alpha!}(-\nu_1)^{\alpha_1} \cdots (-\nu_n)^{\alpha_n}\gamma(x,\nu)$. The discontinuity follows from the fact that these limits depend on ν. For, if the ray defined by the vector ν does not intersect the ball B, then $\gamma(x,\nu) = 0$. On the other hand, there exists ν such that $\gamma(x,\nu) = 0$, otherwise $\int_{\mathbb{R}^n} \omega(z)\, dz = \int_S \left(\int_0^\infty \omega(x + \varrho\nu)\varrho^{n-1}\, d\varrho \right) dS = 0$, where S is the unit sphere in $\mathbb{R}^n$, which contradicts (3.37). For $n = 1$ the discontinuity follows from the formulas for w and w_α given in Remark 11.

and

$$w_l(x,y) = \frac{1}{(l-1)!}\frac{(x-y)^l}{|x-y|^l}w(x,y) = \frac{(\mathrm{sgn}\,(x-y))^{l-1}}{(l-1)!}\Lambda(x,y).$$

Thus, (3.38) takes the form (3.13).

Idea of the proof of Theorem 4. Multiply Taylor's formula (3.36) with $x_0 = y$ and $y \in B$, by $\omega(y)$ and integrate with respect to y over $\mathbb{R}^n$. (We assume that for $y \neq \overline{B}$ $\omega(y)g(y) = 0$ even if $g(y)$ is not defined at the point y.) The left-hand side of (3.36) does not change and the first summand in the right-hand side coincides with the first summand in the right-hand side of (3.38). As for the second summand it takes the form of the second summand in (3.38) after appropriate changes of variables. $\square$

Proof. After multiplying (3.36) with $x_0 = y$ by $\omega(y)$ and integrating with respect to y over $\mathbb{R}^n$ the second summand of the right-hand side of (3.36) takes the form

$$l\sum_{|\alpha|=l}\frac{1}{\alpha!}\int_{\mathbb{R}^n}\omega(y)\left(\int_0^1 (1-t)^{l-1}(D^\alpha f)(y+t(x-y))\,dt\right)dy$$

$$= l\sum_{|\alpha|=l}\frac{1}{\alpha!}\int_0^1 (1-t)^{l-1}\left(\int_{\mathbb{R}^n}(D^\alpha f)(y+t(x-y))\,\omega(y)\,dy\right)dt \equiv l\sum_{|\alpha|=l}\frac{1}{\alpha!}J_\alpha.$$

Changing variables $y + t(x-y) = z$ and taking into account that $(x-y)^\alpha = \frac{(x-z)^\alpha}{(1-t)^l}$, $dy = \frac{dz}{(1-t)^n}$, we establish that

$$J_\alpha = \int_{\mathbb{R}^n}(D^\alpha f)(z)(x-z)^\alpha\left(\int_0^1 \omega\left(\frac{z-tx}{1-t}\right)\frac{dt}{(1-t)^{n+1}}\right)dz.$$

Replacing $\frac{|x-z|}{1-t}$ by ϱ, we have

$$J_\alpha = \int_{\mathbb{R}^n}(D^\alpha f)(z)\frac{(x-z)^\alpha}{|x-z|^l}\left(\int_{|x-z|}^\infty \omega\left(x+\varrho\frac{z-x}{|z-x|}\right)\varrho^{n-1}\,d\varrho\right)dz,$$

which by (3.41) gives (3.38). $\square$

Remark 12 One can replace the ball B in the assumptions of Theorem 4 by some other open set G_x depending in general on $x \in \Omega$ such [18] that $\overline{G_x} \subset \Omega$ and replace the function ω by some function ω_x such that

$$\omega_x \in L_1(\mathbb{R}^n), \quad \operatorname{supp} \omega_x \subset \overline{G_x}, \quad \int_{\mathbb{R}^n} \omega_x(y)\, dy = 1.$$

In this case the same argument as above leads to the integral representation (3.38) in which B, ω and the conic body V_x are replaced by G_x, ω_x, the conic body $V_{x,G_x} = \bigcup_{y \in G_x} (x, y)$ respectively. We shall use this fact in Corollaries $10 - 12$ below.

Remark 13 Set $\omega_{(r)} = r^n \omega(x_0 - rx)$. Then $\omega_{(r)}$ is a kernel of mollification in the sense of Section 1.1 (in general, a non-smooth one since we have only that $\omega_{(r)} \in L_1(\mathbb{R}^n)$). The polynomial $S_{l-1}(x, x_0)$, the first summand in Sobolev's integral representation (3.38), is Taylor's polynomial averaged over the ball $B \equiv B(x_0, r)$ in the following sense:

$$S_{l-1}(x, x_0) = (A_r T_{l-1}(x, \cdot))(x_0).$$

Here $T_{l-1}(x, y)$ is Taylor's polynomial of the function f with respect to the point y, A_δ is the mollifier with the kernel $\omega_{(r)}$ and the mollification is carried out with respect to the variable y. For,

$$(A_r P_{l-1}(x, \cdot))(x_0) = \int_{B(0,1)} P_{l-1}(x, x_0 - rz)\, \omega_{(r)}(z)\, dz$$

$$= \sum_{|\alpha| < l} \frac{1}{\alpha!} \int_{B(0,1)} (D^\alpha f)(x_0 - rz)(x - x_0 + rz)^\alpha r^n \omega(x_0 - rz)\, dz$$

$$= \sum_{|\alpha| < l} \frac{1}{\alpha!} \int_{B(x_0,r)} (D^\alpha f)(y)(x - y)^\alpha \omega(y)\, dy \equiv (S_{l-1} f)(x, x_0).$$

This allows us to characterize Sobolev's integral representation as a "mollified (averaged) Taylor's formula with the remainder written in the form of an integral of potential type" of briefly "averaged Taylor's formula".

---

[18] It is also possible to suppose only that $G_x \subset \Omega$ replacing the assumption $f \in C^l(\Omega)$ by $f \in \overline{C}^l(\Omega)$ in the case in which $\overline{G_x} \cap \partial\Omega \neq \varnothing$. See a detailed Remark 1 for the one-dimensional case.

Theorem 5 *Let $\Omega \subset \mathbb{R}^n$ be a domain star-shaped with respect to the ball $B = B(x_0, r)$ such that $\overline{B} \subset \Omega$,*

$$\omega \in L_\infty(\mathbb{R}^n), \quad \operatorname{supp}\omega \subset \overline{B}, \quad \int_{\mathbb{R}^n} \omega \, dx = 1, \tag{3.44}$$

$l \in \mathbb{N}$ and $f \in (W_1^l)^{loc}(\Omega)$. Then for almost every $x \in \Omega$

$$f(x) = \sum_{|\alpha|<l} \frac{1}{\alpha!} \int_B (D_w^\alpha f)(y)(x-y)^\alpha \omega(y)\, dy + \sum_{|\alpha|=l} \int_{V_x} \frac{(D_w^\alpha f)(y)}{|x-y|^{n-l}} w_\alpha(x,y)\, dy. \tag{3.45}$$

Idea of the proof. For $0 < \delta < \operatorname{dist}(B, \partial\Omega)$, which is such that $-\frac{1}{\delta} < |x_0| - r < |x_0| + r < \frac{1}{\delta}$ set [19] $\Omega_{[\delta]} = \{x \in \Omega : V_x \subset \Omega_\delta \cap B\left(0, \frac{1}{\delta}\right)\} \subset \Omega_\delta$, write (3.38) for $A_\gamma f \in C^\infty(\Omega_{[\delta]})$ where $0 < \gamma \le \delta$ and pass to the limit as $\gamma \to 0+$. $\square$

Proof. For each δ and γ such that $0 < \gamma < \delta$ and $\forall x \in \Omega_{[\delta]}$ we have

$$(A_\gamma f)(x)$$

$$= \sum_{|\alpha|<l} \frac{1}{\alpha!} \int_B (D^\alpha(A_\gamma f))(y)(x-y)^\alpha \omega(y)\, dy + \sum_{|\alpha|=l} \int_{V_x} \frac{(D^\alpha A_\gamma f)(y)}{|x-y|^{n-l}} w_\alpha(x,y)\, dy$$

$$= \sum_{|\alpha|<l} \frac{1}{\alpha!} \int_B (A_\gamma(D_w^\alpha f))(y)(x-y)^\alpha \omega(y)\, dy + \sum_{|\alpha|=l} \int_{V_x} \frac{(A_\gamma(D_w^\alpha f))(y)}{|x-y|^{n-l}} w_\alpha(x,y)\, dy.$$

We have applied Lemma 4 of Chapter 1 and the fact that the weak derivatives $D_w^\alpha f$ where $|\alpha| < l$ exist (Lemma 6 of Chapter 1).

By (1.5)

$$A_\gamma f \to f \quad \text{a.e. on} \quad \Omega_{[\delta]} \tag{3.46}$$

as $\gamma \to 0+$. We shall prove that

$$R_{\alpha,\gamma}(x) \equiv \int_B (A_\gamma(D_w^\alpha f))(y)(x-y)^\alpha \omega(y)\, dy$$

$$\to \int_B (D_w^\alpha f)(y)(x-y)^\alpha \omega(y)\, dy \equiv R_\alpha(x) \quad \text{on} \quad \Omega_{[\delta]} \tag{3.47}$$

---

[19] The necessity of introducing of these more complicated sets than Ω_δ arises in connection with Example 8.

and

$$S_{\alpha,\gamma}(x) \equiv \int\limits_{K_x} \frac{(A_\gamma(D_w^\alpha f))(y)}{|x-y|^{n-l}} w_\alpha(x,y)\,dy \to$$

$$\to \int\limits_{K_x} \frac{(D_w^\alpha f)(y)}{|x-y|^{n-l}} w_\alpha(x,y)\,dy \equiv S_\alpha(x) \quad \text{in} \ \ L_1(\Omega_{[\delta]}). \tag{3.48}$$

The relation (3.47) follows from the estimate

$$|R_{\alpha,\gamma}(x) - R_\alpha(x)| \le M_1 \int\limits_B |(A_\gamma(D_w^\alpha f))(y) - (D_w^\alpha f)(y)|\,dy,$$

where $M_1 = \|(x-y)^\alpha\|_{C(\Omega_{[\delta]}\times B)}\|\,\omega(y)\|_{L_\infty(\mathbb{R}^n)} < \infty$ and the property (1.9).

Set $M_2 = \|w_\alpha(x,y)\|_{L_\infty(\Omega_{[\delta]}\times\mathbb{R}^n)}$. Then by Remark 10 $M_2 < \infty$. Applying the inclusions $V_x \subset \Omega_\delta^* \equiv \Omega_\delta \cap B\left(0,\frac{1}{\delta}\right)$ for $x \in \Omega_{[\delta]}$ and $\Omega_{[\delta]} \subset \Omega_\delta^*$ we obtain

$$\|S_{\alpha,\gamma}(x) - S_\alpha(x)\|_{L_1(\Omega_{[\delta]})} \le M_2 \left\|\ \int\limits_{\Omega_\delta^*} \frac{|(A_\gamma(D_w^\alpha f))(y) - (D_w^\alpha f)(y)|}{|x-y|^{n-l}}\,dy\ \right\|_{L_1(\Omega_\delta^*)}$$

$$= M_2 \int\limits_{\Omega_\delta^*} |A_\gamma(D_w^\alpha f) - D_w^\alpha f|\left(\int\limits_{\Omega_\delta^*} \frac{dx}{|x-y|^{n-l}}\right) dy$$

$$\le M_2 \|\,|z|^{l-n}\|_{L_1(\Omega_\delta^* - \Omega_\delta^*)}\|A_\gamma(D_w^\alpha f) - D_w^\alpha f\|_{L_1(\Omega_\delta^*)}$$

and [20] (1.9) implies (3.48).

From (3.48) it follows that, for some $\gamma_k > 0, k \in \mathbb{N}$, such that $\gamma_k \to 0$ as $k \to \infty$, we have

$$S_{\alpha,\gamma_k}(x) \longrightarrow S_\alpha(x) \quad \text{a.e. on} \ \ \Omega_{[\delta]}. \tag{3.49}$$

Now passing to the limit as $k \to \infty$ in equality for $A_\gamma f$ with γ_k replacing γ, by (3.46), (3.47) and (3.49) we obtain that (3.45) holds almost everywhere on $\Omega_{[\delta]}$. Since $\bigcup\limits_{\delta>0} \Omega_{[\delta]} = \Omega$ we establish that (3.45) is valid almost everywhere on Ω. $\square$

---

[20] In the last inequality in the expression $\Omega_\delta^* - \Omega_\delta^*$ the sign $-$ denotes vector subtraction of sets in $\mathbb{R}^n$, i.e., $A - B = \{z \in \mathbb{R}^n : z = x - y \ \text{where} \ x \in A, y \in B\}$. Clearly,

$$\Omega_\delta^* - \Omega_\delta^* \subset B(0,\delta^{-1}) - B(0,\delta^{-1}) = B(0,2\delta^{-1}).$$

3.5 Corollaries

Corollary 9 *In Theorems* 4, 5 *let conditions* (3.37), (3.44) *respectively, be replaced by*

$$\omega \in C_0^\infty(\Omega), \quad \operatorname{supp}\omega \subset \overline{B}, \quad \int_{\mathbb{R}^n} \omega\, dx = 1. \tag{3.50}$$

Then $\forall f \in C^l(\Omega)$ *for every* $x \in \Omega$ *and* $\forall f \in (W_1^l)^{loc}(\Omega)$ *for almost every* $x \in \Omega$

$$f(x) = \int_B \left(\sum_{|\alpha|<l} \frac{(-1)^{|\alpha|}}{\alpha!} D_y^\alpha[(x-y)^\alpha \omega(y)] \right) f(y)\, dy$$

$$+ \sum_{|\alpha|=l} \int_{V_x} \frac{(D^\alpha f)(y)}{|x-y|^{n-l}} w_\alpha(x, y)\, dy \tag{3.51}$$

with $D_w^\alpha f$ *replacing* $D^\alpha f$ *in the case in which* $f \in (W_1^l)^{loc}(\Omega)$.

Idea of the proof. For $f \in C^l(\Omega)$ — integration by parts in the first summand of the right-hand side of (3.38). For $f \in (W_1^l)^{loc}(\Omega)$ — the same limiting procedure as in the proof of Theorem 5, starting from (3.51) with $A_\gamma f$ replacing f. $\square$

Corollary 10 *In addition to the assumptions of Corollary* 9 *let* $\beta \in \mathbb{N}_0^n$ *and* $0 < |\beta| < l$. *Then* $\forall f \in C^l(\Omega)$ *for every* $x \in \Omega$ *and* $\forall f \in (W_1^l)^{loc}(\Omega)$ *for almost every* $x \in \Omega$

$$(D^\beta f)(x) = \int_B \left(\sum_{|\alpha|<l-|\beta|} \frac{(-1)^{|\alpha|+|\beta|}}{\alpha!} D_y^{\alpha+\beta}[(x-y)^\alpha \omega(y)] \right) f(y)\, dy$$

$$+ \sum_{|\alpha|=l, \alpha \geq \beta} \int_{V_x} \frac{(D^\alpha f)(y)}{|x-y|^{n-l+|\beta|}} w_{\alpha-\beta}(x, y)\, dy \tag{3.52}$$

with $D_w^\beta f$ *replacing* $D^\beta f$ *and* $D_w^\alpha f$ *replacing* $D^\alpha f$ *if* $f \in (W_1^l)^{loc}(\Omega)$.

Idea of the proof. For $f \in C^l(\Omega)$ write equality (3.38) with $D^\beta f \in C^{l-|\beta|}(\Omega)$ replacing f and with $l - |\beta|$ replacing l, integrate additionally by parts in the first summand of the right-hand side, change the multi-index of summation α to $\gamma - \beta$ (then $\sum_{|\alpha|=l-|\beta|} = \sum_{|\gamma|=l, \gamma \geq \beta}$) and write α instead of γ. For $f \in (W_1^l)^{loc}(\Omega)$ apply the limiting procedure from the proof of Theorem 4. $\square$

Remark 14 For $n = 1$ equality (3.52) takes the form (3.17). As in the one-dimesional case the first summand in (3.51) and (3.52) can be rewritten in the form

$$\int_B \Big(\sum_{|\alpha|<l,\,\alpha\geq\beta} \sigma_{\alpha,\beta}(x-y)^{\alpha-\beta}\big(D^\alpha\omega(y)\big)\Big) f(y)\,dy, \qquad (3.53)$$

where $\sigma_{\alpha,\beta}$ depends only on α and β.

Corollary 11 *Let $l \in \mathbb{N}$ and $\Omega \subset \mathbb{R}^n$ be a bounded domain star-shaped with respect to the ball B with the parameters d, D, i.e., $d \leq \operatorname{diam} B \leq \operatorname{diam} \Omega \leq D < \infty$. Then there exists $c_9 > 0$, depending only on n, l, d and D, such that $\forall f \in C^l(\Omega)$ for every $x \in \Omega$ and $\forall f \in (W_1^l)^{loc}(\Omega)$ for almost every $x \in \Omega$*

$$|(D^\beta f)(x)| \leq c_9\Big(\int_B |f|\,dy + \sum_{|\alpha|=l}\int_{V_x} \frac{|(D^\alpha f)(y)|}{|x-y|^{n-l+|\beta|}}\,dy \Big), \qquad (3.54)$$

where $\beta \in \mathbb{N}_0^n$ and $0 < |\beta| < l$, with $D_w^\beta f$ replacing $D^\beta f$ and $D_w^\alpha f$ replacing $D^\alpha f$ in the case in which $f \in (W_1^l)^{loc}(\Omega)$. Moreover,

$$|(D^\beta f)(x)| \leq c_{10}\Big(\Big(\frac{D}{d}\Big)^{l-1} D^{-|\beta|}d^{-n}\int_B |f|\,dy$$

$$+\Big(\frac{D}{d}\Big)^{n-1} \sum_{|\alpha|=l}\int_{V_x} \frac{|(D^\alpha f)(y)|}{|x-y|^{n-l+|\beta|}}\,dy \Big), \qquad (3.55)$$

where $c_{10} > 0$ depends only on n and l.

Idea of the proof. Let μ be a fixed kernel of the mollifier defined by (1.1). In equality (3.52) take $\omega(x) = (\frac{2}{d})^n \mu(\frac{2(x-x_0)}{d})$ and apply (3.53) and (3.43) (see also Corollary 3). $\sqcap$

Proof. First of all $\forall x \in \Omega$ and $\forall y \in B$

$$|(x-y)^{\alpha-\beta}(D^\alpha\omega)(y)| \leq \Big(\frac{2}{d}\Big)^{n+\alpha} |x-y|^{|\alpha|-|\beta|}\Big|(D^\alpha\mu)\Big(\frac{2(x-x_0)}{d}\Big)\Big|$$

$$\leq M_1\|D^\alpha\mu\|_{C(\mathbb{R}^n)}\Big(\frac{D}{d}\Big)^{|\alpha|} D^{-|\beta|}d^{-n} \leq M_2\Big(\frac{D}{d}\Big)^{l-1} D^{-|\beta|}d^{-n},$$

where M_1 and M_2 depend only on n and l (since μ is fixed). Hence

$$\Big|\Big(\sum_{|\alpha|<l,\,\alpha\geq\beta} \sigma_{\alpha,\beta}(x-y)^{\alpha-\beta}(D^\alpha\omega)(y)\Big)\Big| \leq M_3\Big(\frac{D}{d}\Big)^{l-1} D^{-|\beta|}d^{-n}, \qquad (3.56)$$

where M_3 depends only on n and l.

Secondly from (3.43) it follows that $\forall x \in \Omega$ and $y \in V_x$

$$|w_{\alpha-\beta}(x,y)| \leq M_4 \left(\frac{D}{d}\right)^{n-1}, \tag{3.57}$$

where M_4 depends only on n and l.

So (3.53), (3.56) and (3.57) imply (3.55) and hence (3.54). $\square$

Remark 15 For $n = 1$ inequalities (3.54), (3.55) take the form of the first inequality (3.23), the form of (3.21) respectively.

Moreover, if $\Omega = B$ and $\operatorname{diam} B = d$, then (3.55) implies that

$$|(D^\beta f)(x)| \leq M_5 \left(d^{-|\beta|-n} \int_B |f|\, dy + \sum_{|\alpha|=l} \int_B \frac{|(D^\alpha f)(y)|}{|x-y|^{n-l+|\beta|}}\, dy\right), \tag{3.58}$$

where M_5 depends only on n and l, which is a multidimensional analogue of the first inequality (3.22). If, in addition, $l - |\beta| - n \geq 0$, then

$$|(D^\beta f)(x)| \leq M_5 \left(d^{-|\beta|-n} \int_B |f|\, dy + d^{l-|\beta|-n} \sum_{|\alpha|=l} \int_B |(D^\alpha f)(y)|\, dy\right),$$

which is a multidimensional analogue of the second inequality (3.22).

Remark 16 If $l - |\beta| - n \geq 0$ (in the one-dimensional case this condition is always satisfied), then there is no singularity in the integrals of the second summand in the righ-hand side of (3.54). In this case (3.54) implies the inequality

$$|(D^\beta f)(x)| \leq c_{11} \left(\int_B |f|\, dy + \sum_{|\alpha|=l} \int_{V_x} |(D^\alpha f)(y)|\, dy\right)$$

$$\leq c_{11} \left(\int_B |f|\, dy + \sum_{|\alpha|=l} \int_\Omega |(D^\alpha f)(y)|\, dy\right), \tag{3.59}$$

where $c_{11} > 0$ depends only on n, l, d and D.

Applying the same procedure as in the second proof of Corollary 4 one can obtain the related inequality with a small parameter: if $l - |\beta| - n > 0$, then

$$|(D^\beta f)(x)| \leq c_{12} K(\varepsilon) \int_B |f|\, dy + \varepsilon \sum_{|\alpha|=l} \int_B |(D^\alpha f)(y)|\, dy, \tag{3.60}$$

where now $0 < \varepsilon \leq M_5\, d^{l-|\beta|-n}$,

$$K(\varepsilon) = \varepsilon^{-\frac{|\beta|+n}{l-|\beta|-n}} \tag{3.61}$$

and $c_{12} > 0$ depends only on n and l.

Corollary 12 *Let $l \in \mathbb{N}$ and $\Omega \subset \mathbb{R}^n$ be a bounded domain star-shaped with respect to a ball with the parameters d, D. Suppose that $f \in C^l(\Omega)$ (or $f \in (W_1^l)^{loc}(\Omega)$). Then there exists a polynomial $p_{l-1}(x, f)$ of degree less than or equal to $l - 1$ such that $\forall \beta \in \mathbb{N}_0^n$ satisfying $|\beta| < l$ and $\forall x \in \Omega$*

$$|(D^\beta f)(x) - (D^\beta p_{l-1})(x, f)| \le c_{13} \left(\frac{D}{d}\right)^{n-1} \sum_{|\alpha|=l} \int_\Omega \frac{|(D^\alpha f)(y)|}{|x - y|^{n-l+|\beta|}}\, dy, \quad (3.62)$$

where $c_{13} > 0$ depends only on n and l. (If $f \in (W_1^l)^{loc}(\Omega)$, then (3.62) with $D_w^\beta f$ and $D_w^\alpha f$ replacing $D^\beta f$ and $D^\alpha f$ holds for almost every $x \in \Omega$).

Idea of the proof. Set

$$p_{l-1}(x, f) = S_{l-1}(x, f) \equiv \int_B \left(\sum_{|\alpha|<l} \frac{(-1)^{|\alpha|}}{\alpha!} D_y^\alpha[(x - y)^\alpha \omega(y)]\right) f(y)\, dy$$

(this is the first summand in (3.51)), where $B \subset \Omega$ is such that Ω is star-shaped with respect to B and ω is the same as in the proof of Corollary 11. Note that

$$S_{l-|\beta|-1}(x, D^\beta f) = (D^\beta S_{l-1})(x, f) \tag{3.63}$$

and apply (3.51), (3.52) and (3.56). $\square$

Corollary 13 *Let Ω be a bounded convex domain. For $x, y \in \Omega$ ($x \ne y$) denote by $d(x, y)$ the length of the segment of the ray, which goes from the point x through the point y, contained in $\overline{\Omega}$. Then $\forall f \in \overline{C}^l(\Omega)$ and for all $x \in \Omega$ and $\forall f \in (W_1^l)^{loc}(\Omega)$ for almost all $x \in \Omega$*

$$f(x) = \frac{1}{\operatorname{meas}\Omega} \left\{ \sum_{|\alpha|<l} \frac{1}{\alpha!} \int_\Omega (D^\alpha f)(y)(x - y)^\alpha\, dy \right.$$

$$+ \frac{l}{n} \sum_{|\alpha|=l} \frac{1}{\alpha!} \int_\Omega (D^\alpha f)(y) \frac{(x - y)^\alpha}{|x - y|^n} (d(x, y)^n - |x - y|^n)\, dy \tag{3.64}$$

and hence

$$|f(x)| \le \frac{1}{\operatorname{meas}\Omega} \left\{ \sum_{|\alpha|<l} \frac{1}{\alpha!} \int_\Omega d(x, y)^n |(D^\alpha f)(y)|\, dy \right.$$

$$+ \frac{l}{n} \sum_{|\alpha|=l} \frac{1}{\alpha!} \int_\Omega \frac{d(x, y)^n}{|x - y|^{n-l}} |(D^\alpha f)(y)|\, dy. \tag{3.65}$$

In particular,

$$|f(x)| \leq \frac{1}{\operatorname{meas} \Omega} \sum_{|\alpha| \leq n} \frac{D^{|\alpha|}}{\alpha!} \int_\Omega |(D^\alpha f)(y)| \, dy, \qquad (3.66)$$

where $D = \operatorname{diam} \Omega$. (If $f \in (W_1^l)^{loc}(\Omega)$, then $D^\alpha f$ must be replaced by $D_w^\alpha f$.)

Idea of the proof. Suppose that in Theorems 4–5 $\operatorname{supp} \omega \subset \overline{\Omega}$ instead of $\operatorname{supp} \omega \subset B$. Then in (3.38) we have $V_{x,\Omega} = \Omega$ instead of $V_x \equiv V_{x,B}$ – see Remark 12. Take $\omega(x) = (\operatorname{meas} \Omega)^{-1}$ for $x \in \Omega$, then (3.64) follows from (3.38) – (3.40). $\square$

Corollary 14 *Let $\Omega \subset \mathbb{R}^n$ be an open set, $l \in \mathbb{N}$. Then $\forall f \in C_0^l(\Omega)$ for every $x \in \Omega$ and $\forall f \in (W_1^l)_0(\Omega)$ for almost every $x \in \Omega$*

$$f(x) = \frac{l}{\sigma_n} \sum_{|\alpha|=l} \frac{1}{\alpha!} \int_\Omega \frac{(x-y)^\alpha}{|x-y|^n} (D^\alpha f)(y) \, dy \qquad (3.67)$$

and hence

$$|f(x)| \leq \frac{l}{\sigma_n} \sum_{|\alpha|=l} \frac{1}{\alpha!} \int_\Omega \frac{|(D^\alpha f)(y)|}{|x-y|^{n-l}} \, dy \qquad (3.68)$$

with $D_w^\alpha f$ replacing $D^\alpha f$ for $f \in (W_1^l)_0(\Omega)$, where v_n is the n-dimensional measure of the unit ball in $\mathbb{R}^n$ and σ_n is the surface area of the unit sphere in $\mathbb{R}^n$ $(\sigma_n = n v_n)$. [21]

Idea of the proof. Since $\operatorname{supp} f$ is compact in Ω one can assume, without loss of generality, that the function f is defined on $\mathbb{R}^n$ and $f = 0$ outside Ω. Replace in Theorems $4 - 5$, keeping in mind Remark 12, B by $G_x = B(x,r_2) \setminus B(x,r_1)$, where $r_1 < r_2$ are such that $\operatorname{supp} f \subset B(x,r_1)$ and ω by $\omega_x(y) = (\operatorname{meas} G_x)^{-1}$, $y \in G_x$. Then $V_{x,G_x} = B(x,r_2)$.

Moreover, from (3.40) it follows that $\forall y \in \operatorname{supp} f$

$$w(x,y) = \frac{1}{v_n(r_2^n - r_1^n)} \int_{r_1}^{r_2} \varrho^{n-1} \, d\varrho = \frac{1}{\sigma_n}.$$

Since $f = 0$ on G_x equality (3.67) follows from (3.28). Furthermore, inequality (3.68) follows from (3.67) because $\alpha! \geq \frac{n}{l}$ for $|\alpha| = l$. $\square$

---

[21] We recall that $v_n = \frac{\pi^{\frac{n}{2}}}{\Gamma(\frac{n}{2}+1)}$, where $\Gamma(u) = \int_0^\infty x^{u-1} e^{-x} \, dx$, $u > 0$, is the gamma-function.

Remark 17 For $n = 1$, $l = 1$ and $\Omega = (a, b)$ equality (3.65) reduces to the obvious equality: $\forall f \in C_0^1(a, b)$ and $\forall x \in (a, b)$

$$f(x) = \frac{1}{2} \int\limits_a^b \operatorname{sgn}(x - y)\, f'(y)\, dy \tag{3.69}$$

(see (3.7)). Starting from this equality it is possible to give another proof of equality (3.67).

Remark 18 Consider the following particular case of (3.67)

$$f(x) = \frac{1}{\sigma_n} \int \frac{(\nabla f)(y) \cdot (x - y)}{|x - y|^n}\, dy = \frac{1}{\sigma_n} \sum_{j=1}^n \frac{x_j}{|x|^n} * \nabla f. \tag{3.70}$$

We note that

$$\frac{x_j}{|x|^2} = \frac{\partial}{\partial x_j}(\ln |x|),\ n = 2, \quad \frac{x_j}{|x|^n} = -\frac{1}{n-2}\frac{\partial}{\partial x_j}(|x|^{2-n}),\ n \geq 3,$$

and for $\varphi \in (W_1^1)^{loc}(\mathbb{R}^n)$, $\psi \in C_0^1(\mathbb{R}^n)$ we have $\frac{\partial}{\partial x_j}(\varphi * \psi) = \frac{\partial \varphi}{\partial x_j} * \psi = \varphi * \frac{\partial \psi}{\partial x_j}$. Consequently, $\forall f \in C_0^2(\mathbb{R}^n)$

$$f(x) = \frac{1}{2\pi} \ln \frac{1}{|x|} * \Delta f, \quad n = 2 \tag{3.71}$$

(*logarithmic potential*) and

$$f(x) = -\frac{1}{(n-2)\sigma_n} |x|^{2-n} * \Delta f, \quad n \geq 3 \tag{3.72}$$

(*the Newton potential*).

Corollary 15 *Let $\Omega \subset \mathbb{R}^n$ be an open set satisfying the cone condition with the parameters $r > 0$ and $h > 0$ and K_x be the cone of that condition. Suppose that $\forall x \in \Omega$*

$$\omega_x \in C_0^\infty(\mathbb{R}^n), \quad \operatorname{supp}\omega_x \subset \overline{K_x}, \quad \int\limits_{\mathbb{R}^n} \omega_x(y)\, dy = 1, \tag{3.73}$$

$l \in \mathbb{N}$, $\beta \in \mathbb{N}_0^n$ and $|\alpha| < l$. Then $\forall f \in C^l(\Omega)$ for every $x \in \Omega$ and $\forall f \in (W_1^l)^{loc}(\Omega)$ for almost every $x \in \Omega$

$$(D^\beta f)(x) = \int\limits_{K_x} \left(\sum_{|\alpha|<l-|\beta|} \frac{(-1)^{|\alpha|+|\beta|}}{\alpha!} D_y^{\alpha+\beta}[(x-y)^\alpha \omega_x(y)] \right) f(y)\, dy$$

$$+ \sum_{|\alpha|=l, \ \alpha \geq \beta} \int_{K_x} \frac{(D^\alpha f)(y)}{|x-y|^{n-l+|\beta|}} w_{\alpha-\beta,x}(x,y) \, dy \qquad (3.74)$$

(with $D_w^\beta f$ and $D_w^\alpha f$ replacing $D^\beta f$ and $D^\alpha f$ for $f \in (W_1^l)^{loc}(\Omega))$, where

$$w_{\alpha,x}(x,y) = \frac{|\alpha|}{\alpha!} \frac{(x-y)^\alpha}{|x-y|^n} \int_{|x-y|}^{\sqrt{h^2+r^2}} \omega_x \left(x + \varrho \frac{y-x}{|y-x|} \right) \varrho^{n-1} \, d\varrho. \qquad (3.75)$$

Remark 19 In contrast to other integral representations the first summand in (3.74) is no more a polynomial.

Idea of the proof. Apply Remark 12 with $G_x = K_x$ and ω_x replacing ω. Note that $V_{x,K_x} = K_x$ and $d_2 \leq \sqrt{h^2+r^2}$ (d_2 is defined in Remark 8). $\square$

Corollary 16 *Let $\Omega \subset \mathbb{R}^n$ be an open set satisfying the cone condition with the parameters $r > 0$ and $h > 0$, $l \in \mathbb{N}$, $\beta \in \mathbb{N}_0^n$ and $|\beta| < l$. Then there exists $c_{14} > 0$, depending only on n, l, r and h, such that $\forall f \in C^l(\Omega)$ for every $x \in \Omega$ and $\forall f \in (W_1^l)^{loc}(\Omega)$ for almost every $x \in \Omega$*

$$|(D^\beta f)(x)| \leq c_{14} \left(\int_\Omega |f| \, dy + \sum_{|\alpha|=l} \int_\Omega \frac{|(D^\alpha f)(y)|}{|x-y|^{n-l+|\beta|}} \, dy \right) \qquad (3.76)$$

(with $D_w^\beta f$ and $D_w^\alpha f$ replacing $D^\beta f$, $D^\alpha f$ respectively, for $f \in (W_1^l)^{loc}(\Omega))$.

Moreover,

$$|(D^\beta f)(x)| \leq c_{15} \left(\left(\frac{h}{r_1} \right)^{-|\beta|} r_1^{-n} \int_{K_x} |f| \, dy + \left(\frac{h}{r_1} \right)^{n-1} \sum_{|\alpha|=l} \int_{K_x} \frac{|(D^\alpha f)(y)|}{|x-y|^{n-l+|\beta|}} \, dy \right),$$
$$(3.77)$$

where $r_1 = \min\{r, h\}$ and $c_{15} > 0$ depends only on n and l.

Remark 20 Compared with (3.54) inequality (3.76) is valid for a wider class of open sets satisfying the cone condition. On the other hand, (3.54) is a sharper version of (3.76) (for in the right-hand side of (3.54) $\int_B |f| \, dx$ replaces $\int_\Omega |f| \, dx$) for a narrower class of domains star-shaped with respect to the ball B.

Idea of the proof. Let $K = K(r, h)$ if $h \geq r$ and $K = K(h, h)$ if $h < r$, and let $B(x_0, r_2)$ be the ball inscribed into the cone K. Here $r_2 = rh(\sqrt{r^2+h^2}+r)^{-1} \geq$

$r(1+\sqrt{2})^{-1}$ if $h \geq r$ and $r_2 = h(1+\sqrt{2})^{-1}$ if $h < r$. Hence $B(x_0, r_1(1+\sqrt{2})^{-1}) \subset K$. Moreover, let ω be a fixed function defined by (3.50). Suppose that in (3.74) ω_x is defined by: $\omega_x(y) = (\frac{r_1}{1+\sqrt{2}})^{-n}\omega((\frac{r_1}{1+\sqrt{2}})^{-1}(y - \xi_x))$, where $\xi_x = A_x(x_0)$ and A_x is a linear transformation such that $K_x = A_x(K)$. Following the proof of estimates (3.56) and (3.57), establish that $\forall x \in \Omega$ and $\forall y \in K_x$

$$\left| \sum_{|\alpha| < l - |\beta|} \frac{(-1)^{|\alpha|+|\beta|}}{\alpha!} D_y^{\alpha+\beta}[(x - y)^\alpha \omega_x(y)] \right| \leq M_1 \left(\frac{h}{r_1} \right)^{l-1} h^{-|\beta|} r_1^{-n} \qquad (3.78)$$

and

$$|w_{\alpha-\beta,x}(x, y)| \leq M_2 \left(\frac{h}{r_1} \right)^{n-1}, \qquad (3.79)$$

where M_1 and M_2 depend only on n and l. Estimates (3.77) and hence (3.76) follow from (3.74), (3.78) and (3.79). $\square$

Remark 21 From (3.77) it also follows that in (3.76) $\int_\Omega |f|\, dx$ can be replaced by $\|f\|_{L_p(\Omega)}$ for any $p \in [1, \infty]$.

Chapter 4

Embedding theorems

The main aim of this chapter is to prove various inequalities related to those
of the form

$$\|D_w^\alpha f\|_{L_q(\Omega)} \le c_1 \|f\|_{W_p^l(\Omega)},$$

where $\alpha \in \mathbb{N}_0^n$, $|\alpha| < l$ $(D_w^0 f \equiv f)$ and $c_1 > 0$ does not depend on f.

These inequalities may be also presented in an equivalent form as the so-
called embedding theorems.

4.1 Embeddings and inequalities

We start with the consideration of the notion of a continuous embedding as it
relates to the general theory of function spaces, which are, in the framework of
this book, normed or semi-normed vector spaces.

First let Z_1 and Z_2 be normed vector spaces. We say that Z_1 is *embedded*
in Z_2 if

$$Z_1 \subset Z_2. \tag{4.1}$$

The identity operator I, considered as an operator acting from Z_1 in Z_2:

$$\forall f \in Z_1 \quad If = f, \quad I: \; Z_1 \to Z_2, \tag{4.2}$$

which is possible because of (4.1), will be called the *embedding operator* corresponding to the embedding (4.1).

Definition 1 *Let Z_1 and Z_2 be normed vector spaces. We say that Z_1 is continuously embedded in Z_2 and write*

$$Z_1 \hookrightarrow Z_2 \tag{4.3}$$

if, in addition to (4.1), the corresponding embedding operator is continuous, i.e., there exists $c_2 > 0$ such that $\forall f \in Z_1$

$$\|f\|_{Z_2} \le c_2 \|f\|_{Z_1}. \tag{4.4}$$

In the cases we are interested in relations (4.1) and (4.3) are equivalent, which follows from the next statement.

Lemma 1 *Let Z_1 and Z_2 be Banach spaces such that $Z_1 \subset Z_2$. Suppose that the corresponding embedding operator is closed, i.e., for any $f_k \in Z_1$ where $k \in \mathbb{N}, g_1 \in Z_1$ and $g_2 \in Z_2$*

$$\lim_{k \to \infty} f_k = g_1 \ \ \text{in} \ \ Z_1, \quad \lim_{k \to \infty} f_k = g_2 \ \ \text{in} \ \ Z_2 \implies g_1 = g_2. \tag{4.5}$$

Then (4.4) is satisfied and, hence, $Z_1 \hookrightarrow Z_2$.

Idea of the proof. Since the embedding operator $I : Z_1 \to Z_2$ is defined on the whole Z_1 and is closed, by the Banach closed graph theorem the operator I is bounded. $\square$

Remark 1 Let us introduce for Banach spaces Z_1 and Z_2 such that $Z_1 \subset Z_2$ one more norm on Z_1, namely, $\forall f \in Z_1$

$$\|f\|_{Z_{12}} = \|f\|_{Z_1} + \|f\|_{Z_2}.$$

It is a norm on Z_1 considered as the intersection of Z_1 and Z_2 . Condition (4.5) is equivalent to the following: Z_1 is complete with respect to the norm $\|\cdot\|_{Z_{12}}$. Indeed, if $\{f_k\}_{k \in \mathbb{N}}$ is a Cauchy sequence with respect to $\|\cdot\|_{Z_{12}}$, then

$$\lim_{k,m \to \infty} \|f_k - f_m\|_{Z_1} = \lim_{k,m \to \infty} \|f_k - f_m\|_{Z_2} = 0.$$

Consequently, by the completeness of Z_1 and Z_2 elements $g_1 \in Z_1$ and $g_2 \in Z_2$ exist such that $\lim_{k \to \infty} f_k = g_1$ in Z_1 and $\lim_{k \to \infty} f_k = g_2$ in Z_2. By (4.5) $g_1 = g_2$ and, hence, $\lim_{k \to \infty} \|f_k - g_1\|_{Z_{12}} = 0$. Conversely, let the above relations be satisfied. Then $\{f_k\}_{k \in \mathbb{N}}$ is a Cauchy sequence with respect to the norm $\|\cdot\|_{Z_{12}}$. By the completeness of Z_{12} an element $g \in Z_1$ exists such that

$$\lim_{k \to \infty} (\|f_k - g\|_{Z_1} + \|f_k - g\|_{Z_2}) = 0.$$

From the uniqueness of limits in Z_1 and Z_2 it follows that $g_1 = g_2(= g)$.

We note also that the closedness of the embedding operator I is a necessary and sufficient condition for the equivalence of (4.1) and (4.3). The sufficiency is proved in Lemma 1, while the necessity is obvious, since the boundedness of I implies its closedness.

Now let Z_1 and Z_2 be semi-normed vector spaces and

$$\theta_k = \{f \in Z_k : \|f\|_{Z_k} = 0\}, \quad k = 1, 2.$$

Definition 2 *Let Z_1 and Z_2 be semi-normed vector spaces, which are subsets of a linear space Z. We say that Z_1 is continuously embedded in Z_2 and write*

$$Z_1 \hookrightarrow Z_2 \tag{4.6}$$

if [1]

$$Z_1 \subset Z_2 + \theta_1,$$

and the corresponding embedding operator $I : Z_1 \to Z_2$, defined by $If = g$, is bounded. [2]

Remark 2 This means that $\forall f \in Z_1$ there exists $g \in Z_2$ such that g is equivalent to f in Z_1, i.e., $g - f \in \theta_1$, and there exists $c_3 > 0$ such that $\forall f \in Z_1$

$$\|g\|_{Z_2} \leq c_3 \|f\|_{Z_1}. \tag{4.7}$$

Remark 3 If $\theta_1 \subset \theta_2$ (in particular, if Z_1 and Z_2 are normed vector spaces), then $Z_2 + \theta_1 = Z_2$, $\|g\|_{Z_2} = \|f\|_{Z_2}$ and Definition 2 has the same form as Definition 1.

Remark 4 Assume that the semi-normed vector spaces Z_1 and Z_2 possess the following property:

the semi-norm $\|f\|_{Z_2}$ makes sense for each $f \in Z_1$ with $\|f\|_{Z_2} < \infty$ or $\|f\|_{Z_2} = \infty$ and $\forall f \in Z_1$ there exists $g \in Z_2$ such that

$$\inf_{h \in \theta_1} \|f - h\|_{Z_2} = \|g\|_{Z_2}. \tag{4.8}$$

In this case Definition 2 is equivalent to:
there exists $c_3 > 0$ such that $\forall f \in Z_1$

$$\inf_{h \in \theta_1} \|f - h\|_{Z_2} \leq c_3 \|f\|_{Z_1}. \tag{4.9}$$

---

[1] The sign + denotes the vector sum of sets.

[2] In this case, in general, the embedding operator is not unique. However, one may easily verify that for different embedding operators, say I_1 and I_2, $\forall f \in Z_1$ we have $\|I_1 f\|_{Z_2} = \|I_2 f\|_{Z_2}$.

Lemma 2 *Let Z_1 and Z_2 be semi-Banach spaces such that $Z_1 \subset Z_2$. Suppose that for any $f_k \in Z_1, k \in \mathbb{N}, g_1 \in Z_1$ and $g_2 \in Z_2$*

$$\lim_{k \to \infty} f_k = g_1 \text{ in } Z_1, \quad \lim_{k \to \infty} f_k = g_2 \text{ in } Z_2 \Longrightarrow g_1 - g_2 \in \theta_1. \tag{4.10}$$

Then (4.9) is satisfied.

Idea of the proof. Apply the Banach closed graph theorem to the factor spaces $\widetilde{Z}_1 = Z_1/\theta_1$ and $\widetilde{Z}_2 = Z_2/\theta_1$ and [3] the embedding operator $\tilde{I} : \widetilde{Z}_1 \to \widetilde{Z}_2$. $\square$

Proof. We recall that $\widetilde{Z}_1$ is a Banach space and $\forall \tilde{f} \in \widetilde{Z}_1$

$$\|\tilde{f}\|_{\widetilde{Z}_1} = \|f\|_{Z_1}, \tag{4.11}$$

where f is an arbitrary element in $\tilde{f}$. (If $f_1, f_2 \in \tilde{f}$, then $\|f_1\| = \|f_2\|$.) As for $\widetilde{Z}_2$ it is, in general, a semi-Banach space and

$$\|\tilde{f}\|_{\widetilde{Z}_2} = \inf_{h \in \theta_1} \|f - h\|_{Z_2} \tag{4.12}$$

where $f \in \tilde{f}$. (The right-hand side does not depend on the choice of $f \in \tilde{f}$.)

From $Z_1 \subset Z_2$ it follows that $\widetilde{Z}_1 \subset \widetilde{Z}_2$ and by (4.10) the corresponding embedding operator $\tilde{I}$ is closed. For, let $\tilde{f}_k \in \widetilde{Z}_1, k \in \mathbb{N}, \tilde{g}_1 \in \widetilde{Z}_1, \tilde{g}_2 \in \widetilde{Z}_2$ and

$$\lim_{k \to \infty} \tilde{f}_k = \tilde{g}_1 \text{ in } \widetilde{Z}_1, \quad \lim_{k \to \infty} \tilde{I} \tilde{f}_k = \lim_{k \to \infty} \tilde{f}_k = \tilde{g}_2 \text{ in } \widetilde{Z}_2.$$

Suppose that $f_k \in \tilde{f}_k$, $g_1 \in \tilde{g}_1$ and $g_2 \in \tilde{g}_2$. Then $f_k - g_1 \in \tilde{f}_k - \tilde{g}_1$, $f_k - g_2 \in \tilde{f}_k - \tilde{g}_2$ and

$$\lim_{k \to \infty} \|f_k - g_1\|_{Z_1} = 0, \quad \lim_{k \to \infty} \big(\inf_{h \in \theta_1} \|f_k - g_2 - h\|_{Z_2} \big) = 0.$$

Therefore, $\forall k \in \mathbb{N}$ there exists $h_k \in \theta_1$ such that $\lim_{k \to \infty} \|f_k - g_2 - h_k\|_{Z_2} = 0$. Thus $f_k - h_k \to g_2$ in Z_2 as $k \to \infty$. Moreover, since $\|f_k - h_k - g_1\|_{Z_1} = \|f_k - g_1\|_{Z_1}$, we also have that $f_k - h_k \to g_1$ in Z_1 as $k \to \infty$. By (4.10) $g_1 - g_2 \in \theta_1$ and, hence, $\tilde{g}_1 = \tilde{g}_2$.

Now by the Banach closed graph theorem the operator $\tilde{I}$ is bounded: for some $c_4 > 0$ we have $\forall \tilde{f} \in \widetilde{Z}_1$

$$\|\tilde{f}\|_{\widetilde{Z}_2} = \|\tilde{I} \tilde{f}\|_{\widetilde{Z}_2} \leq c_4 \|\tilde{f}\|_{\widetilde{Z}_1}.$$

Consequently, by (4.10) and (4.12) the desired inequality (4.9) follows. $\square$

---

[3] The spaces $\widetilde{Z}_1$ and $\widetilde{Z}_2$ consist of nonintersecting classes $\tilde{f} \subset Z_1$, $\tilde{f} \subset Z_2$ respectively, such that $f_1, f_2 \in \tilde{f} \Longleftrightarrow f_1 - f_2 \in \theta_1$, $f_1 - f_2 \in \theta_2$ respectively.

Corollary 1 *If, in addition to the assumptions of Lemma 2, (4.8) is satisfied, then $Z_1 \subset Z_2$ is equivalent to $Z_1 \hookrightarrow Z_2$.*

Idea of the proof. Apply Remark 4. $\square$

Corollary 2 *In addition to the assumptions of Lemma 2, let*

$$\theta_1 \subset \theta_2. \tag{4.13}$$

Then there exists $c_5 > 0$ such that $\forall f \in Z_1$

$$\|f\|_{Z_2} \leq c_5 \|f\|_{Z_1}. \tag{4.14}$$

Idea of the proof. Since $\theta_1 \subset \theta_2$, we have $\|h\|_{Z_2} = 0$ for each $h \in \theta_1$. Furthermore, $\forall f \in Z_1$ we also have $\|f - h\|_{Z_2} = \|f\|_{Z_2}$ and (4.9) coincides with (4.14). $\square$

Corollary 3 *Let Z be a semi-normed vector space, equipped with two semi-norms $\|\cdot\|_1$ and $\|\cdot\|_2$ and complete with respect to both of them. Moreover, suppose that for any $f_k \in Z$, $k \in \mathbb{N}$, $g_1, g_2 \in Z$*

$$\lim_{k \to \infty} \|f_k - g_1\|_1 = 0, \quad \lim_{k \to \infty} \|f_k - g_2\|_2 = 0 \implies g_1 - g_2 \in \theta_1 \cap \theta_2. \tag{4.15}$$

Then the semi-norms $\|\cdot\|_1$ and $\|\cdot\|_2$ are equivalent [4] if, and only if,

$$\theta_1 = \theta_2. \tag{4.17}$$

Suppose, in particular, that Z is a normed vector space, equipped with two norms $\|\cdot\|_1$ and $\|\cdot\|_2$ and complete with respect to both of them. If for any $f_k \in Z$, $k \in \mathbb{N}$, $g_1, g_2 \in Z$

$$\lim_{k \to \infty} \|f_k - g_1\|_1 = 0, \quad \lim_{k \to \infty} \|f_k - g_2\|_2 = 0 \implies g_1 = g_2, \tag{4.18}$$

then the norms $\|\cdot\|_1$ and $\|\cdot\|_2$ are equivalent.

---

[4] I.e., there exist $c_6, c_7 > 0$ such that $\forall f \in Z$

$$c_6 \|f\|_2 \leq \|f\|_1 \leq c_7 \|f\|_2. \tag{4.16}$$

Idea of the proof. Necessity of (4.17) follows directly from (4.16). To prove sufficiency apply Corollary 1 to the semi-Banach spaces Z_1 and Z_2, which are the same set Z, equipped with the semi-norms $\|\cdot\|_1$ and $\|\cdot\|_2$. $\square$

Now we pass to the case of function spaces. Let $\Omega \subset \mathbb{R}^n$ be an open set. Moreover, suppose that $Z(\Omega)$ is a semi-normed vector space of functions defined on Ω. Denote

$$\theta_{Z(\Omega)} = \{f \in Z(\Omega) : \; \|f\|_{Z(\Omega)} = 0\}$$

and

$$\theta(\Omega) = \{f : \; f(x) = 0 \text{ for almost every } x \in \Omega\}.$$

All function spaces $Z(\Omega)$, which are considered in this book, possess the following property:

$$Z(\Omega) \hookrightarrow L_1^{loc}(\Omega), \tag{4.19}$$

i.e., $Z(\Omega) \subset L_1^{loc}(\Omega)$ and for each compact $K \subset \Omega$ there exists $c_8(K) > 0$ such that $\forall f \in Z(\Omega)$

$$\inf_{h \in \theta_{Z(\Omega)}} \|f - h\|_{L_1(K)} \leq c_8(K) \|f\|_{Z(\Omega)}. \tag{4.20}$$

For many of the function spaces considered

$$\theta_{Z(\Omega)} = \theta(\Omega). \tag{4.21}$$

If this property is satisfied, then (4.20) takes the form

$$\|f\|_{L_1(K)} \leq c_8(K) \|f\|_{Z(\Omega)}. \tag{4.22}$$

Remark 5 If two semi-normed vector spaces $Z_1(\Omega)$ and $Z_2(\Omega)$ satisfy (4.19) and $\theta_{Z_1(\Omega)}, \theta_{Z_2(\Omega)} \subset \theta(\Omega)$, then for any $f_k \in Z_1(\Omega) \cap Z_2(\Omega)$, $k \in \mathbb{N}$, $g_1 \in Z_1(\Omega)$ and $g_2 \in Z_2(\Omega)$

$$\lim_{k\to\infty} f_k = g_1 \text{ in } Z_1(\Omega), \lim_{k\to\infty} f_k = g_2 \text{ in } Z_2(\Omega) \implies g_1 \sim g_2 \text{ on } \Omega,$$

i.e., $g_1 - g_2 \in \theta(\Omega)$.

Lemma 3 *Let $\Omega \subset \mathbb{R}^n$ be an open set and let $Z_1(\Omega)$ and $Z_2(\Omega)$ be semi-Banach function spaces satisfying (4.19) and (4.21). If*

$$Z_1(\Omega) \subset Z_2(\Omega) \tag{4.23}$$

then there exists $c_9 > 0$ such that $\forall f \in Z_1(\Omega)$

$$\|f\|_{Z_2(\Omega)} \leq c_9 \|f\|_{Z_1(\Omega)} \tag{4.24}$$

and, hence, (4.23) is equivalent to

$$Z_1(\Omega) \hookrightarrow Z_2(\Omega). \tag{4.25}$$

Idea of the proof. Apply Corollary 2. $\square$

Proof. If $f_k \in Z_1(\Omega)$, $k \in \mathbb{N}$, $g_1 \in Z_1(\Omega)$, $g_2 \in Z_2(\Omega)$, and

$$\lim_{k \to \infty} f_k = g_1 \text{ in } Z_1(\Omega), \quad \lim_{k \to \infty} f_k = g_2 \text{ in } Z_2(\Omega), \tag{4.26}$$

then by (4.22)

$$\lim_{k \to \infty} f_k = g_1 \text{ in } L_1^{loc}(\Omega), \quad \lim_{k \to \infty} f_k = g_2 \text{ in } L_1^{loc}(\Omega) \tag{4.27}$$

and $g_1 - g_2 \in \theta(\Omega) = \theta_{Z_1(\Omega)}$. Hence, (4.24) follows from (4.13)

From (4.21) it follows that $Z_2(\Omega) + \theta_{Z_1(\Omega)} = Z_2(\Omega)$. Moreover, for each $g \in Z_2(\Omega)$, for which $g - f \in \theta_{Z_1(\Omega)}$ we have $\|g\|_{Z_2(\Omega)} = \|f\|_{Z_2(\Omega)}$. Consequently, (4.23) coincides with (4.6), (4.24) coincides with (4.7) and, hence, (4.23) is equivalent to (4.25). $\square$

Corollary 4 *Let $\Omega \subset R^n$ be an open set and $Z(\Omega)$ a semi-normed vector space equipped with two semi-norms and complete with respect to both of them. Moreover, suppose that conditions (4.19) and (4.21) are satisfied. Then these semi-norms are equivalent.*

Idea of the proof. Apply Lemma 3 to the semi-normed vector spaces $Z_1(\Omega)$ and $Z_2(\Omega)$, which are the same set $Z(\Omega)$, equipped with the given semi-norms. $\square$

Finally, we collect together all the statements about equivalence of inequalities, embeddings and continuous embeddings in the case of Sobolev spaces.

Theorem 1 *Let $l \in \mathbb{N}, m \in \mathbb{N}_0, m < l, 1 \le p, q \le \infty$ and let $\Omega \subset \mathbb{R}^n$ be an open set.*

1. *The continuous embedding*

$$W_p^l(\Omega) \hookrightarrow W_q^m(\Omega), \tag{4.28}$$

i.e., the inequality

$$\|f\|_{W_q^m(\Omega)} \le c_{10} \|f\|_{W_p^l(\Omega)}, \tag{4.29}$$

where $c_{10} > 0$ does not depend on f, is equivalent to the embedding

$$W_p^l(\Omega) \subset W_q^m(\Omega). \tag{4.30}$$

2. *The continuous embedding*

$$W_p^l(\Omega) \hookrightarrow C_b^m(\Omega), \tag{4.31}$$

i.e., the statement: $\forall f \in W_p^l(\Omega)$ *there exists a function* $g \in C_b^m(\Omega)$ *such that* $g \sim f$ *on* Ω *and*

$$\|g\|_{C^m(\Omega)} \leq c_{11} \|f\|_{W_p^l(\Omega)}, \tag{4.32}$$

where $c_{11} > 0$ *does not depend on* f, *is equivalent to inequality* (4.29) *and embedding* (4.30), *where* $q = \infty$, *and in* (4.29) $c_{10}|_{q=\infty} = c_{11}$.

3. *If inequality* (4.29) *holds for all* $f \in W_p^l(\Omega) \cap C^\infty(\Omega)$, *then it holds for all* $f \in W_p^l(\Omega)$.

Idea of the proof. Apply Definition 2 and Lemmas 2 and 3. To prove the equivalence of (4.32) and (4.29) when $q = \infty$ apply Theorem 1 of Chapter 2 and the completeness of the spaces under consideration. If $m > 0$ apply also the closedness of the differentiation operator D^α where $|\alpha| = m$ in $C_b(\Omega)$. To prove the third statement of the theorem again apply Theorem 1 of Chapter 2, the completeness of $W_q^m(\Omega)$ and, for $m > 0$, the closedness of the weak differentiation operator D_w^α where $|\alpha| = m$ in $L_q(\Omega)$. $\square$

Proof. 1. Clearly, (4.30) follows from (4.28). As for the converse, it is a direct corollary of Lemma 3, because $\theta_{W_p^l(\Omega)} = \theta_{W_q^m(\Omega)} = \theta(\Omega)$.

2. Furthermore, (4.31) implies (4.29) with $q = \infty$, which, by the first statement of the theorem, is equivalent to (4.30) with $q = \infty$.

Let us prove that (4.29) with $q = \infty$ implies (4.32) where $c_{11} = c_{10}|_{q=\infty}$. First suppose that $m = 0$ and $1 \leq p < \infty$. Then $\forall f \in W_p^l(\Omega)$ there exist $f_k \in C^\infty(\Omega) \cap W_p^l(\Omega)$, $k \in \mathbb{N}$, such that $f_k \to f$ in $W_p^l(\Omega)$ – see Theorem 1 in Chapter 2. By (4.29) with $q = \infty$, $f_k \in C_b(\Omega), k \in \mathbb{N}$, and $\forall k, s \in \mathbb{N}$

$$\|f_k - f_s\|_{C(\Omega)} = \|f_k - f_s\|_{L_\infty(\Omega)} \leq c_{11} \|f_k - f_s\|_{W_p^l(\Omega)}$$

(see footnote 4 on page 12). Hence, $\{f_k\}_{k\in\mathbb{N}}$ is a Cauchy sequence in $C_b(\Omega)$. Since $C_b(\Omega)$ is complete, there exists $g \in C_b(\Omega)$ such that $f_k \to g$ in $C_b(\Omega)$ as $k \to \infty$. Since both $W_p^l(\Omega)$ and $C_b(\Omega)$ are continuosly embedded into $L_1^{loc}(\Omega)$, it follows that $g \sim f$ on Ω – see Remark 5.

If $m = 0$ and $p = \infty$, then $\forall f \in W_\infty^l(\Omega)$ there exist $f_k \in C^\infty(\Omega) \cap W_\infty^l(\Omega), k \in \mathbb{N}$, such that $f_k \to f$ in $W_\infty^r(\Omega)$ for $r = 1, ..., l - 1$ and $\|f_k\|_{W_\infty^l(\Omega)} \to \|f\|_{W_\infty^l(\Omega)}$ as $k \to \infty$ (see Theorem 1 in Chapter 2). Since $\|f_k - f_s\|_{C(\Omega)} = \|f_k - f_s\|_{L_\infty(\Omega)}$, $\{f_k\}_{k\in\mathbb{N}}$ is again a Cauchy sequence in $C_b(\Omega)$. The rest is the same as for the case in which $1 \leq p < \infty$.

If $m > 0$ and $1 \leq p \leq \infty$, then the same argument as above shows that there exist $h \in C_b(\Omega)$ and $h_\alpha \in C_b(\Omega)$ where $|\alpha| = m$ such that $f_k \to h$ in $C_b(\Omega)$ and $D^\alpha f_k \to h_\alpha$ in $C_b(\Omega)$. Since the differentiation operator D^α is closed in $C_b(\Omega)$, it follows that $h_\alpha = D^\alpha h$. Hence $h \in C_b^m(\Omega)$ and $f_k \to g$ in $C_b^m(\Omega)$.

Finally, for all $m \geq 0$ and $1 \leq p \leq \infty$, we take $f = f_k$ in (4.29) where $q = \infty$ and passing to the limit as $k \to \infty$ we get (4.32) since

$$\|g\|_{C^m(\Omega)} = \lim_{k \to \infty} \|f_k\|_{C^m(\Omega)} = \lim_{k \to \infty} \|f_k\|_{W_\infty^m(\Omega)}$$

$$\leq c_{10} \lim_{k \to \infty} \|f_k\|_{W_p^l(\Omega)} = c_{10} \|f\|_{W_p^l(\Omega)}.$$

3. The proof of the third statement of the theorem is analogous. $\square$

4.2 The one-dimensional case

We start with inequalities for intermediate derivatives.

Theorem 2 *Let* $-\infty \leq a < b \leq \infty$, $l \in N$ *and* $1 \leq p \leq \infty$.
   1. For each function $f \in W_p^l(a,b)$ *and* $m = 1, ..., l-1$

$$\|f_w^{(m)}\|_{L_p(a,b)} \leq c_{12} \|f\|_{W_p^l(a,b)}, \tag{4.33}$$

where $c_{12} > 0$ *depends only on* l *and* $b - a$.
   2. If $-\infty < a \leq \alpha < \beta \leq b < \infty$, *then for* $m = 1, ..., l-1$

$$\|f_w^{(m)}\|_{L_p(a,b)} \leq c_{13} (\|f\|_{L_1(\alpha,\beta)} + \|f_w^{(l)}\|_{L_p(a,b)}), \tag{4.34}$$

where $c_{13} > 0$ *depends only on* $l, b - a, \beta - \alpha$ *and* f *is such that the right-hand side is finite.*

Remark 6 If $b - a = \infty$ and $\beta - \alpha < \infty$, then inequality (4.34) does not hold. This follows by setting $f(x) = x^k$ where $m \leq k < l$.

Idea of the proof. Apply inequality (3.21) and Remark 5 of Chapter 3. $\square$
Proof. Let $-\infty < a < b < \infty$. From (3.21), by Hölder's inequality and Remark 5 of Chapter 3, it follows that

$$\|f_w^{(m)}\|_{L_p(a,b)} \leq M_1 (b - a)^{l-m-\frac{1}{p'}} \left((\beta - \alpha)^{-l}\|f\|_{L_1(\alpha,\beta)} + (b - a)^{\frac{1}{p'}} \|f_w^{(l)}\|_{L_p(a,b)} \right), \tag{4.35}$$

where M_1 depends only on n. This inequality implies (4.34).

   Now let $b - a = \infty$. Say, for example, $-\infty < a < \infty$ and $b = \infty$. If $1 \leq p < \infty$, then by (4.33)

$$\|f_w^{(m)}\|_{L_p(a,\infty)} = \left(\sum_{k=1}^{\infty} \|f_w^{(m)}\|_{L_p(a+k-1,a+k)}^p \right)^{\frac{1}{p}}$$

$$\leq M_2 \Big(\sum_{k=1}^{\infty} (\|f\|_{L_p(a+k-1,a+k)}^p + \|f_w^{(l)}\|_{L_p(a+k-1,a+k)}^p) \Big)^{\frac{1}{p}}$$

$$= M_2 (\|f\|_{L_p(a,\infty)}^p + \|f_w^{(l)}\|_{L_p(a,\infty)}^p)^{\frac{1}{p}} \leq M_2 (\|f\|_{L_p(a,\infty)} + \|f_w^{(l)}\|_{L_p(a,\infty)}).$$

Here M_2 is the constant c_{12} in (4.33) for the case in which $b - a = 1$.

If $p = \infty$, then

$$\|f_w^{(m)}\|_{L_\infty(a,\infty)} = \sup_{k \in \mathbb{N}} \|f_w^{(m)}\|_{L_\infty(a+k-1,a+k)}$$

$$\leq M_2 \sup_{k \in \mathbb{N}} (\|f\|_{L_\infty(a+k-1,a+k)} + \|f_w^{(l)}\|_{L_\infty(a+k-1,a+k)})$$

$$\leq M_2 (\|f\|_{L_\infty(a,\infty)} + \|f_w^{(l)}\|_{L_\infty(a,\infty)}). \quad \Box$$

Corollary 5 *Let* $1 \leq p \leq \infty$. *The norm*

$$\sum_{m=0}^{l} \|f_w^{(m)}\|_{L_p(a,b)} \tag{4.36}$$

is equivalent to $\|f\|_{W_p^l(a,b)}$ *for any interval* $(a,b) \subset \mathbb{R}$. *The norm*

$$\|f\|_{L_1(\alpha,\beta)} + \|f_w^{(l)}\|_{L_p(a,b)} \tag{4.37}$$

is equivalent to $\|f\|_{W_p^l(a,b)}$ *if* $-\infty < a \leq \alpha < \beta \leq b < \infty$.

Idea of the proof. Apply inequality (4.33) and inequality (4.34) with $m = 0$. $\Box$

Corollary 6 *Let* $-\infty < a < b < \infty$, $l \in \mathbb{N}$ *and* $1 \leq p \leq \infty$. *Then*

$$w_p^l(a,b) = W_p^l(a,b). \tag{4.38}$$

(*If* $b - a = \infty$, *then* $1 \in w_p^l(a,b) \setminus W_p^l(a,b)$. *Thus, the embedding* $W_p^l(a,b) \subset w_p^l(a,b)$ *is strict.*)

Idea of the proof. If $f \in w_p^l(a,b)$, then $f \in L_1^{loc}(a,b)$, and Corollary 5 implies (4.38). $\Box$

Remark 7 Equality (4.38) is an equality of sets of functions. Since $\theta_{w_p^l(a,b)} \neq \theta_{W_p^l(a,b)}$, the semi-norms $\| \cdot \|_{w_p^l(a,b)}$ and $\| \cdot \|_{W_p^l(a,b)}$ are not equivalent. (See Corollary 3 of Section 4.1.)

Corollary 7 *Let $l, m \in N$, $m < l$ and $1 \leq p \leq \infty$.*

1. If $-\infty < a < b < \infty$, then the validity of inequality (4.33) with some $c_{12} > 0$ independent of f is equivalent to the validity of

$$\|f_w^{(m)}\|_{L_p(a,b)} \leq c_{14}\left((b-a)^{-m}\|f\|_{L_p(a,b)} + (b-a)^{l-m}\|f_w^{(l)}\|_{L_p(a,b)}\right), \qquad (4.39)$$

$$\|f_w^{(m)}\|_{L_p(a,b)} \leq c_{15}\,\varepsilon^{-\frac{m}{l-m}}\,\|f\|_{L_p(a,b)} + \varepsilon\,\|f_w^{(l)}\|_{L_p(a,b)} \qquad (4.40)$$

for [5] $0 < \varepsilon \leq c_{14}(b-a)^{l-m}$ and

$$\|f_w^{(m)}\|_{L_p(a,b)} \leq c_{16}\,\|f\|_{L_p(a,b)}^{1-\frac{m}{l}}\,\|f\|_{W_p^l(a,b)}^{\frac{m}{l}} \qquad (4.41)$$

with some $c_{14}, c_{15} > 0$ independent of f, a and b and some $c_{16} > 0$ independent of f.

2. If $b - a = \infty$, then inequality (4.33) is equivalent to

$$\|f_w^{(m)}\|_{L_p(a,b)} \leq c_{12}^{\frac{l}{l-m}}\,\varepsilon^{-\frac{m}{l-m}}\,\|f\|_{L_p(a,b)} + \varepsilon\,\|f_w^{(l)}\|_{L_p(a,b)} \qquad (4.42)$$

for $0 < \varepsilon < \infty$,

$$\|f_w^{(m)}\|_{L_p(a,b)} \leq c_{12}\left[\left(\frac{m}{l}\right)^{\frac{m}{l}}\left(1 - \frac{m}{l}\right)^{1-\frac{m}{l}}\right]^{-1}\|f\|_{L_p(a,b)}^{1-\frac{m}{l}}\,\|f_w^{(l)}\|_{L_p(a,b)}^{\frac{m}{l}}, \qquad (4.43)$$

$$\int_a^b |f_w^{(m)}|^p\,dx \leq c_{12}^p\left[\left(\frac{m}{l}\right)^{\frac{m}{l}}\left(1 - \frac{m}{l}\right)^{1-\frac{m}{l}}\right]^{1-p}\int_a^b\left(|f|^p + |f_w^{(l)}|^p\right)dx \qquad (4.44)$$

if $1 \leq p < \infty$ and

$$\int_a^b |f_w^{(m)}|^p\,dx \leq c_{12}^{\frac{pl}{l-m}}\left[\left(\frac{m}{l}\right)^{\frac{m}{l}}\left(1 - \frac{m}{l}\right)^{1-\frac{m}{l}}\right]^{\frac{(1-p)l}{l-m}}\varepsilon^{-\frac{m}{l-m}}\int_a^b |f|^p\,dx + \varepsilon\int_a^b |f_w^{(l)}|^p\,dx \qquad (4.45)$$

for $0 < \varepsilon < \infty$ if $1 \leq p < \infty$.

---

[5] One may consider $0 < \varepsilon \leq \varepsilon_0$, where ε_0 is an arbitrary positive number. In this case $c_{15} > 0$ depends on ε_0 as well. It also follows that $\forall \varepsilon > 0$ there exists $C(\varepsilon)$ such that

$$\|f_w^{(m)}\|_{L_p(a,b)} \leq C(\varepsilon)\,\|f\|_{L_p(a,b)} + \varepsilon\,\|f_w^{(l)}\|_{L_p(a,b)}.$$

Note that one cannot replace here $C(\varepsilon)$ by ε and ε by $C(\varepsilon)$. (If it were so, then taking $f(x) = x^m$ and passing to the limit as $\varepsilon \to 0+$ would give a contradiction.)

From (4.40) it also follows that

$$\|f_w^{(m)}\|_{L_p(a,b)} \leq M((b-a)^{-m} + 1)\,\|f\|_{W_p^l(a,b)},$$

where $M > 0$ is independent of f, a and b.

Idea of the proof. 1. Changing variables reduce the case of an arbitrary interval (a, b) to the case of the interval $(0, 1)$ and deduce (4.39) from (4.33). For $0 < \delta < b - a$ set $N = [\frac{b-a}{\delta}] + 1$, $\delta_1 = \frac{b-a}{N}$, $a_k = a + \delta_1(k - 1)$, $k = 1, ..., N + 1$. Apply the equality [6]

$$\|f\|_{L_p(a,b)} = \Big(\sum_{k=1}^{N} \|f\|_{L_p(a_k,a_{k+1})}^p\Big)^{\frac{1}{p}} \tag{4.46}$$

and the inequality $\frac{\delta}{2} \leq \delta_1 \leq \delta$ to deduce (4.40) from (4.39). Minimize the right-hand side of (4.40) with respect to ε in order to prove (4.41). Finally, note that inequality (4.39) follows from (4.41) and the inequality

$$x^\alpha y^{1-\alpha} \leq \alpha^\alpha (1 - \alpha)^{1-\alpha}(x + y), \tag{4.47}$$

where $x, y \geq 0$, $0 < \alpha < 1$.

2. Apply (4.33) to $f(a + \delta(x - a))$ if $b = \infty$, to $f(b - \delta(b - x))$ if $a = -\infty$ or to $f(\delta x)$ if $a = -\infty, b = \infty$ where $\delta > 0$, and deduce (4.42). Minimize the right-hand side of (4.42) to get (4.43). Note that (4.33) follows from (4.43) and (4.47). Raise (4.43) to the power p and apply (4.47) to establish (4.44). Deduce, by applying dilations once more, (4.45) from (4.44). Minimizing the right-hand side of (4.45), verify that (4.45) implies (4.43). $\square$

Proof. 1. Setting $y = \frac{x-a}{b-a}$ we get

$$\|f_w^{(m)}\|_{L_p(a,b)} = (b - a)^{\frac{1}{p}-m} \|(f(a + y(b - a)))_w^{(m)}\|_{L_p(0,1)}$$

$$\leq M_1(b - a)^{\frac{1}{p}-m}(\|f(a + y(b - a))\|_{L_p(0,1)} + \|(f(a + y(b - a)))_w^{(l)}\|_{L_p(0,1)})$$

$$= M_1((b - a)^{-m}\|f\|_{L_p(a,b)} + (b - a)^{l-m}\|f_w^{(l)}\|_{L_p(a,b)}),$$

where M_1 is the constant c_{12} in (4.33) for the case in which $(a, b) = (0, 1)$.

Moreover, for $0 < \delta \leq b - a$ and $1 \leq p \leq \infty$ from (4.46), (4.39) and Minkowski's inequality it follows that

$$\|f_w^{(m)}\|_{L_p(a,b)} \leq c_{14}\Big(\sum_{k=1}^{N}(\delta_1^{-m}\|f\|_{L_p(a_k,a_{k+1})} + \delta_1^{l-m}\|f_w^{(l)}\|_{L_p(a_k,a_{k+1})})^p\Big)^{\frac{1}{p}}$$

$$\leq c_{14}\Big(\delta_1^{-m}\Big(\sum_{k=1}^{N}\|f\|_{L_p(a_k,a_{k+1})}^p\Big)^{\frac{1}{p}} + \delta_1^{l-m}\Big(\sum_{k=1}^{N}\|f_w^{(l)}\|_{L_p(a_k,a_{k+1})}^p\Big)^{\frac{1}{p}}\Big)$$

---

[6] If $p = \infty$ this means that $\|f\|_{L_\infty(a,b)} = \max\limits_{k=1,...,N} \|f\|_{L_\infty(a_k,a_{k+1})}$.

$$\leq c_{14}(2^m \delta^{-m} \|f\|_{L_p(a,b)} + \delta^{l-m} \|f_w^{(l)}\|_{L_p(a,b)}).$$

Setting $\varepsilon = c_{14}\delta^{l-m}$ we establish (4.40). The minimum of the right-hand side of (4.40) is equal to

$$c_{15}^{1-\frac{m}{l}} \left(\left(\tfrac{m}{l}\right)^{\frac{m}{l}} (1 - \tfrac{m}{l})^{1-\frac{m}{l}} \right)^{-1} \|f\|_{L_p(a,b)}^{1-\frac{m}{l}} \|f_w^{(l)}\|_{L_p(a,b)}^{\frac{m}{l}}$$

and is achieved for $\varepsilon = \varepsilon_1$, where

$$\varepsilon_1 = \left(\tfrac{m}{l-m} c_{15} \|f\|_{L_p(a,b)} \|f_w^{(l)}\|_{L_p(a,b)}^{-1} \right)^{1-\frac{m}{l}}.$$

If $\varepsilon_1 \leq (b-a)^{l-m}$, then, setting $\varepsilon = \varepsilon_1$ in (4.40) we get (4.41). Now let $\varepsilon_1 \geq (b-a)^{l-m}$. This is equivalent to

$$\|f_w^{(l)}\|_{L_p(a,b)} \leq \tfrac{m}{l-m} c_{15} (b-a)^{-l} \|f\|_{L_p(a,b)}.$$

Since

$$\|f\|_{L_p(a,b)} \leq \|f\|_{L_p(a,b)}^{1-\frac{m}{l}} \|f\|_{W_p^l(a,b)}^{\frac{m}{l}}, \qquad \|f_w^{(l)}\|_{L_p(a,b)} \leq \|f_w^{(l)}\|_{L_p(a,b)}^{1-\frac{m}{l}} \|f\|_{W_p^l(a,b)}^{\frac{m}{l}},$$

inequality (4.39) with $\varepsilon = c_{14}(b-a)^{l-m}$ implies that

$$\|f_w^{(m)}\|_{L_p(a,b)} \leq M_2((b-a)^{-m} + 1) \|f\|_{L_p(a,b)}^{1-\frac{m}{l}} \|f\|_{W_p^l(a,b)}^{\frac{m}{l}},$$

where M_2 depends only on l, and (4.41) follows. In its turn (4.41) and (4.47) imply (4.33).

2. Let $b - a = \infty$, say $a = -\infty, b = \infty$. Given a function $f \in W_p^l(-\infty, \infty)$ and $\delta > 0$, by (4.33) we have

$$\|f_w^{(m)}\|_{L_p(-\infty,\infty)} = \delta^{-m+\frac{1}{p}} \|(f(\delta x))_w^{(m)}\|_{L_p(-\infty,\infty)}$$

$$\leq c_{12}\, \delta^{-m+\frac{1}{p}} \left(\|f(\delta x)\|_{L_p(-\infty,\infty)} + \|(f(\delta x))_w^{(l)}\|_{L_p(-\infty,\infty)} \right)$$

$$= c_{12} \left(\delta^{-m} \|f\|_{L_p(-\infty,\infty)} + \delta^{l-m} \|f_w^{(l)}\|_{L_p(-\infty,\infty)} \right).$$

Setting $c_{12}\delta^{l-m} = \varepsilon$, we get (4.42).

The rest of the proof is as in step 1. $\square$

Corollary 8 *Let $l, m \in \mathbb{N}, m < l$. Then*

$$\|f_w^{(m)}\|_{L_2(-\infty,\infty)} \leq \|f\|_{L_2(-\infty,\infty)}^{1-\frac{m}{l}} \|f_w^{(l)}\|_{L_2(-\infty,\infty)}^{\frac{m}{l}} \tag{4.48}$$

and the constant 1 in this inequality is sharp.

Idea of the proof. Prove (4.44) when $p = 2$, $a = -\infty$, $b = \infty$ by using Fourier transforms and Parseval's inequality, and apply Corollary 7. $\square$

Proof. By Parseval's equality and inequality (4.47)

$$\|f_w^{(m)}\|_{L_2(-\infty,\infty)}^2 = \|F(f_w^{(m)})\|_{L_2(-\infty,\infty)}^2 = \int_{-\infty}^{\infty} \xi^{2m} |(Ff)(\xi)|^2 \, d\xi$$

$$\leq \left(\tfrac{m}{l}\right)^{\frac{m}{l}} \left(1 - \tfrac{m}{l}\right)^{1-\frac{m}{l}} \int_{-\infty}^{\infty} (1 + \xi^{2l}) |(Ff)(\xi)|^2 \, d\xi$$

$$= \left(\tfrac{m}{l}\right)^{\frac{m}{l}} \left(1 - \tfrac{m}{l}\right)^{1-\frac{m}{l}} \left(\|Ff\|_{L_2(-\infty,\infty)}^2 + \|F(f_w^{(l)})\|_{L_2(-\infty,\infty)}^2 \right)$$

$$= \left(\tfrac{m}{l}\right)^{\frac{m}{l}} \left(1 - \tfrac{m}{l}\right)^{1-\frac{m}{l}} \left(\|f\|_{L_2(-\infty,\infty)}^2 + \|f_w^{(l)}\|_{L_2(-\infty,\infty)}^2 \right).$$

Since $\xi^{2m} = \left(\tfrac{m}{l}\right)^{\frac{m}{l}} \left(1 - \tfrac{m}{l}\right)^{1-\frac{m}{l}} (1 + \xi^{2l})$ if, and only if, $|\xi| = \left(\tfrac{m}{l-m}\right)^{\frac{1}{2m}} \equiv \xi_0$ we set $f = f_\varepsilon$, where $f_\varepsilon(x) = (F^{-1}(\chi_{(\xi_0-\varepsilon,\xi_0+\varepsilon)}))(x) = \sqrt{\tfrac{2}{\pi}} \, \tfrac{\sin \varepsilon x}{x} \, e^{-i\xi_0 x}$. Passing to the limit as $\varepsilon \to 0+$ we obtain that $\left(\tfrac{m}{l}\right)^{\frac{m}{l}} \left(1 - \tfrac{m}{l}\right)^{1-\frac{m}{l}}$ is a sharp constant. $\square$

Remark 8 Let $l, m \in \mathbb{N}$, $m < l$ and $1 \leq p \leq \infty$. The value of the sharp constant $c_{m,l,p}$ in the inequality

$$\|f_w^{(m)}\|_{L_p(-\infty,\infty)} \leq c_{m,l,p} \|f_w\|_{L_p(-\infty,\infty)}^{1-\frac{m}{l}} \|f_w^{(l)}\|_{L_p(-\infty,\infty)}^{\frac{m}{l}} \tag{4.49}$$

is also known in the cases $p = \infty$ and $p = 1$:

$$c_{m,l,1} = c_{m,l,\infty} = \frac{K_{l-m}}{K_l^{l-m}},$$

where for $j \in \mathbb{N}$

$$K_{2j-1} = \frac{4}{\pi} \sum_{i=0}^{\infty} \frac{1}{(2i+1)^{2j}} \, , \qquad K_{2j} = \frac{4}{\pi} \sum_{i=0}^{\infty} \frac{(-1)^i}{(2i+1)^{2j+1}} \, .$$

The following inequality holds

$$1 = c_{m,l,2} \leq c_{m,l,p} \leq c_{m,l,1} = c_{m,l,\infty} \leq \frac{\pi}{2}.$$

Thus, $\forall f \in W_p^l(-\infty, \infty)$ for each $1 \leq p \leq \infty$

$$\|f_w^{(m)}\|_{L_p(-\infty,\infty)} \leq \frac{\pi}{2} \|f\|_{L_p(-\infty,\infty)}^{1-\frac{m}{l}} \|f_w^{(l)}\|_{L_p(-\infty,\infty)}^{\frac{m}{l}}. \tag{4.50}$$

Remark 9 We note a simple particular case of (4.49):

$$\|f'_w\|_{L_\infty(-\infty,\infty)} \leq \sqrt{2}\,\|f\|^{\frac{1}{2}}_{L_\infty(-\infty,\infty)}\|f''_w\|^{\frac{1}{2}}_{L_\infty(-\infty,\infty)}\,, \tag{4.51}$$

where $\sqrt{2}$ is a sharp constant. One can easily prove this inequality by applying the integral representation (3.27) with $a = x + \varepsilon, b = x - \varepsilon, \varepsilon > 0$. It follows that for each function f whose derivative f' is locally absolutely continuous

$$|f'(x)| \leq \tfrac{1}{\varepsilon}\|f\|_{L_\infty(-\infty,\infty)} + \tfrac{\varepsilon}{2}\|f''\|_{L_\infty(-\infty,\infty)}.$$

We get the desired inequality by minimizing with respect to $\varepsilon > 0$ and applying Definition 4 of Chapter 1.

In the sequel we shall need a more general inequality: for $1 \leq p \leq \infty$

$$\|f'_w\|_{L_\infty(-\infty,\infty)} \leq 2\,\|f\|^{\frac{1}{2}(1-\frac{1}{p})}_{L_p(-\infty,\infty)}\,\|f''_w\|^{\frac{1}{2}(1+\frac{1}{p})}_{L_p(-\infty,\infty)}. \tag{4.52}$$

To prove it we apply Hölder's inequality to the integral representation (3.24) and get

$$|f'(x)| \leq \|\omega'\|_{L_{p'}(a,b)}\|f\|_{L_p(a,b)} + \|\Lambda(x,\cdot)\|_{L_{p'}(a,b)}\|f''_w\|_{L_p(a,b)}$$

almost everywhere on (a,b). Choosing ω in such a way that $\|\omega'\|_{L_{p'}(a,b)}$ is minimal, we establish that [7]

$$\omega(x) = (1 + \tfrac{1}{p})\frac{(b-a)^p - |2x - (a+b)|^p}{(b-a)^{p+1}} \leq (1 + \tfrac{1}{p})\frac{1}{b-a}$$

and

$$\|\omega'\|_{L_{p'}(a,b)} = 2\,(p+1)^{\frac{1}{p}}.$$

Moreover,

$$\left|\Lambda\!\left(\frac{a+b}{2}, y\right)\right| \leq \begin{cases} \displaystyle\int_a^b \omega(u)\,du \leq (1 + \tfrac{1}{p})\frac{y-a}{b-a}, & \text{for } a \leq y \leq \frac{a+b}{2}, \\[2ex] \displaystyle\int_y^b \omega(u)\,du \leq (1 + \tfrac{1}{p})\frac{b-y}{b-a}, & \text{for } \frac{a+b}{2} \leq y \leq b, \end{cases}$$

---

[7] Euler's equation for the extremal problem $\int_a^b |\omega'(x)|^{p'}\,dx \to \min, \int_a^b \omega(x)\,dx = 1$ where $1 < p < \infty$ has the form $(|\omega'(x)|^{p'}\mathrm{sgn}\,\omega'(x))' = \lambda$. So, $\omega'(x) = |\lambda_1 x + \lambda_2|^{p-1}\mathrm{sgn}\,(\lambda_1 x + \lambda_2)$. Here $\lambda, \lambda_1, \lambda_2$ are some constants.

and

$$\left\|\Lambda\left(\frac{a+b}{2},\cdot\right)\right\|_{L_{p'}(a,b)} \le \tfrac{1}{2}\left(1+\tfrac{1}{p}\right)(p'+1)^{-\frac{1}{p'}}(b-a)^{\frac{1}{p'}}.$$

Taking $a = x + \varepsilon, b = x - \varepsilon$, we get

$$|f'(x)| \le \left(\tfrac{p+1}{2}\right)^{\frac{1}{p}}\varepsilon^{-1-\frac{1}{p}}\|f\|_{L_p(-\infty,\infty)}$$

$$+2^{-\frac{1}{p}}\left(1+\tfrac{1}{p}\right)(p'+1)^{-\frac{1}{p'}}\varepsilon^{1-\frac{1}{p}}\|f''_w\|_{L_p(-\infty,\infty)}$$

almost everywhere on $(-\infty,\infty)$. By minimizing with respect to $\varepsilon > 0$ and applying Definition 4 of Chapter 1 we have

$$\|f'_w\|_{L_\infty(-\infty,\infty)} \le A_p \|f\|_{L_p(-\infty,\infty)}^{\frac{1}{2}\left(1-\frac{1}{p}\right)} \|f''_w\|_{L_p(-\infty,\infty)}^{\frac{1}{2}\left(1+\frac{1}{p}\right)},$$

where [8]

$$A_p = \left(\frac{4p'}{p'+1}\left(\frac{p+1}{p'+1}\right)^{\frac{1}{p}}\right)^{\frac{1}{2p'}} \le \sqrt{2\,e^{\frac{1}{e}}} < 2.$$

Remark 10 By (4.52) and (4.50) it follows that $\forall l \in \mathbb{N}, l \ge 2$,

$$\|f_w^{(l-1)}\|_{L_\infty(-\infty,\infty)} \le 2\,\|f_w^{(l-2)}\|_{L_p(-\infty,\infty)}^{\frac{1}{2}\left(1-\frac{1}{p}\right)} \|f_w^{(l)}\|_{L_p(-\infty,\infty)}^{\frac{1}{2}\left(1+\frac{1}{p}\right)}$$

$$\le 2\left(\frac{\pi}{2}\,\|f\|_{L_p(-\infty,\infty)}^{\frac{2}{l}}\,\|f_w^{(l)}\|_{L_p(-\infty,\infty)}^{1-\frac{2}{l}}\right)^{\frac{1}{2}\left(1-\frac{1}{p}\right)} \|f_w^{(l)}\|_{L_p(-\infty,\infty)}^{\frac{1}{2}\left(1+\frac{1}{p}\right)}$$

$$\le \sqrt{2\,\pi}\,\|f\|_{L_p(-\infty,\infty)}^{\frac{1}{l}\left(1-\frac{1}{p}\right)}\|f_w^{(l)}\|_{L_p(-\infty,\infty)}^{\frac{1}{l}\left(l-1+\frac{1}{p}\right)}. \tag{4.53}$$

Remark 11 Let $l, m \in \mathbb{N}, m < l$ and $1 \le p \le \infty$. Then $\forall f \in W_p^l(0,\infty)$

$$\|f_w^{(m)}\|_{L_p(0,\infty)} \le \frac{\pi}{2}\,8^l\,\|f\|_{L_p(0,\infty)}^{1-\frac{m}{l}}\|f_w^{(l)}\|_{L_p(0,\infty)}^{\frac{m}{l}}. \tag{4.54}$$

This can be proved with the help of the extension operator T_2, constructed in Section 6.1 of Chapter 6 (see Remark 1):

$$\|f_w^{(m)}\|_{L_p(0,\infty)} \le \|(T_2 f)_w^{(m)}\|_{L_p(-\infty,\infty)} \le \frac{\pi}{2}\,\|T_2 f\|_{L_p(-\infty,\infty)}^{1-\frac{m}{l}}\|(T_2 f)_w^{(l)}\|_{L_p(-\infty,\infty)}^{\frac{m}{l}}$$

---

[8] This inequality is equivalent to $\frac{p^2-1}{2p-1} \le \left(\tfrac{1}{2}\nu^{\frac{2p}{p-1}}\right)^p\left(1-\tfrac{1}{2p}\right)^p$, where $\nu = \sqrt{2\,e^{\frac{1}{e}}}$, which is clear since for $p > 1$

$$\frac{p^2-1}{2p-1} \le \frac{p}{2} \le \frac{1}{2}\,e^{\frac{2}{e}} \le e^{\frac{2}{e}}\left(1-\frac{1}{2p}\right)^p \le \left(\frac{1}{2}\nu^{\frac{2p}{p-1}}\right)^p\left(1-\frac{1}{2p}\right)^p.$$

$$\le \frac{\pi}{2} 8^l \, \|f\|_{L_p(0,\infty)}^{1-\frac{m}{l}} \, \|f_w^{(l)}\|_{L_p(0,\infty)}^{\frac{m}{l}}.$$

It can be proved that, in contrast to the case of the whole line, the best constant in the case of the half-line for fixed m tends to ∞ as $l \to \infty$. To verify this one may consider the function f defined by $f(x) = (\sqrt[l]{l!} - x)^l$ for $0 \le x \le \sqrt[l]{l!}$, and $f(x) = 0$ for $x \ge \sqrt[l]{l!}$.

Corollary 9 *Let* $-\infty \le a < b \le \infty$, $l, m \in \mathbb{N}$, $m < l$ *and* $1 \le p \le \infty$. *If a sequence* $\{f_k\}_{k \in \mathbb{N}}$ *is bounded in* $W_p^l(a,b)$ *and converges in* $L_p(a, b)$ *to a function* f, *then it converges to* f *in* $W_p^m(a,b)$.

Idea of the proof. Apply inequalities with a parameter (4.40) and (4.42) or multiplicative inequalities (4.41) and (4.43). $\square$

Proof. Let $\|f_k\|_{W_p^l(a,b)} \le K$ for each $k \in \mathbb{N}$ and $f_k \to f$ in $L_p(a,b)$ as $k \to \infty$.

1. By footnote 5 it follows that $\forall \varepsilon > 0$ there exists $C(\varepsilon)$ such that

$$\|f_w^{(m)}\|_{L_p(a,b)} \le C(\varepsilon) \, \|f\|_{L_p(a,b)} + \varepsilon \, \|f_w^{(l)}\|_{L_p(a,b)}.$$

for each $f \in W_p^l(a,b)$. Consequently, $\forall k, s \in \mathbb{N}$

$$\|(f_k)_w^{(m)} - (f_s)_w^{(m)}\|_{L_p(a,b)} \le C(\varepsilon) \, \|f_k - f_s\|_{L_p(a,b)} + 2\varepsilon K.$$

Given $\delta > 0$ we choose ε in such a way that $2\varepsilon K < \delta$. Since f_k is a Cauchy sequence in $L_p(a,b)$, there exists $N \in \mathbb{N}$ such that $C(\varepsilon) \, \|f_k - f_s\|_{L_p(a,b)} < \frac{\delta}{2}$ if $k, s \ge N$. Hence, $\forall k, s \ge N$ we have $\|(f_k)_w^{(m)} - (f_s)_w^{(m)}\|_{L_p(a,b)} < \delta$, i.e., the sequence $(f_k)_w^{(m)}$ is Cauchy in $L_p(a,b)$. Because of the completeness of $L_p(a,b)$ there exists $g \in L_p(a,b)$ such that $(f_k)_w^{(m)} \to g$ as $k \to \infty$ in $L_p(a,b)$. Since the weak differentiation operator is closed (see Section 1.2), g is a weak derivative of order m of f on (a,b). Consequently, $f_k \to f$ in $W_p^m(a,b)$ as $k \to \infty$.

2. By (4.41) and (4.43) it follows that $\forall k, s \in \mathbb{N}$

$$\|(f_k)_w^{(m)} - (f_s)_w^{(m)}\|_{L_p^m(a,b)} \le M \, \|f_k - f_s\|_{L_p(a,b)}^{1-\frac{m}{l}} \, \|f_k - f_s\|_{W_p^l(a,b)}^{\frac{m}{l}}$$

$$\le M(2K)^{\frac{m}{l}} \, \|f_k - f_s\|_{L_p(a,b)}^{1-\frac{m}{l}} \, ,$$

where M depends only on l. Consequently, we can again state that $(f_k)_w^{(m)}$ is a Cauchy sequence in $L_p(a,b)$. The rest is the same as in step 1. $\square$

Theorem 3 *Let* $-\infty < a < b < \infty, l \in \mathbb{N}, m \in \mathbb{N}_0, m < l$ *and* $1 \le p \le \infty$. *Then the embedding*

$$W_p^l(a,b) \hookrightarrow W_p^m(a,b)$$

is compact, i.e., the embedding operator $I : W_p^l(a,b) \to W_p^m(a,b)$ is compact. [9]

Idea of the proof. Let S be an arbitrary bounded set in $W_p^l(a,b)$. In the case $m = 0$ consider any bounded extension operator $T : W_p^l(a,b) \to W_p^l(-\infty,\infty)$. (See Section 6.1.) For $\delta > 0$ set $f_\delta = A_\delta(Tf)$, where A_δ is a mollifier with a nonnegative kernel. Prove that there exists $M_1 > 0$ such that $\forall f \in S$ and $\forall \delta > 0$

$$\|f - f_\delta\|_{L_p(a,b)} \leq M_1\,\delta. \tag{4.55}$$

Moreover, prove that there exists $M_2(\delta) > 0$ such that $\forall f \in S$ and $\forall x, y \in [a,b]$

$$|f_\delta(x)| \leq M_2(\delta), \quad |f_\delta(x) - f_\delta(y)| \leq M_2(\delta)\,|x - y|. \tag{4.56}$$

Finally, apply the criterion of compactness in terms of ε-nets and Arzela's theorem. [10] In the case $m > 0$ apply Corollary 9. $\square$

Proof. By (1.8) we have

$$\|f - f_\delta\|_{L_p(a,b)} \leq \|A_\delta(Tf) - Tf\|_{L_p(-\infty,\infty)}$$

$$\leq \sup_{|h| \leq \delta} \|(Tf)(x + h) - (Tf)(x)\|_{L_p(-\infty,\infty)}.$$

By Corollary 7 of Chapter 3 and inequality (4.33)

$$\|(Tf)(x + h) - (Tf)(x)\|_{L_p(-\infty,\infty)} \leq |h|\,\|(Tf)'_w\|_{L_p(-\infty,\infty)}$$

$$\leq M_3\,|h|\,\|Tf\|_{W_p^l(-\infty,\infty)} \leq M_4\,|h|\,\|f\|_{W_p^l(a,b)},$$

where M_3 and M_4 are independent of f.

Since S is bounded in $W_p^l(a,b)$, say $\|f\|_{W_p^l(a,b)} \leq K$ for each $f \in S$, inequality (4.55) follows. Furthermore, by Hölder's inequality $\forall x \in [a,b]$

$$|f_\delta(x)| \leq \frac{1}{\delta} \int_{x-\delta}^{x+\delta} \omega\left(\frac{x - y}{\delta}\right) |(Tf)(y)|\,dy \leq \frac{M_5}{\delta}\,(2\,\delta)^{\frac{1}{p'}}\,\|Tf\|_{L_p(\mathbb{R})}$$

$$\leq M_6(\delta)\,\|f\|_{W_p^l(a,b)} \leq K\,M_6(\delta)$$

---

[9] This means that each set bounded in $W_p^l(a,b)$ is compact in $W_p^m(a,b)$ ($\equiv$ precompact), i.e., each of its infinite subsets contains a sequence convergent in $W_p^m(a,b)$.

[10] Let $\Omega \subset R^n$ be a compact. A set $S \subset C(\Omega)$ is compact in $C(\Omega)$ ($\equiv$ precompact) if, and only if, S is bounded and equicontinuous, i.e., $\forall \varepsilon > 0\ \exists \delta > 0$ such that $\forall f \in S$ and $\forall x, y \in \Omega$ satisfying $|x - y| < \delta$ the inequality $|f(x) - f(y)| < \varepsilon$ holds.

and the first inequality (4.56) follows. Here $M_5 = \max\limits_{|z| \le 1} |\omega(z)|$ and $M_6(\delta)$ is independent of f.

Moreover, $\forall x, y \in [a, b]$

$$|f_\delta(x) - f_\delta(y)| = \frac{1}{\delta} \left| \int\limits_{\mathbb{R}} \left(\omega\left(\frac{x-y}{\delta}\right) - \omega\left(\frac{y-u}{\delta}\right) \right) (Tf)(u)\, du \right|$$

$$\le \frac{1}{\delta} \int\limits_{(x-\delta,x+\delta)\cup(y-\delta,y+\delta)} \left| \omega\left(\frac{x-u}{\delta}\right) - \omega\left(\frac{y-u}{\delta}\right) \right| |(Tf)(u)|\, du$$

$$\le \frac{M_7 |x - y|}{\delta^2} (4\delta)^{\frac{1}{p'}} \|Tf\|_{L_p(\mathbb{R})} \le M_8(\delta)\, |x - y|\, \|f\|_{W_p^l(a,b)} \le K\, M_8(\delta)\, |x - y|$$

and the second inequality (4.56) follows. Here $M_7 = \max\limits_{|z| \le 1} |\omega'(z)|$ and $M_8(\delta)$ is independent of f.

Given $\varepsilon > 0$ by (4.55) there exists $\delta > 0$ such that $\|f - f_\delta\|_{L_p(a,b)} < \frac{\varepsilon}{2}$ for each $f \in S$. Then an $\frac{\varepsilon}{2}$-net for the set $S_\delta = \{f_\delta : f \in S\}$ will be an ε-net for S. If we now establish the compactness of S_δ in $L_p(a, b)$ it will imply the existence of a finite $\frac{\varepsilon}{2}$-net for S_δ. This means that we may construct finite ε-nets for S for an arbitrary $\varepsilon > 0$, which implies that S is compact in $L_p(a, b)$.

Finally, it is enough to note that from (4.56) it follows that the set S_δ is bounded and equicontinuous in $C[a, b]$. Hence, by Arzela's theorem S_δ is compact in $C[a, b]$ and consequently in $L_p[a, b]$ since convergence in $C[a, b]$ implies convergence in $L_p[a, b]$.

2. Let $m > 0$. By step 1 each infinite subset of S contains a sequence $\{f_k\}_{k \in \mathbb{N}}$ convergent to a function f in $L_p(a, b)$. By Corollary 9 $f \in W_p^m(a, b)$ and $f_k \to f$ as $k \to \infty$ in $W_p^m(a, b)$. $\square$

Example 1 If $b - a = \infty$, then Theorem 3 does not hold. Let, for example, $(a, b) = (0, \infty)$. Suppose, that $\varphi \in C_0^\infty(-\infty, \infty)$ is such that $\mathrm{supp}\, \varphi \subset [0, 1]$ and $\varphi \not\equiv 0$. Then the set $S = \{\varphi(x - k)\}_{k \in \mathbb{N}}$ is bounded in $W_p^l(0, \infty)$ since $\|\varphi(x - k)\|_{W_p^l(0,\infty)} = \|\varphi\|_{W_p^l(0,\infty)}$. However, it is not compact in $W_p^m(0, \infty)$ because for each $k, m \in \mathbb{N}, k \ne m$

$$\|\varphi(x - k) - \varphi(x - m)\|_{W_p^m(0,\infty)} \ge \|\varphi(x - k) - \varphi(x - m)\|_{L_p(0,\infty)} = 2^{\frac{1}{p}}.$$

(Consequently, any sequence in S, i.e., $\{\varphi(x - k_s)\}_{s \in \mathbb{N}}$, is not convergent in $W_p^m(0, \infty)$.)

Next we pass to the embedding theorems in the simplest case of Sobolev spaces $W_p^1(a,b)$. In this case it is possible to evaluate sharp constants in many of the relevant inequalities.

Theorem 4 *Let* $-\infty \le a < b \le \infty$ *and* $1 \le p < \infty$. *Then each function* $f \in W_p^1(a,b)$ *is equivalent to a function* $h \in \overline{C}(a,b)$. *Moreover,*
   1) *if* $-\infty < a < b < \infty$, *then*

$$\|f\|_{L_\infty(a,b)} \le (b-a)^{-\frac{1}{p}}\|f\|_{L_p(a,b)} + (p'+1)^{-\frac{1}{p'}}(b-a)^{\frac{1}{p'}}\|f'_w\|_{L_p(a,b)} \qquad (4.57)$$

and, consequently,
$$\|f\|_{L_\infty(a,b)} \le c_{17}\,\|f\|_{W_p^1(a,b)}\,, \qquad (4.58)$$

where
$$c_{17} = \max\{(b-a)^{-\frac{1}{p}}, (p'+1)^{-\frac{1}{p'}}(b-a)^{\frac{1}{p'}}\};$$

   2) *if* $-\infty < a < b < \infty$, *then*

$$\left\|f(x) - \frac{1}{b-a}\int_a^b f(y)\,dy\right\|_{L_\infty(a,b)} \le (p'+1)^{-\frac{1}{p'}}(b-a)^{\frac{1}{p'}}\,\|f'_w\|_{L_p(a,b)}; \qquad (4.59)$$

   3) *if* $-\infty < a < b = \infty$, *then* $\lim\limits_{x\to+\infty} h(x) = 0$ *and*

$$\|f\|_{L_\infty(a,\infty)} \le (p')^{-\frac{1}{p'}}\|f\|_{W_p^1(a,\infty)}; \qquad (4.60)$$

   4) *if* $(a,b) = (-\infty, \infty)$, *then* $\lim\limits_{x\to-\infty} h(x) = \lim\limits_{x\to+\infty} h(x) = 0$ *and*

$$\|f\|_{L_\infty(-\infty,\infty)} \le 2^{-\frac{1}{p}}(p')^{-\frac{1}{p'}}\|f\|_{W_p^1(-\infty,\infty)}. \qquad (4.61)$$

All the constants in the inequalities (4.57), (4.59)−(4.61) *are sharp. The constant* c_{17} *in inequality* (4.58) *is sharp if* $b - a \le (p'+1)^{\frac{1}{p'}}$.

Remark 12 For $p = 1$ inequality (4.57) takes the form

$$\|f\|_{L_\infty(a,b)} \le (b-a)^{-1}\|f\|_{L_1(a,b)} + \|f'_w\|_{L_1(a,b)}. \qquad (4.62)$$

We also note that for $p = 1$ inequalities (4.60) and (4.61) are equivalent to

$$\|f\|_{L_\infty(a,\infty)} \le \|f'_w\|_{L_1(a,\infty)} \qquad (4.63)$$

and

$$\|f\|_{L_\infty(-\infty,\infty)} \le \tfrac{1}{2}\|f'_w\|_{L_1(-\infty,\infty)} \qquad (4.64)$$

respectively.

Idea of the proof. Apply Definition 4 and Remark 6 of Chapter 1. In order to prove inequality (4.59) apply the integral representation (3.6). Deduce (4.57) and (4.58) from (4.59). Inequality (4.63) follows from (4.62) and implies inequality (4.64). Apply (4.63) and (4.64) to $|f|^p$ and deduce (4.60) and, respectively, (4.61). Set $f \equiv 1$ to prove sharpness of c_{17} in (4.58) and of the constant multiplying $\|f\|_{L_p(a,b)}$ in (4.57). Set $f(x) = (x-a)^{p'}$ to prove sharpness of the constant in (4.59). Set $f(x) = f_\varepsilon(x) \equiv 1 + \varepsilon(x-a)^{p'}$ and pass to the limit as $\varepsilon \to 0+$ to prove sharpness of the constant multiplying $\|f'_w\|_{L_p(a,b)}$ in (4.57). Set $f(x) = e^{-\mu(x-a)}$ in (4.60), $f(x) = e^{-\mu|x|}$ in (4.61) respectively, and choose an appropriate μ to prove sharpness of the constants in those inequalities. $\square$

Proof. 1. Let $f \in W_p^1(a,b)$. By Definition 4 of Chapter 1 there is a function h equivalent to f on (a,b), which is locally absolutely continuous on (a,b). Moreover, if $a > -\infty$ or $b < \infty$, then the limits $\lim_{x\to a+} h(x)$ and $\lim_{x\to b-} h(x)$ exist. If $a = -\infty$ or $b = \infty$, then as will be proved in steps 3-4 $\lim_{x\to-\infty} h(x) = 0$, $\lim_{x\to\infty} h(x) = 0$ respectively. Hence h is bounded and uniformly continuous on (a,b), i.e., $h \in \overline{C}(a,b)$ for all $-\infty \le a \le b \le \infty$.

2. By applying Hölder's inequality to (3.6) we obtain that for almost every $x \in (a,b)$

$$\left| f(x) - \frac{1}{b-a} \int_a^b f(y)\, dy \right|$$

$$\le (b-a)^{-1} \left(\int_a^x (y-a)^{p'}\, dy + \int_x^b (b-y)^{p'}\, dy \right)^{\frac{1}{p'}} \|f'_w\|_{L_p(a,b)}$$

$$= (p'+1)^{-\frac{1}{p'}} (b-a)^{-1} [(x-a)^{p'+1} + (b-x)^{p'+1}]^{\frac{1}{p'}} \|f'_w\|_{L_p(a,b)}. \tag{4.65}$$

Since $\max_{a \le x \le b} [(x-a)^{p'+1} + (b-x)^{p'+1}]^{\frac{1}{p'}} = (b-a)^{1+\frac{1}{p'}}$ we have established (4.59). Inequality (4.57) and, hence, (4.58) follow since

$$|f(x)| \le \left| \frac{1}{b-a} \int_a^b f(y)\, dy \right| + \left| f(x) - \frac{1}{b-a} \int_a^b f(y)\, dy \right|$$

$$\le (b-a)^{-\frac{1}{p}} \|f\|_{L_p(a,b)} + \left| f(x) - \frac{1}{b-a} \int_a^b f(y)\, dy \right|.$$

3. By letting $b \to +\infty$ in (4.62) we obtain (4.63). Moreover, $\forall x \in (a,\infty)$

$$|h(x)| \le \|h\|_{C(x,\infty)} = \|f\|_{L_\infty(x,\infty)} \le \int_x^\infty |f'_w(y)|\, dy$$

and it follows that $\lim\limits_{x\to+\infty} h(x) = 0$.

4. If $f \in W_1^1(-\infty, +\infty)$, then

$$|h(x)| \leq \int\limits_{-\infty}^{x} |f_w'|\, dy$$

and $\lim\limits_{x\to-\infty} h(x) = 0$ as well. Adding the last inequality and the previous one, we get

$$|h(x)| \leq \tfrac{1}{2}\, \|f_w'\|_{L_1(-\infty,\infty)}$$

and (4.64) follows since $\|f\|_{L_\infty(-\infty,\infty)} = \|h\|_{C(-\infty,\infty)}$.

5. If $p > 1$, then by (4.63) and Hölder's inequality

$$\|f\|_{L_\infty(a,\infty)} = \| \, |f|^p\|_{L_\infty(a,\infty)}^{\frac{1}{p}} \leq \|(|f|^p)_w'\|_{L_1(a,\infty)}^{\frac{1}{p}} = p^{\frac{1}{p}} \| \, |f|^{p-1} f_w'\|_{L_1(a,\infty)}^{\frac{1}{p}}$$

$$\leq p^{\frac{1}{p}} \|f\|_{L_p(a,\infty)}^{\frac{1}{p'}} \|f_w'\|_{L_p(a,\infty)}^{\frac{1}{p}}. \tag{4.66}$$

We establish (4.60) by applying inequality (4.47). Inequality (4.61) is proved in a similar way.

6. Setting $f \equiv 1$ we obtain that the constant $(b-a)^{-\frac{1}{p}}$ multiplying $\|f\|_{L_p(a,b)}$ in (4.57) and the constant c_{17} in (4.58), if $b - a \leq (p'+1)^{\frac{1}{p'}}$, cannot be diminished. If $f(x) = (x-a)^{p'}$, then one can easily verify that there is equality in (4.59).

Now let us consider the inequality

$$\|f\|_{L_\infty(a,b)} \leq (b-a)^{-\frac{1}{p}} \|f\|_{L_p(a,b)} + A \|f_w'\|_{L_p(a,b)}.$$

We prove that $A \geq (p'+1)^{-\frac{1}{p}}(b-a)^{\frac{1}{p'}}$, which means that the constant multiplying $\|f_w'\|_{L_p(a,b)}$ in (4.57) is sharp. Indeed, set $f(x) = f_\varepsilon(x) \equiv 1 + c(x-a)^{p'}$, where $\varepsilon > 0$, then $\|f_\varepsilon\|_{L_\infty(a,b)} = 1 + \varepsilon(b-a)^{p'}$, $\|(f_\varepsilon)'\|_{L_p(a,b)} = \varepsilon p'(p'+1)^{-\frac{1}{p}}(b-a)^{p'-\frac{1}{p'}}$ and $\|f_\varepsilon\|_{L_p(a,b)} = (b-a)^{\frac{1}{p}}(1+\varepsilon(p'+1)^{-1}(b-a)^{p'} + o(\varepsilon))$ as $\varepsilon \to 0+$. Consequently,

$$A \geq \lim_{\varepsilon\to0+} \frac{\|f_\varepsilon\|_{L_\infty(a,b)} - (b-a)^{-\frac{1}{p}}\|f_\varepsilon\|_{L_p(a,b)}}{\|(f_\varepsilon)'\|_{L_p(a,b)}} = (p'+1)^{-\frac{1}{p'}}(b-a)^{\frac{1}{p'}}.$$

Finally, for $f(x) = e^{-\mu(x-a)}$ inequality (4.60) with $1 < p < \infty$ is equivalent to the inequality $1 \leq (p')^{-\frac{1}{p'}} p^{-\frac{1}{p}}(\mu^{-\frac{1}{p}} + \mu^{\frac{1}{p'}})$. For $\mu = \frac{1}{p-1}$ the quantity $\mu^{-\frac{1}{p}} + \mu^{\frac{1}{p'}}$ is minimal and this inequality becomes an equality. Hence, for $f(x) = e^{-\frac{x-a}{p-1}}$

there is equality in (4.60). Analogously for $f(x) = e^{-\frac{|x|}{p-1}}$ there is equality in (4.61).

If $p = 1$, then equality in (4.60) and (4.61) holds if, and only if, f is equivalent to 0. This follows from inequalities (4.63), (4.64) respectively. However, the constants 1 in (4.60) and $\frac{1}{2}$ in (4.61) are sharp, which easily follows by setting $f(x) = e^{-\mu(x-a)}$, $f(x) = e^{-\mu|x|}$ respectively, and passing to the limit as $\mu \to +\infty$. $\square$

Remark 13 We note that for a function h, which is equivalent to f on (a, b) and which is absolutely continuous on $[a, b]$, inequality (4.63) may be rewritten as

$$\max_{a \leq x < \infty} |h(x)| \leq \operatorname*{Var}_{[a,\infty)} h.$$

The maximum exists since $h(x) \to 0$ as $x \to +\infty$. The inequality is clear since for each function h of bounded variation $\forall x \in [a, \infty)$ we have

$$|h(x)| = \lim_{y \to +\infty} |h(x) - h(y)| \leq \operatorname*{Var}_{[a,\infty)} h.$$

It is also clear that for $f \in W_p^1(a, \infty)$ equality is achieved in (4.63) if, and only if, f is equivalent to a nonnegative and nonincreasing function or a nonpositive and nondecreasing one on $[a, \infty)$. Similarly for $f \in W_p^1(-\infty, \infty)$ equality is achieved in (4.64) if, and only if, f is equivalent to a function, which is nonnegative, nondecreasing on $(-\infty, x_0]$ and nonincreasing on $[x_0, \infty)$ for some x_0 or nonpositive, nonincreasing on $(-\infty, x_0]$ and nondecreasing on $[x_0, \infty)$.

Remark 14 Analysis of the cases, in which there is equality in Hölder's inequality, [11] suggests the choice of test-functions, which allows one to state the sharpness of the constants. In the case of inequality (4.59) we take $x = b$ in (4.65). If $(f')^p = M_1(x - a)^{p'}$ on (a, b) for some $M_1 > 0$, and, in particular, $f(x) = (x - a)^{p'}$, then there is equality in inequality (4.65) and, consequently, in (4.59). Let $f > 0$ and $f' < 0$ on (a, ∞) in the case of inequality (4.60). Then by Remark 13 there is equality in the first inequality (4.66). Furthermore, if $(-f')^p = M_2(f^{p-1})^{p'}$ on (a, ∞), $M_2 > 0$, then there is equality in the second

---

[11] Let f and g be measurable on (a, b) and $1 < p < \infty$. The equality

$$\|fg\|_{L_1(a,b)} = \|f\|_{L_p(a,b)} \|g\|_{L_{p'}(a,b)}$$

holds if, and only if, $A|f|^p = B|g|^{p'}$ almost everywhere on (a, b) for some nonnegative A and B.

inequality (4.66). All solutions $f \in L_p(a, \infty)$ of this equation have the from $f(x) = e^{-\mu(x-a)}$ for some $\mu > 0$.

A more sophisticated argument of similar type explains the choice of test-functions $f(x) = 1 + \varepsilon(y - a)^{p'}$ in the case of inequality (4.57).

Corollary 10 (inequalities with a small parameter multiplying the norm of a derivative) *Let* $-\infty \le a < b \le \infty, 1 < p < \infty$.

1) *If* $-\infty < a < b < \infty$, *then* $\forall \varepsilon \in (0, \varepsilon_0]$, *where* $\varepsilon_0 = ((p' + 1)^{-1}(b - a))^{\frac{1}{p'}}$,

$$\|f\|_{L_\infty(a,b)} \le (p' + 1)^{-\frac{1}{p}} \varepsilon^{-\frac{1}{p-1}} \|f\|_{L_p(a,b)} + \varepsilon \|f'_w\|_{L_p(a,b)}. \tag{4.67}$$

2) *If* $-\infty < a < b = \infty$, *then* $\forall \varepsilon \in (0, \infty)$

$$\|f\|_{L_\infty(a,\infty)} \le (p')^{-1} \varepsilon^{-\frac{1}{p-1}} \|f\|_{L_p(a,\infty)} + \varepsilon \|f'_w\|_{L_p(a,b)}. \tag{4.68}$$

3) *If* $(a, b) = (-\infty, \infty)$, *then* $\forall \varepsilon \in (0, \infty)$

$$\|f\|_{L_\infty(-\infty,\infty)} \le (p')^{-1}(2\varepsilon)^{-\frac{1}{p-1}} \|f\|_{L_p(-\infty,\infty)} + \varepsilon \|f'_w\|_{L_p(-\infty,\infty)}. \tag{4.69}$$

The constants in inequalities (4.68) and (4.69) are sharp.

Idea of the proof. Apply the proofs of inequalities (4.40) and (4.42). Verify that there is equality in (4.68) for $f(x) = \exp\left(-\frac{\varepsilon^{-p'}(x-a)}{p}\right)$ and in (4.69) for $f(x) = \exp\left(-\frac{(\varepsilon 2^{\frac{1}{p}})^{-p'}}{p} |x|\right)$. See also Remarks 15–16 and 18 below. $\square$

Corollary 11 (multiplicative inequalities) *Let* $-\infty \le a < b \le \infty, 1 \le p \le \infty$.
1) *If* $-\infty < a < b < \infty$, *then*

$$\|f\|_{L_\infty(a,b)} \le c_{18} \|f\|_{L_p(a,b)}^{\frac{1}{p'}} \|f\|_{W_p^1(a,b)}^{\frac{1}{p}}, \tag{4.70}$$

where $c_{18} = (b - a)^{-\frac{1}{p}} + p^{\frac{1}{p}}(p')^{\frac{1}{p'}}(p' + 1)^{-\frac{1}{pp'}} \le (b - a)^{-\frac{1}{p}} + 2$.
2) *If* $-\infty < a < b = \infty$, *then*

$$\|f\|_{L_\infty(a,\infty)} \le p^{\frac{1}{p}} \|f\|_{L_p(a,\infty)}^{\frac{1}{p'}} \|f'_w\|_{L_p(a,\infty)}^{\frac{1}{p}}. \tag{4.71}$$

3) *If* $(a, b) = (-\infty, \infty)$, *then*

$$\|f\|_{L_\infty(-\infty,\infty)} \le (\tfrac{p}{2})^{\frac{1}{p}} \|f\|_{L_p(-\infty,\infty)}^{\frac{1}{p'}} \|f'_w\|_{L_p(a,b)}^{\frac{1}{p}}. \tag{4.72}$$

The constants in inequalities (4.71) and (4.72) are sharp.

Idea of the proof. Inequalities (4.71) and (4.72) have already been established in the proof of Theorem 4. If $f(x) = \exp\left(-(x-a)\right)$ or $f(x) = \exp\left(-|x|\right)$, then inequalities (4.71) and (4.72) become equalities. Apply the proof of inequality (4.41) to prove (4.70). See also Remarks 15–16 and 18 below. $\square$

Remark 15 We note that for $1 < p < \infty$ the additive inequality (4.60), the inequality with a parameter (4.61) and the multiplicative inequality (4.71) are equivalent. Indeed, (4.68) was derived from (4.60) with the help of dilations and (4.60) was derived from (4.71) with the help of inequality (4.47). Finally (4.68) implies (4.71) by minimizing the right-hand side of (4.68) with respect to a parameter.

These inequalities are also equivalent to the following ones:

$$\|f\|_{L_\infty(a,\infty)} \le (p-1)^{\frac{1}{pp'}}\left(\int\limits_a^\infty \left(|f|^p + |f'_w|^p\right)dx\right)^{\frac{1}{p}} \tag{4.73}$$

and, $\forall \varepsilon > 0$,

$$\|f\|_{L_\infty(a,\infty)} \le \left((p-1)^{\frac{1}{p}}\varepsilon^{-\frac{1}{p-1}}\int_a^\infty |f|^p\,dx + \varepsilon \int_a^\infty |f'_w|^p\,dx\right)^{\frac{1}{p}}. \tag{4.74}$$

For, (4.73) follows from inequality (4.71) raised to the power p and (4.47), (4.74) follows from (4.73) with the help of dilations and (4.71) follows from (4.74) by minimizing its right-hand side.

For the same reasons inequalities (4.61), (4.69), (4.72) and the inequalities

$$\|f\|_{L_\infty(-\infty,\infty)} \le 2^{-\frac{1}{p}}(p-1)^{\frac{1}{pp'}}\left(\int\limits_{-\infty}^\infty \left(|f|^p\,dx + |f'_w|^p\right)^{\frac{1}{p}}, \tag{4.75}$$

and

$$\|f\|_{L_\infty(-\infty,\infty)} \le \left(\tfrac{1}{2}(p-1)^{\frac{1}{p}}\varepsilon^{-\frac{1}{p-1}} + \varepsilon\int\limits_{-\infty}^\infty |f'_w|^p\,dx\right)^{\frac{1}{p}} \tag{4.76}$$

with an arbitrary $\varepsilon > 0$, are equivalent as well. Equalities in (4.73)–(4.74), (4.75)–(4.76) respectively, hold for $f(x) = \exp\left(-\mu(x-a)\right)$, $f(x) = \exp\left(-\mu|x|\right)$ respectively, with appropriate choice of μ. For example, in the case of inequality (4.75) $\mu = (p-1)^{-\frac{1}{p}}$.

Moreover, the listed inequalities for the halfline and for the whole line are also equivalent. This follows from the equivalence of (4.73) and (4.75). For, if

(4.73) holds, then, replacing x by $2a - x$, we have also that

$$\|f\|_{L_\infty(-\infty,a)} \le (p-1)^{\frac{1}{pp'}} \left(\int_{-\infty}^{a} (|f|^p + |f'_w|^p) \, dx \right)^{\frac{1}{p}}.$$

Consequently,

$$\|f\|_{L_\infty(-\infty,\infty)}^p = \tfrac{1}{2} \left(\|f\|_{L_\infty(-\infty,a)}^p + \|f\|_{L_\infty(a,\infty)}^p \right) = \frac{(p-1)^{\frac{1}{p'}}}{2} \int_{-\infty}^{\infty} (|f|^p + |f'_w|^p) \, dx$$

and (4.75) follows. Conversely, if (4.75) holds and $f \in W_p^1(a,\infty)$ we apply (4.75) to the even extension F of f: $F(x) = f(x)$, if $x > a$, and $F(x) = f(2a - x)$, if $x < a$. Then

$$\|f\|_{L_\infty(a,\infty)} = \|F\|_{L_\infty(-\infty,\infty)} \le 2^{-\frac{1}{p}} (p-1)^{\frac{1}{pp'}} \left(\int_{-\infty}^{\infty} (|F|^p + |F'_w|^p) \, dx \right)^{\frac{1}{p}}$$

$$= (p-1)^{\frac{1}{pp'}} \left(\int_{a}^{\infty} (|f|^p + |f'_w|^p) \, dx \right)^{\frac{1}{p}}.$$

Remark 16 For $p = 2$ all the inequalities discussed in Remark 15 may be deduced by taking Fourier transforms since by Parseval's equality

$$\|f\|_{L_\infty(-\infty,\infty)} = \|F^{-1}Ff\|_{L_\infty(-\infty,\infty)} = \frac{1}{\sqrt{2\pi}} \left\| \int_{-\infty}^{\infty} e^{ix\xi} (Ff)(\xi) \, d\xi \right\|_{L_\infty(-\infty,\infty)}$$

$$\le \frac{1}{\sqrt{2\pi}} \|Ff\|_{L_1(-\infty,\infty)} = \frac{1}{\sqrt{2\pi}} \|(1+\xi^2)^{-\frac{1}{2}} (1+\xi^2)^{\frac{1}{2}} (Ff)(\xi)\|_{L_1(-\infty,\infty)}$$

$$\le \frac{1}{\sqrt{2\pi}} \left(\int_{-\infty}^{\infty} (1+\xi^2)^{-1} \, d\xi \right)^{\frac{1}{2}} \left(\int_{-\infty}^{\infty} (|(Ff)(\xi)|^2 + |\xi(Ff)(\xi)|^2 \, d\xi \right)^{\frac{1}{2}}$$

$$= \frac{1}{\sqrt{2}} \left(\int_{-\infty}^{\infty} (|Ff|^2 + |Ff'_w|^2) \, d\xi \right)^{\frac{1}{2}} = \frac{1}{\sqrt{2}} \left(\int_{-\infty}^{\infty} (|f|^2 + |f'_w|^2) \, dx \right)^{\frac{1}{2}}.$$

Thus we obtain (4.75) for $p = 2$.

There is an alternative way of using Fourier transforms, which leads to two other inequalities [12]

$$\|f\|_{L_\infty(-\infty,\infty)} \le \frac{1}{\sqrt{2}} \left(\int\limits_{-\infty}^{\infty} |f \pm f_w'|^2 \, dx \right)^{\frac{1}{2}},$$

which hold for each [13] $f \in W_2^1(-\infty,\infty)$. The constant $\frac{1}{\sqrt{2}}$ is sharp. Indeed, by the properties of Fourier transforms and by the Cauchy-Schwartz inequality we have

$$\|f\|_{L_\infty(-\infty,\infty)} = \left\| F^{-1}\left(\frac{1}{1 \pm i\xi} F(f \pm f_w') \right) \right\|_{L_\infty(-\infty,\infty)}$$

$$= \frac{1}{\sqrt{2\pi}} \left\| F^{-1}\left(\frac{1}{1 \pm i\xi} \right) * (f \pm f_w') \right\|_{L_\infty(-\infty,\infty)}$$

$$= \frac{1}{\sqrt{2\pi}} \left\| \int\limits_{-\infty}^{\infty} \left(F^{-1}\left(\frac{1}{1 \pm i\xi} \right) \right)(x - y) \, (f(y) \pm f_w'(y)) \, dy \right\|_{L_\infty(-\infty,\infty)}$$

$$\le \frac{1}{\sqrt{2\pi}} \left\| F^{-1}\left(\frac{1}{1 \pm i\xi} \right) \right\|_{L_2(-\infty,\infty)} \|f \pm f_w'\|_{L_2(-\infty,\infty)}$$

$$= \frac{1}{\sqrt{2\pi}} \left\| \frac{1}{1 \pm i\xi} \right\|_{L_2(-\infty,\infty)} \|f \pm f_w'\|_{L_2(-\infty,\infty)}$$

and the desired inequality follows. If $f(x) = e^{-|x|}$, then all three inequalities under consideration become equalities.

The second approach is applicable to the case $1 \le p \le \infty$ as well and leads to the inequalities

$$\|f\|_{L_\infty(-\infty,\infty)} \le (p')^{-\frac{1}{p'}} \left(\int\limits_{-\infty}^{\infty} |f \pm f_w'|^p \, dx \right)^{\frac{1}{p}}$$

for each $f \in W_p^1(-\infty,\infty)$ since, for example,

$$\left\| F^{(-1)}\left(\frac{1}{1 + i\xi} \right) \right\|_{L_{p'}(-\infty,\infty)} = \sqrt{2\pi} \|e^{-x}\|_{L_{p'}(0,\infty)} = \sqrt{2\pi} (p')^{-\frac{1}{p'}}.$$

If $1 < p < \infty$, $f(x) = e^{-x}$ for $x \ge 0$ and $f(x) = e^{\frac{x}{p-1}}$ for $x < 0$, then these inequalities become equalities.

---

[12] If we square and add them, we obtain the previous inequality.

[13] We note that this inequality does not hold for each function f, which is such that the right-hand side is finite. (It does not hold, say, for $f(x) = e^{\mp x}$.)

Remark 17 We note two corollaries of (4.65) under the supposition that f is absolutely continuous on $[a,b]$:

$$|f(a)| \leq (b-a)^{-\frac{1}{p}}\|f\|_{L_p(a,b)} + (p'+1)^{-\frac{1}{p}}(b-a)^{\frac{1}{p'}}\|f'\|_{L_p(a,b)} \qquad (4.77)$$

(for $p=1$ (4.77) coincides with (3.8)) and

$$\left|f\left(\frac{a+b}{2}\right)\right| \leq (b-a)^{-\frac{1}{p}}\|f\|_{L_p(a,b)} + \frac{1}{2}(p'+1)^{-\frac{1}{p}}(b-a)^{\frac{1}{p'}}\|f'\|_{L_p(a,b)}. \qquad (4.78)$$

Both constants in (4.77) are the same as in (4.57) and are sharp. This is proved by using the same test-functions as in the case of inequality (4.57).

In inequality (4.78) both constants are also sharp. The test-function $f \equiv 1$ shows that the constant multiplying $\|f\|_{L_p(a,b)}$ is sharp. Moreover, the test-functions $f = f_\varepsilon$, where $\varepsilon > 0$ and f_ε is defined by $f_\varepsilon(x) = 1 + \varepsilon(x-a)^{p'}$, if $a \leq x \leq \frac{a+b}{2}$, and $f_\varepsilon(x) = 1 + \varepsilon(b-a)^{p'}$, if $\frac{a+b}{2} \leq x \leq b$, show by passing to the limit as $\varepsilon \to 0+$ that the constant multiplying $\|f'\|_{L_p(a,b)}$ is sharp.

Corollary 12 *Let $l \in \mathbb{N}, 1 \leq p \leq \infty$ and $-\infty \leq a < b \leq \infty$. Then each function $f \in W_p^l(a,b)$ is equivalent to a function $h \in \overline{C}^{\,l-1}(a,b)$ and*

$$\|g^{(m)}\|_{C(a,b)} \leq c_{19}\|f\|_{W_p^l(a,b)}, \quad m = 0, ..., l-1, \qquad (4.79)$$

where $c_{19} > 0$ is independent of f, i.e., $W_p^l(a,b) \hookrightarrow \overline{C}^{\,m}(a,b)$.

If $a = -\infty$, then $\lim\limits_{x \to -\infty} h^{(m)}(x) = 0$, if $b = \infty$, then $\lim\limits_{x \to \infty} h^{(m)}(x) = 0$, where $m = 0, ..., l-1$.

Idea of the proof. Apply Remark 6 of Section 1.3 and Theorems 4 and 2. $\square$

Theorem 5 *Let $l \in \mathbb{N}, m \in \mathbb{N}_0, m < l, 1 \leq p, q \leq \infty$ and $-\infty < a < b \leq \infty$. Then the embedding*

$$W_p^l(a,b) \hookrightarrow W_q^m(a,b) \qquad (4.80)$$

holds if, and only if, $b - a < \infty$, or $b - a = \infty$ and $p \leq q$. Moreover, this embedding is compact if, and only if, $b - a < \infty$ and the equalities $m = l-1, p=1$ and $q = \infty$ are not satisfied simultaneously.

Remark 18 As in the simplest case discussed in Corollary 7, from the inequality, accompanying embedding (4.80),

$$\|f_w^{(m)}\|_{L_q(a,b)} \leq M\|f\|_{W_p^l(a,b)}, \qquad (4.81)$$

where $M > 0$ is independent of f, it follows that, for $b - a < \infty$,

$$\|f_w^{(m)}\|_{L_q(a,b)} \leq c_{20} (b-a)^{\frac{1}{q}-\frac{1}{p}} \left((b-a)^{-m}\|f\|_{L_p(a,b)} + (b-a)^{l-m}\|f_w^{(l)}\|_{L_p(a,b)} \right), \tag{4.82}$$

where $c_{20} > 0$ is independent of f, a and b.

If $q \geq p$, then, excluding the case in which $m = l - 1, p = 1$ and $q = \infty$, it also follows that

$$\|f_w^{(m)}\|_{L_q(a,b)} \leq c_{21}\, \varepsilon^{-\frac{m+\frac{1}{p}-\frac{1}{q}}{l-m-\frac{1}{p}+\frac{1}{q}}}\, \|f\|_{L_p(a,b)} + \varepsilon\|f_w^{(l)}\|_{L_p(a,b)}\,, \tag{4.83}$$

where $0 < \varepsilon \leq c_{22}(b-a)^{l-m-\frac{1}{p}+\frac{1}{q}}$ and $c_{21}, c_{22} > 0$ are independent of f, a and b, and

$$\|f_w^{(m)}\|_{L_q(a,b)} \leq c_{23} \|f\|_{L_p(a,b)}^{\frac{1}{l}(l-m+\frac{1}{p}-\frac{1}{q})} \|f\|_{W_p^l(a,b)}^{\frac{1}{l}(m-\frac{1}{p}+\frac{1}{q})}\,, \tag{4.84}$$

where $c_{23} > 0$ is independent of f. Moreover, inequalities $(4.81)-(4.84)$ are equivalent.

The proof is similar to the proof of Corollary 7. One should notice, in addition, that since $q \geq p$, by Jensen's inequality

$$\left(\sum_{k=1}^{N} \|g\|_{L_p(a_k,a_{k+1})}^q\right)^{\frac{1}{q}} \leq \left(\sum_{k=1}^{N} \|g\|_{L_p(a_k,a_{k+1})}^p\right)^{\frac{1}{p}} = \|g\|_{L_p(a,b)}.$$

If $b - a = \infty$, then inequality (4.83) holds $\forall \varepsilon > 0$ and in inequality (4.84) $\|f\|_{W_p^l(a,b)}$ can be replaced by $\|f_w^{(l)}\|_{L_p(a,b)}$.

Idea of the proof. To prove (4.81) apply Corollary 12 and Hölder's inequality if $b - a < \infty$ and the inequality

$$\|f\|_{L_q(a,b)} \leq \|f\|_{L_p(a,b)}^{\frac{p}{q}}\|f\|_{L_\infty(a,b)}^{1-\frac{p}{q}}\,,$$

where if $b - a = \infty$ and $p < q$. If $b - a = \infty$ and $q < p$, set $f(x) = (1 + x^2)^{-\frac{2}{q}}$, or

$$f(x) = \sum_{k \in \mathbb{Z}:(k,k+1)\subset(a,b)} |k|^{-\frac{1}{q}}\varphi(x - k), \tag{4.85}$$

where $\varphi \in C_0^\infty(\mathbb{R}), \varphi \not\equiv 0$ and $\operatorname{supp}\varphi \subset [0,1]$, to verify that (4.81) does not hold.

To prove the compactness apply Theorem 3 and inequality (4.83) or (4.84). If $b - a < \infty, m = l - 1, p = 1$ and $q = \infty$, consider the sequence

$$f_k(x) = k^{1-l}\eta\left(\frac{a+b}{2} + k\left(x - \frac{a+b}{2}\right)\right), \tag{4.86}$$

where $k \in \mathbb{N}, \eta \in C_0^\infty(-\infty, \infty)$, $\operatorname{supp} \eta \subset (a, b)$ and $\eta^{(l-1)}(\frac{a+b}{2}) = 1$. Finally, if $b - a = \infty$, apply Example 1. $\square$

Proof. The proof of the statements concerning embedding (4.80) being clear, we pass to the proof of the statements concerning the compactness.

1. Let $b - a < \infty$ and $f_k, k \in \mathbb{N}$, be a sequence bounded in $W_p^l(a, b)$. Then by Theorem 3 there exists a subsequence f_{k_s}, $s \in \mathbb{N}$, and a function $f \in L_p(a, b)$ such that $f_{k_s} \to f$ in $L_p(a, b)$. If $l - m - \frac{1}{p} + \frac{1}{q} > 0$, then from (4.83) or (4.84) it follows that $f_{k_s} \to f$ in $W_q^m(a, b)$.

2. If $b - a < \infty$ and $l - m - \frac{1}{p} + \frac{1}{q} = 0$, i.e., $m = l - 1, p = 1$ and $q = \infty$, then for the functions f_k defined by (4.86) we have

$$\|f_k\|_{W_1^l(a,b)} = k^{-l} \|\eta\|_{L_1(a,b)} + \|\eta^{(l)}\|_{L_1(a,b)} \leq \|\eta\|_{W_1^l(a,b)}$$

and $\lim\limits_{k \to \infty} f_k^{(l-1)}(x) = h(x)$, where $h(0) = 1$ and $h(x) = 0$ for $x \neq 0$. Consequently, the sequence f_k, $k \in \mathbb{N}$, is bounded in $W_1^l(a, b)$, but none of its subsequences f_{k_s}, $s \in \mathbb{N}$, converges in $L_\infty(a, b)$. Otherwise, for some subsequence f_{k_s}, $\lim\limits_{s,\sigma \to \infty} \|f_{k_s} - f_{k_\sigma}\|_{C[a,b]} = \lim\limits_{s,\sigma \to \infty} \|f_{k_s} - f_{k_\sigma}\|_{L_\infty(a,b)} = 0$. Hence, f_{k_s} convergers uniformly on $[a, b]$ to h, which contradicts the discontinuity of the function h.

3. If $b - a = \infty$, then Example 1 shows that embedding (4.80) is not compact for any admissible values of the parameters. $\square$

4.3 Open sets with quasi-resolvable, quasi-continuous, smooth and Lipschitz boundaries

We say that a domain $\Omega \subset \mathbb{R}^n$ is a *bounded elementary domain with a resolved boundary* with the parameters d, D, satisfying $0 < d \leq D < \infty$, if

$$\Omega = \{\, x \in \mathbb{R}^n : a_n < x_n < \varphi(\bar{x}), \ \bar{x} \in W \,\}, \tag{4.87}$$

where[14] $\operatorname{diam} \Omega \leq D$, $\bar{x} = (x_1, ..., x_{n-1})$, $W = \{\bar{x} \in \mathbb{R}^{n-1} : \ a_i < x_i < b_i, \ i = 1, ...n - 1\}$, $-\infty \leq a_i < b_i \leq \infty$, and

$$a_n + d \leq \varphi(\bar{x}), \quad \bar{x} \in W. \tag{4.88}$$

---

[14] Since Ω is a domain, hence measurable, by Fubini's theorem the function φ is measurable on W and $\operatorname{meas} \Omega = \int\limits_W (\varphi(\bar{x}) - a_n) \, d\bar{x}$.

If, in addition, $\varphi \in C(\overline{W})$ or $\varphi \in C^l(\overline{W})$ for some $l \in \mathbb{N}$ and $\|D^\alpha \varphi\|_{C(\overline{W})} \leq M$ if $1 \leq |\alpha| \leq l$ where $0 \leq M < \infty$ or φ satisfies the Lipschitz conditon

$$|\varphi(\bar{x}) - \varphi(\bar{y})| \leq M\, |\bar{x} - \bar{y}|, \quad \bar{x}, \bar{y} \in \overline{W}, \tag{4.89}$$

then we say that Ω is a *bounded elementary domain with a continuous boundary* with the parameters d, D, *with a C^l-boundary* with the parameters d, D, M, or *with a Lipschitz boundary* with the parameters d, D, M respectively.

Moreover, we say that an open set $\Omega \subset \mathbb{R}^n$ has a *resolved boundary* with the parameters d, $0 < d < \infty$, D, $0 < D \leq \infty$ and $\varkappa \in \mathbb{N}$ if there exist open parallelepipeds V_j, $j = \overline{1, s}$, where $s \in \mathbb{N}$ for bounded Ω and $s = \infty$ for unbounded Ω such that

1) $(V_j)_d \cap \Omega \neq \varnothing$ and $\operatorname{diam} V_j \leq D$,

2) $\displaystyle \Omega \subset \bigcup_{j=1}^{s} (V_j)_d$,

3) the multiplicity of the covering $\{V_j\}_{j=1}^{s}$ does not exceed $\varkappa$,

4) there exist maps λ_j, $j = \overline{1, s}$, which are compositions of rotations, reflections and translations and are such that

$$\lambda_j(V_j) = \big\{ x \in \mathbb{R}^n : \ a_{ij} < x_i < b_{ij}, \ i = 1, ..., n \big\}$$

and

$$\lambda_j(\Omega \cap V_j) = \{ x \in \mathbb{R}^n : \ a_{nj} < x_n < \varphi_j(\bar{x}), \ \bar{x} \in W_j \}, \tag{4.90}$$

where $\bar{x} = (x_1, ..., x_{n-1})$, $W_j = \{ \bar{x} \in \mathbb{R}^{n-1} : \ a_{ij} < x_i < b_{ij}, \ i = 1, ..., n - 1 \}$, and

$$a_{nj} + d \leq \varphi_j(\bar{x}) \leq b_{nj} - d, \quad \bar{x} \in W_j, \tag{4.91}$$

if $V_j \cap \partial \Omega \neq \varnothing$. If $V_j \subset \Omega$, then $\varphi_j(\bar{x}) \equiv b_{nj}$. (The left inequality (4.91) is satisfied autimatically since by 1) $b_{nj} - a_{nj} \geq 2\, d$.)

We note that $\lambda_j(\Omega \cap V_j)$ and, if $V_j \cap \partial \Omega \neq \varnothing$, also $\lambda_j^-((^c\overline{\Omega}) \cap V_j)$ are bounded elementary domains with a resolved boundary with the parameters d, D, where $\lambda_j^-(x) = (\lambda_{j,1}(x), ..., \lambda_{j,n-1}(x), -\lambda_{j,n}(x))$.

Since by 1) $b_{ij} - a_{i,j} \geq 2d$, $i = 1, ..., n-1$, by 4) it follows that $\operatorname{meas}(\Omega \cap V_j) \geq d^n$, $j = \overline{1, s}$. So by 3), for unbounded Ω, $\operatorname{meas}\Omega = \infty$, because by (2.60) $\displaystyle \sum_{j=1}^{\infty} \operatorname{meas}(\Omega \cap V_j) \leq \varkappa \operatorname{meas}\Omega$.

If an open set $\Omega \subset \mathbb{R}^n$ has a resolved boundary with the parameters d, D, $\varkappa$ and, in addition, for some $l \in \mathbb{N}$ all functions $\varphi_j \in C^l(\overline{W}_j)$ and $\|D^\alpha \varphi_j\|_{C(\overline{W}_j)} \leq M$ if $1 \leq |\alpha| \leq l$ where $0 \leq M < \infty$ and is independent of j or all functions φ_j satisfy the Lipschitz condition

$$|\varphi_j(\bar{x}) - \varphi_j(\bar{y})| \leq M\, |\bar{x} - \bar{y}|, \ \bar{x}, \bar{y} \in \overline{W}_j, \tag{4.92}$$

where M is independent of $\bar{x}, \bar{y}$ and j, then we say that Ω has a C^l-*boundary* (briefly $\partial\Omega \in C^l$) with the parameters d, D, $\varkappa$, M, or a *Lipschitz boundary* (briefly $\partial\Omega \in \mathrm{Lip1}$) with the parameters d, D, $\varkappa$, M respectively.

If all functions φ_j are continuous on $\overline{W}$ we say that Ω has a *continuous boundary* with the parameters d, D, $\varkappa$.

Furthermore, an open set $\Omega \subset \mathbb{R}^n$ has a *quasi-resolved* (*quasi-continuous*) *boundary* with the parameters d, D, $\varkappa$ if $\Omega = \bigcup_{k=1}^{s} \Omega_k$, where $s \in \mathbb{N}$ or $s = \infty$, and Ω_k, $k = \overline{1, s}$, are open sets, which have a resolved (continuous) boundary with the parameters d, D, $\varkappa$, and the multiplicity of the covering $\{\Omega_k\}_{k=1}^{s}$ does not exceed $\varkappa$. (We note that if Ω is bounded, then $s \in \mathbb{N}$.)

Finally, we say that an open set $\Omega \subset \mathbb{R}^n$ has a *resolved* (*quasi-resolved, continuous, quasi-continuous*) *boundary* if for some $d, D, \varkappa$, satisfying $0 < d \leq D < \infty$ and $\varkappa \in \mathbb{N}$, it has a resolved (quasi-resolved, continuous, quasi-continuous) boundary with the parameters $d, D, \varkappa$). Respectively an open set $\Omega \subset \mathbb{R}^n$ has a C^l- (Lipschitz) boundary if for some $d, D, \varkappa, M$, satisfying $0 < d \leq D < \infty, \varkappa \in \mathbb{N}$ and $0 \leq M < \infty$, it has a C^l- (Lipschitz) boundary with the parameters $d, D, \varkappa, M$.

Example 2 Suppose that $\Omega = \{(x_1, x_2) \in \mathbb{R}^2 : -1 < x_2 < 1$ if $-1 < x_1 < 0, -1 < x_2 < x_1^\gamma$ if $0 \leq x_1 < 1\}$ where $0 < \gamma < 1$. Then Ω is a bounded elementary domain with a resolved boundary, which is not a quasi-continuous boundary.

Example 3 Let $\Omega = \{(x_1, x_2) \in \mathbb{R}^2 : 0 < x_1 < 1, x_1^\gamma < x_2 < 2\,x_1^\gamma\}$ where $0 < \gamma < \infty, \gamma \neq 1$. Then $\partial\Omega$ is not a quasi-resolved boundary while $^c\overline{\Omega}$ satisfies the cone condition.

Example 4 For the elementary domain Ω defined by (4.87) the Lipschitz condition (4.89) means geometrically that $\forall x \in \partial\Omega$ the cones

$$K_x^+ = \{y \in \mathbb{R}^n : y_n < \varphi(\bar{x}) - M|\bar{x} - \bar{y}|\}, \quad K_x^- = \{y \in \mathbb{R}^n : \varphi(\bar{x}) + M|\bar{x} - \bar{y}| < y_n\}$$

are such that

$$K_x^+ \cap \widehat{W} \subset \Omega, \quad K_x^- \cap \widehat{W} \subset {}^c\overline{\Omega}, \tag{4.93}$$

where $\widehat{W} = \{x \in \mathbb{R}^n : \bar{x} \in W, a_n < x_n < \infty\}$.

For, if (4.89) is satisfied and $y \in K_x^+ \cap \widehat{W}$, then $y_n < \varphi(\bar{x}) - M|\bar{x} - \bar{y}| \leq \varphi(\bar{y})$ and $y \in \Omega$. Similarly, $K_x^- \cap \widehat{W} \subset {}^c\overline{\Omega}$. Suppose that (4.93) is satisfied. Since $(\bar{y}, \varphi(\bar{y})) \notin \Omega$ the inclusion $K_x^- \cap \widehat{W} \subset \Omega$ implies that $\varphi(\bar{y}) \geq \varphi(\bar{x}) - M|\bar{x} - \bar{y}|$.

(For, if $\varphi(\bar{y}) < \varphi(\bar{x}) - M|\bar{x} - \bar{y}|$, then $(\bar{y}, \varphi(\bar{y})) \in \Omega$.) Similarly, $K_x^- \cap \widehat{W} \subset {}^c\overline{\Omega}$ implies that $\varphi(\bar{y}) \le \varphi(\bar{x}) + M|\bar{x} - \bar{y}|$ and (4.89) follows.

We note also that the tangent of the angle at the common vertex of both cones K_x^+ and K_x^- is equal to $\frac{1}{M}$.

Example 5 Let $\Omega = \{(x_1, x_2) \in \mathbb{R}^2 : x_2 < \varphi(x_1)\}$, where $\varphi(x_1) = -|x_1|^\gamma$ if $x_1 \le 0$, $\varphi(x_1) = x_1^\gamma$ if $x_1 \ge 0$ and $\gamma > 0$. Then the function φ satisfies a Lipschitz condition on $\mathbb{R}$ if, and only if, $\gamma \le 1$, while Ω has a Lipschitz boundary in the sense of the above definition for each $\gamma > 0$.

Example 6 Let $\gamma > 0$. Both the domain $\Omega_1 = \{x \in \mathbb{R}^n : |\bar{x}|^\gamma < x_n < 1, |\bar{x}| < 1\}$ and the domain $\Omega_2 = \{x \in \mathbb{R}^n : -1 < x_n < |\bar{x}|^\gamma, |\bar{x}| < 1\}$ have a Lipschitz boundary if, and only if, $\gamma \ge 1$. (Compare with Examples 6 and 7 of Chapter 3.)

Lemma 4 *If an open set $\Omega \subset \mathbb{R}^n$ has a Lipschitz boundary with the parameters d, D, $\varkappa$ and M, then both Ω and ${}^c\overline{\Omega}$ satisfy the cone condition with the parameters r, h depending only on d, M and n.*

Idea of the proof. Let $z \in V_j \cap \partial\Omega$ and $x = \lambda_j(z)$. Consider the cones K_x^+ and K_x^- defined in Example 4, where φ is replaced by φ_j. $\square$

Proof. By Example 4 we have

$$K_x^+ \cap \lambda_j(V_j) \subset \lambda_j(V_j \cap \Omega), \quad K_x^- \cap \lambda_j(V_j) \subset \lambda_j(V_j \cap {}^c\overline{\Omega}).$$

By 1) $b_{ij} - a_{ij} \ge 2d$ and (4.90) implies that there exist $r, h > 0$ depending only on d, M and n such that $\lambda_j(V_j \cap \Omega)$ and $\lambda_j(V_j \cap {}^c\overline{\Omega})$ satisfy the cone condition with the parameters r and h. (The cone condition is satisfied for the largest cone with vertex the origin, which is contained in the intersection of the cone $K(d, \frac{d}{M})$ defined by (3.34) and the infinite rectangular block $x_1, ..., x_{n-1} > 0$). Since λ_j is a composition of rotations, reflections and translations, the sets $V_j \cap \Omega$, $V_j \cap {}^c\overline{\Omega}$ and, hence, the sets Ω and ${}^c\overline{\Omega}$ also satisfy the cone condition with the parameters r and h. $\square$

Example 7 Let $\Omega = Q_1 \bigcup Q_2$, where Q_1 and Q_2 are open cubes such that the intersection $\overline{Q}_1 \cap \overline{Q}_2$ consists of just one point. Then both Ω and ${}^c\overline{\Omega}$ satisfy the cone condition, but the boundary of Ω is not Lipschitz. (It is not even resolvable.)

Lemma 5 *A bounded domain $\Omega \subset \mathbb{R}^n$ star-shaped with respect to the ball $B \subset \Omega$ has a Lipschitz boundary with the parameters depending only on $\operatorname{diam} B$, $\operatorname{diam} \Omega$ and n.*

Idea of the proof. Apply the proof of Lemma 2 of Chapter 3 and Example 1. $\square$

Proof. Let Ω be star-shaped with respect to the ball $B(x_0, r)$ and $z \in \partial\Omega$. We consider the conic body $V_z = \bigcup_{y \in B(x_0,r)} (y, z)$ and the supplementary infinite cone

$\widehat{V}_z = \bigcup_{y \in B(x_0,r)} (y, z)\widehat{\ }$, where $(y, z)\widehat{\ } = \{z + \varrho(z - y) : 0 < \varrho < \infty\}$ is an open ray

that goes from the point z in the direction of the vector $\overrightarrow{yz}$. Then $V_z \subset \Omega$ and by the proof of Lemma 2 of Chapter 3 $\widehat{V}_z \subset^c \overline{\Omega}$.

Without loss of generality we assume that the vector $\overrightarrow{x_0z}$ is parallel to the axis Ox_n, hence, $z = (\bar{x}_0, z_n)$, where $\bar{x}_0 = (x_{0,1}, ..., x_{0,n-1})$ and $z_n > x_{0,n}$, and consider the parallelepiped $U_z = \{y \in \mathbb{R}^n : x_{0,n} < y_n < 2z_n - x_{0,n}, \bar{y} \in U_z^*\}$, where $U_z^* = \{\bar{y} \in \mathbb{R}^{n-1} : |y_i - x_{0,i}| < \frac{r}{2}, i = 1, ..., n-1\}$. Then $\forall \bar{y} \in U_z^*$ the ray that goes from the point $(\bar{y}, x_{0,n})$ in the direction of the vector $\overrightarrow{x_0z}$ intersects the boundary $\partial\Omega$ at a single[15] point, which we denote by $y = (\bar{y}, \varphi(\bar{y}))$. In particular, $\varphi(\bar{z}) = z_n$.

Since the tangent of the angle at the common vertex of V_z and $\widehat{V}_z$ is greater than or equal to $\frac{r}{R_2}$, where $R_2 = \max_{y \in \partial\Omega} |x_0 - y|$, it follows (see Example 1) that

$|\varphi(\bar{z}) - \varphi(\bar{y})| \leq \frac{R_2}{r} |\bar{z} - \bar{y}|$, $\bar{y} \in U_z^*$. We note that if $\bar{y} \in U_z^*$, then the conic body V_y contains the cone K_y with the point y as a vertex, whose axis is parallel to Ox_n and which is congruent to the cone defined by (3.34) with the parameters $\frac{r}{2}$, $\varphi(\bar{y}) - x_{0,n}$. Moreover, $\widehat{V}_y$ contains the supplementary infinite cone $\widehat{K}_y$. The tangent of the angle at the common vertex of these cones is equal to

$$\frac{r}{2(\varphi(\bar{y}) - x_{0,n})} \geq \frac{r}{2(\varphi(\bar{z}) - x_{0,n} + \frac{R_2}{r}|\bar{z} - \bar{y}|)} \geq \frac{r}{4R_2}.$$

Consequently (see Example 1),

$$|\varphi(\bar{x}) - \varphi(\bar{y})| \leq \frac{4R_2}{r} |\bar{x} - \bar{y}|, \quad \bar{x}, \bar{y} \in U_z^*.$$

Moreover, since $V_z \subset \Omega$ and $\widehat{V}_z \subset^c \overline{\Omega}$, we have $x_{o,n} + \frac{r}{2} < \varphi(\bar{x}) < 2z_n - x_{0,n} - \frac{r}{2}$, $\bar{x} \in U_z^*$. We note also that

$$B(z, \tfrac{r}{2}) \subset U_z \subset B(z, (R_2^2 + (n-1)^2(\tfrac{r}{2})^2)^{\frac{1}{2}}). \tag{4.94}$$

Finally, we consider a minimal covering of $\mathbb{R}^n$ by open balls of radius $\frac{r}{6}$. (Its multiplicity is less than or equal to 2^n.) Denote by $B_1, ..., B_s$ a collection of those of them, which covers the $\frac{r}{6}$-neigbourhood of the boundary $\partial\Omega$. Each of

---

[15] Suppose that $\eta \in \partial\Omega, \eta \neq y$ and $\bar{\eta} = \bar{y}$. If $\eta_n > y_n$, then $y \in V_\eta \subset \Omega$. If $\eta_n < y_n$, then $y \in \widehat{V}_z \subset^c \overline{\Omega}$. In both cases we arrive at a contadiction since $y \in \partial\Omega$.

these balls is contained in a ball of the radius $\frac{r}{2}$ centered at a point of $\partial\Omega$. Since $\forall z \in \partial\Omega$ we have $U_z \supset B(z, \frac{r}{2})$, we can choose $U_{z_1}, ..., U_{z_s}$, where $z_k \in \partial\Omega$ in such a way that $U_{z_k} \supset B_k$. Consequently, the parallelepipeds $U_{z_1}, ..., U_{z_s}$ cover the $\frac{r}{6}$-neigbourhoods of $\partial\Omega$. From (4.94) it follows that the multiplicity of this covering does not exceed $\varkappa = 2^n \left(1 + \frac{2}{r}\left(R_2^2 + (n-1)^2\left(\frac{r}{2}\right)^2\right)^{\frac{1}{2}}\right)^n$. (See footnote 15 of Chapter 3.)

Thus, Ω has a Lipschitz boundary with the parameters $d = \frac{r}{6}$, $D = \operatorname{diam}\Omega$, $M = \frac{4R_2}{r} \le \frac{4D}{r}$ and $\varkappa$.

Lemma 6 1. *A bounded open set $\Omega \subset \mathbb{R}^n$ satisfies the cone condition if, and only if, there exist $s \in \mathbb{N}$ and elementary bounded domains Ω_k, $k = 1, ..., s$, with Lipschitz boundaries with the same parameters such that $\Omega = \bigcup\limits_{k=1}^{s} \Omega_k$.*

2. An unbounded open set $\Omega \subset \mathbb{R}^n$ satisfies the cone condition if, and only if, there exist elementary bounded domains Ω_k, $k \in \mathbb{N}$, with Lipschitz boundaries with the same parameters such that

1) $\Omega = \bigcup\limits_{k=1}^{\infty} \Omega_k$,

and

2) the multiplicity of the covering $\varkappa\left(\{\Omega_k\}_{k=1}^{\infty}\right)$ is finite.

Idea of the proof. To prove the necessity combine Lemma 4 of Chapter 3 and Lemma 5. Note that if the boundaries of the elementary domains Ω_k, $k = \overline{1,s}$ are Lipschitz with the parameters d_k, D_k and M_k then they are Lipschitz with the parameters $d = \inf\limits_{k=\overline{1,s}} d_k, D = \sup\limits_{k=\overline{1,s}} D_k, M = \sup\limits_{k=\overline{1,s}} M_k$ as well if $d > 0$, $D < \infty$ and $M < \infty$. To prove the sufficiency apply Lemma 4 and Example 5 of Chapter 3. $\square$

Remark 19 If in Lemma 6 Ω_k are elementary bounded domains with Lipschitz boundaries with the same parameters d, D, M, then Ω satisfies the cone condition with the parameters r, h depending only on d and M.

Remark 20 If we introduce the notion of an open set with a quasi-Lipschitz boundary in the same manner as in the case of a quasi-continuous boundary, then by Lemma 6 this notion coincides with the notion of an open set satisfying the cone condition. If we define an open set satisfying the quasi-cone condition, then this notion again coincides with the notion of an open set satisfying the cone condition.

Lemma 4 of Chapter 3 and Lemma 6 allow us to reduce the proofs of embedding theorems for open sets satisfying the cone condition to the case of bounded domains star-shaped with respect to a ball or to the case of elementary bounded domains having Lipschitz boundaries. To do this we need the following lemmas about addition of inequalities for the norms of functions.

Lemma 7 *Let $m_0 \in \mathbb{N}, 1 \leq p_1, ...p_m, q \leq \infty$ and let $\Omega = \bigcup\limits_{k=1}^{s} \Omega_k$, where $\Omega_k \subset$ $\mathbb{R}^n$ are measurable sets, $s \in \mathbb{N}$ for $q < \max\limits_{m=1,...,m_0} p_m$ and $s \in \mathbb{N}$ or $s = \infty$ otherwise. Moreover, if $s = \infty$ and $q < \infty$, suppose that the multiplicity of the covering $\varkappa \equiv \varkappa(\{\Omega_k\}_{k=1}^{s})$ is finite. Furthermore, let $f_m, m = 1, ..., m_0$, and g be functions measurable on Ω.*

Suppose that for some $\sigma_m > 0, m = 1, ..., m_0$, for each k

$$\|g\|_{L_q(\Omega_k)} \leq \sum_{m=1}^{m_0} \sigma_m \|f_m\|_{L_{p_m}(\Omega_k)}. \tag{4.95}$$

Then

$$\|g\|_{L_q(\Omega)} \leq A^{\frac{1}{q}} \sum_{m=1}^{m_0} \sigma_m \|f_m\|_{L_{p_m}(\Omega)}, \tag{4.96}$$

where $A = s$ if $q < \max\limits_{m=1,...,m_0} p_m$ and $A = \varkappa$ otherwise.

Idea of the proof. If $p_1 = ... = p_m = q = 1$ add inequalities (4.95) and apply inequality (2.59). In the general case apply Minkowski's or Hölder's inequalities for sums (for $q > p_m$, for $q < p_m$ respectively) and inequality (2.59). $\square$

Proof. Let $q < \infty$. [16] By (4.95) and Minkowski's inequality it follows, that

$$\|g\|_{L_q(\Omega)} \leq \left(\sum_{k=1}^{s} \|g\|_{L_q(\Omega_k)}^{q}\right)^{\frac{1}{q}} \leq \left(\sum_{k=1}^{s}\left(\sum_{m=1}^{m_0} \sigma_m \|f_m\|_{L_{p_m}(\Omega_k)}\right)^{q}\right)^{\frac{1}{q}}$$

$$\leq \sum_{m=1}^{m_0}\left(\sum_{k=1}^{s}(\sigma_m \|f_m\|_{L_p(\Omega_k)})^{q}\right)^{\frac{1}{q}} = \sum_{m=1}^{m_0} \sigma_m \left(\sum_{k=1}^{s} \|f_m\|_{L_{p_m}(\Omega_k)}^{q}\right)^{\frac{1}{q}}.$$

---

[16] The case $q = \infty$ is trivial and the statement holds for $\Omega = \bigcup\limits_{i \in I} \Omega_i$, where I is an arbitrary set of indices:

$$\|g\|_{L_\infty(\Omega)} = \sup_{i \in I} \|g\|_{L_\infty(\Omega_i)} \leq \sum_{m=1}^{m_0} \sigma_m \|f_m\|_{L_{p_m}(\Omega)}.$$

If $q \geq p_m$, denote by χ_k the characteristic function of Ω_k. Since $\sum_{k=1}^{s} \chi_k(x) \leq \varkappa$, by Minkowski's inequality we have

$$\left(\sum_{k=1}^{s} \|f_m\|_{L_{pm}(\Omega_k)}^q \right)^{\frac{1}{q}} = \left(\sum_{k=1}^{s} \|f_m \chi_k\|_{L_{pm}(\Omega)}^q \right)^{\frac{1}{q}}$$

$$= \left(\left(\sum_{k=1}^{s} \left(\int_{\Omega} |f_m(x)|^{p_m} \chi_k(x)\, dx \right)^{\frac{q}{p_m}} \right)^{\frac{p_m}{q}} \right)^{\frac{1}{p_m}}$$

$$\leq \left(\int_{\Omega} \left(\sum_{k=1}^{s} |f_m(x)|^q \chi_k(x) \right)^{\frac{p_m}{q}} dx \right)^{\frac{1}{p_m}}$$

$$= \left(\int_{\Omega} |f_m(x)|^{p_m} \left(\sum_{k=1}^{s} \chi_k(x) \right)^{\frac{p_m}{q}} dx \right)^{\frac{1}{p_m}} \leq \varkappa^{\frac{1}{q}} \|f_m\|_{L_{pm}(\Omega)}.$$

If $q < p_m < \infty$, then by Hölder's inequality with the exponent $\frac{p_m}{q} > 1$ and (2.59)

$$\left(\sum_{k=1}^{s} \|f_m\|_{L_{pm}(\Omega_k)}^q \right)^{\frac{1}{q}} \leq s^{\frac{1}{q} - \frac{1}{p_m}} \left(\sum_{k=1}^{s} \|f_m\|_{L_{pm}(\Omega_k)}^{p_m} \right)^{\frac{1}{p_m}}$$

$$\leq s^{\frac{1}{q} - \frac{1}{p_m}} \varkappa^{\frac{1}{p_m}} \|f\|_{L_{pm}(\Omega)} \leq s^{\frac{1}{q}} \|f\|_{L_{pm}(\Omega)}$$

and inequality (4.96) follows.

The case in which some $p_m = \infty$ is treated in a similar way with suprema replacing sums. $\square$

Corollary 13 *Let $l \in \mathbb{N}, \beta \in \mathbb{N}_0^n$ satisfy $|\beta| < l, 1 \leq p_0, p, q \leq \infty, \Omega = \bigcup_{k=1}^{s} \Omega_k$, where $\Omega_k \subset \mathbb{R}^n$ are open sets, $s \in \mathbb{N}$ if $q < p_0$ or $q < p$ and $s \in \mathbb{N}$ or $s = \infty$ if $q \geq p_0, p$. Moreover, if $s = \infty$ and $q < \infty$, suppose that the multiplicity of the covering $\varkappa = \varkappa\left(\{\Omega_k\}_{k=1}^{s} \right)$ is finite.*

Suppose that $f \in L_{p_0}(\Omega) \cap w_p^l(\Omega), c_{25}, c_{26} > 0$ and

$$\|D_w^\beta f\|_{L_q(\Omega_k)} \leq c_{25} \|f\|_{L_{p_0}(\Omega_k)} + c_{26} \|f\|_{w_p^l(\Omega_k)}, \quad k = \overline{1, s}. \tag{4.97}$$

Then

$$\|D_w^\beta f\|_{L_q(\Omega)} \leq A^{\frac{1}{q}} \left(c_{25} \|f\|_{L_{p_0}(\Omega)} + c_{26} \|f\|_{w_p^l(\Omega)} \right). \tag{4.98}$$

Idea of the proof. Direct application of Lemma 7. $\square$

Lemma 8 *Let $l \in \mathbb{N}, m \in \mathbb{N}_0, m < l, 1 \le p, q \le \infty, \Omega = \bigcup_{k=1}^{s} \Omega_k$, where $s \in \mathbb{N}$ and $\Omega_k \subset \mathbb{R}^n$ are open sets such that*

$$W_p^l(\Omega_k) \hookrightarrow W_q^m(\Omega_k), \quad k = 1, ..., s. \tag{4.99}$$

Then

$$W_p^l(\Omega) \hookrightarrow W_q^m(\Omega). \tag{4.100}$$

Moreover, if embeddings (4.99) are compact, then embedding (4.100) is also compact.

Idea of the proof. Apply Theorem 1 and Lemma 7 to prove embedding (4.100). To prove its compactness consider a sequence of functions bounded in $W_p^l(\Omega)$ and, applying successively the compactness of embeddings (4.99), get a subsequence convergent in $W_q^m(\Omega)$.

Proof. 1. By Theorem 1 (4.99) is equivalent to the inequality

$$\|f\|_{W_q^m(\Omega_k)} \le M_k \|f\|_{W_p^l(\Omega_k)},$$

where $k = 1, ..., s$ and M_k are independent of f. By Lemma 7 it follows that

$$\|f\|_{W_q^m(\Omega)} \le M_0 \max_{k=1,...,s} M_k \|f\|_{W_p^l(\Omega)},$$

where M_0 depends only on n, m, p, q, and (4.100) follows.

2. Let $M > 0$ and $\|f_i\|_{W_p^l(\Omega)} \le M$ for each $i \in \mathbb{N}$. Then , in particular, $\|f_i\|_{W_p^l(\Omega_1)} \le M$. Consequently, there exist a function $g_1 \in W_q^m(\Omega_1)$ and a subsequence $f_{i_j^{(1)}} \to g_1$ in $W_q^m(\Omega_1)$ as $j \to \infty$. Furthermore, $\|f_{i_j^{(1)}}\|_{W_p^l(\Omega_2)} \le M$ and, hence, there exist $g_2 \in W_q^m(\Omega_2)$ and a subsequence $f_{i_j^{(2)}}$ of $f_{i_j^{(1)}}$ such that $f_{i_j^{(2)}} \to g_2$ in $W_q^m(\Omega_2)$. Moreover, $f_{i_j^{(2)}} \to g_1$ in $W_q^m(\Omega_1)$. Repeating this procedure $s - 2$ times, we get functions $g_k \in W_q^m(\Omega_k)$, $k = 1, ..., s$ and a subsequence f_{i_j} such that $f_{i_j} \to g_k$ in $W_q^m(\Omega_k)$ as $j \to \infty$. We note that g_k is equivalent to g_σ on $\Omega_k \cap \Omega_\sigma$. Hence, there exists a function g, defined on Ω, such that $g \sim g_k$ on Ω_k, $k = 1, ..., s$. By the properties of weak derivatives (see Section 1.2) $g \in W_q^m(\Omega)$ and

$$\|f_{i_j} - g\|_{W_q^m(\Omega)} \le \sum_{k=1}^{s} \|f_{i_j} - g_k\|_{W_q^m(\Omega_k)} \to 0$$

as $j \to \infty$. $\square$

Lemma 9 *Let $l \in \mathbb{N}$, $m \in \mathbb{N}_0$, $m < l$ and $1 \leq p,q \leq \infty$. Suppose that for each bounded elementary domain $G \subset \mathbb{R}^n$ with a resolved (continuous) boundary there exists $c_{26} > 0$ such that for each $\beta \in \mathbb{N}_0^n$ satisfying $|\beta| = m$ and $\forall f \in W_p^l(G)$*

$$\|D_w^\beta f\|_{L_q(G)} \leq c_{26} \|f\|_{W_p^l(G)}. \tag{4.101}$$

Then for each bounded open set $\Omega \subset \mathbb{R}^n$ having a quasi-resolved (respectively, quasi-continuous) boundary there exists $c_{27} > 0$ such that

$$\|D_w^\beta f\|_{L_q(\Omega)} \leq c_{27} \|f\|_{W_p^l(\Omega)} \tag{4.102}$$

for each $\beta \in \mathbb{N}_0^n$ satisfying $|\beta| = m$ and $\forall f \in W_p^l(\Omega)$.

If $p \leq q$ and c_{26} depends only on n, l, m, p, q and the parameters d and D of a bounded elementary domain with a resolved (continuous) boundary, then for each unbounded open set $\Omega \subset \mathbb{R}^n$ having a quasi-resolved (quasi-continuous) boundary there exists $c_{27} > 0$ such that inequality (4.102) holds.

Idea of the proof. Apply (4.101), where G, f are replaced by $\lambda_j(\Omega \cap V_j)$, $f_j = f(\lambda_j)$ respectively, and the parallelepipeds V_j and the maps λ_j are as in the definition of a resolved (continuous) boundary. Change the variables, setting $y = \lambda_j^{(-1)}(x)$, and obtain (4.101) where $G = \Omega \cap V_j$. Apply Corollary 13 twice to prove (4.102) succesively for open sets Ω with a resolved (continuous) and quasi-resolved (quasi-continuous) boundary. $\square$

Proof. First suppose that Ω has a resolved boundary. We notice that $\lambda_j(x) = A_j x + b_j$, $\lambda_j^{(-1)}(x) = A_j^{-1}x - A_j^{-1}b_j$, where $b_j \in \mathbb{R}^n$, $A_j = \left(a_{ik}^{(j)}\right)_{i,k=1}^n$, $A_j^{-1} = \left(b_{ij}^{(j)}\right)_{i,k=1}^n$, $|a_{ik}^{(j)}|$, $|b_{ij}^{(j)}| \leq 1$ and $|\det A_j| = |\det A_j^{-1}| = 1$. Consequently, we have

$$\left|(D_w^\beta f)(x)\right| = \left|D_w^\beta(f_j(\lambda_j^{(-1)}(x)))\right| = \left|\left(\frac{\partial}{\partial x_{i_1}}\right)_w \cdots \left(\frac{\partial}{\partial x_{i_m}}\right)_w \left(f_j\left(\lambda_j^{(-1)}(x)\right)\right)\right|$$

$$= \left|\sum_{k_1,\ldots,k_m=1}^n b_{k_1 i_1}^{(j)} \cdots b_{k_m i_m}^{(j)} \left(\left(\frac{\partial}{\partial x_{k_1}}\right)_w \cdots \left(\frac{\partial}{\partial x_m}\right)_w f_j\right)\left(\lambda_j^{(-1)}(x)\right)\right|$$

$$\leq \sum_{|\gamma|=m} \frac{m!}{\gamma!}\left|\left(D_w^\gamma f_j\right)(\lambda_j^{(-1)}(x))\right| \leq n^m \sum_{|\gamma|=m}\left|\left(D_w^\gamma f_j\right)(\lambda_j^{(-1)}(x))\right|.$$

Setting $y = \lambda_j^{(-1)}(x)$ we establish that

$$\|D_w^\beta f\|_{L_q(\Omega \cap V_j)} \leq n^m \sum_{|\gamma|=m} \|D_w^\gamma f_j\|_{L_q(\lambda_j(\Omega \cap V_j))}.$$

Similarly, for $\alpha \in \mathbb{N}_0^n$ satisfying $|\alpha| = l$,

$$\|D_w^\alpha f_j\|_{L_p(\lambda_j(\Omega \cap V_j))} \leq n^l \sum_{|\gamma|=l} \|D_w^\gamma f_j\|_{L_p(\Omega \cap V_j)}.$$

Hence, inequality (4.101) implies that

$$\|D_w^\beta f\|_{L_q(\Omega \cap V_j)} \leq n^{m+l} c_{26}(\lambda_j(\Omega \cap V_j)) \|f\|_{W_p^l(\Omega \cap V_j)}.$$

If Ω is bounded, then the number of parallelepipeds V_j is finite, say, s. Hence, by Corollary 13,

$$\|D_w^\beta f\|_{L_q(\Omega)} \leq n^{m+l} s^{\frac{1}{q}} \max_{j=1,\ldots,s} c_{26}(\lambda_j(\Omega \cap V_j)) \|f\|_{W_p^l(\Omega)}.$$

Let Ω be unbounded, then the set of parallelepipeds V_j is denumerable. Suppose that n, l, m, p, q are fixed ($p \leq q$). Then in (4.101) $c_{26}(G) = c_{26}^*(d, D)$. Hence $\forall j \in \mathbb{N}$

$$\|D_w^\beta f\|_{L_q(\Omega \cap V_j)} \leq n^{m+l} c_{26}^*(d, D) \|f\|_{W_p^l(\Omega \cap V_j)}.$$

By Corollary 13 it follows that

$$\|D_w^\beta f\|_{L_q(\Omega)} \leq n^{m+l} \varkappa^{\frac{1}{q}} c_{26}^*(d, D) \|f\|_{W_p^l(\Omega)} = c_{26}^{**}(d, D, \varkappa) \|f\|_{W_p^l(\Omega)}.$$

Thus, (4.102) is proved for an Ω with a resolved boundary. If Ω has a quasi-resolved boundary one needs to apply Corollary 13 once more, in a similar way. The case of Ω having a quasi-continuous boundary is similar. $\square$

Lemma 10 *Let $l \in \mathbb{N}$, $m \in \mathbb{N}_0$, $m < l$ and $1 \leq p, q \leq \infty$. Suppose that for each bounded domain $\Omega \subset \mathbb{R}^n$ star-shaped with respect to a ball there exists $c_{26} > 0$ such that $\forall f \in W_p^l(G)$ inequality (4.101) holds.*

Then for each open set $\Omega \subset \mathbb{R}^n$ satisfying the cone condition there exists $c_{27} > 0$ such that $\forall f \in W_p^l(\Omega)$ inequality (4.102) holds.

If $p \leq q$ and c_{26} depends only on n, l, m, p, q and the parameters d and D of a domain star-shaped with respect to a ball, then for each unbounded open set $\Omega \subset \mathbb{R}^n$ satisfying the cone condition there exists $c_{27} > 0$ such that inequality (4.102) holds.

Idea of the proof. Apply Lemma 4 and, if Ω is unbounded, Remark 7 of Chapter 3 and Corollary 13. $\square$

Proof. Let Ω satisfy the cone condition with the parameters r, h. By Lemma 4 and Remark 7 of Chapter 3, $\Omega = \bigcup_{k=1}^{s} \Omega_k$, where $s \in \mathbb{N}$ for bounded Ω, $s = \infty$ for unbounded Ω, and Ω_k are bounded domains star-shaped with respect to the balls $B_k \subset \overline{B}_k \subset \Omega_k$. Moreover, $0 < M_1 \leq \operatorname{diam} B_k \leq \operatorname{diam} \Omega_k \leq M_2 < \infty$ and $\varkappa(\{\Omega_k\}_{k=1}^{s}) \leq M_3 < \infty$, where M_1, M_2 and M_3 depend only on n, r and h.

If Ω is bounded, then

$$\|D_w^{\beta} f\|_{L_q(\Omega_k)} \leq c_{26}(\Omega_k) \|f\|_{W_p^l(\Omega_k)}, \quad k = 1, ..., s. \tag{4.103}$$

Hence, by Corollary 13,

$$\|D_w^{\beta} f\|_{L_q(\Omega)} \leq s^{\frac{1}{q}} \max_{k=1,...,s} c_{26}(\Omega_k) \|f\|_{W_p^l(\Omega)}. \tag{4.104}$$

Suppose that Ω is unbounded. Denote by $A(d, D)$ the set of all domains, whose diameters do not exceed D and which are star-shaped with respect to balls whose diameters are greater than or equal to d and set $c_{26}^{*}(d, D) = \sup_{G \in A(d,D)} c_{26}(G)$. Clearly $A(d, D) \subset A(d_1, D_1)$ if $0 < d_1 \leq d \leq D \leq D_1 < \infty$.
Then $\forall k \in \mathbb{N}$

$$\|D_w^{\beta} f\|_{L_q(\Omega_k)} \leq c_{26}^{*}(d_k, D_k) \|f\|_{W_p^l(\Omega_k)} \leq c_{26}^{*}(M_1, M_2) \|f\|_{W_p^l(\Omega_k)}$$

and, by Corollary 13,

$$\|D_w^{\beta} f\|_{L_q(\Omega)} \leq M_3^{\frac{1}{q}} c_{26}^{*}(M_1, M_2) \|f\|_{W_p^l(\Omega)}. \quad \square$$

Lemma 11 *Let $l \in \mathbb{N}$, $m \in \mathbb{N}_0$, $m < l$ and $1 \leq p, q \leq \infty$. Suppose that for each bounded elementary domain $G \subset \mathbb{R}^n$ with a Lipschitz boundary there exists $c_{26} > 0$ such that for each $\beta \in \mathbb{N}_0^n$ satisfying $|\beta| = m$ and $\forall f \in W_p^l(G)$ inequality (4.101) holds.*

Then for each bounded open set $\Omega \subset \mathbb{R}^n$ satisfying the cone condition there exists $c_{27} > 0$ such that for each $\beta \in \mathbb{N}_0^n$ satisfying $|\beta| = m$ and $\forall f \in W_p^l(\Omega)$ inequality (4.102) holds.

If $p \leq q$ and c_{26} depends only on n, l, m, p, q and the parameters d, D and M of a bounded elementary domain with a Lipschitz boundary, then for each unbounded open set $\Omega \subset \mathbb{R}^n$ satisfying the cone condition there exists $c_{27} > 0$ such that inequality (4.102) holds.

Idea of the proof. Apply Lemma 6, Remark 19 and the proof of Lemma 9. $\square$

Proof. Let Ω satisfy the cone condition with the parameters r, h. By Lemma 6 and Remark 19, $\Omega = \bigcup_{k=1}^{s} \Omega_k$, where $s \in \mathbb{N}$ for bounded Ω and $s = \infty$ for unbounded Ω. Here Ω_k are bounded elementary domains with Lipschitz boundaries with the same parameters d, D, M depending only on n, r and h. Moreover, $\varkappa(\{\Omega_k\}_{k=1}^{s}) \leq M_3$, where M_3 also depends only on n, r and h. If Ω is bounded , then as in the proof of Lemma 10 we have inequalities (4.103) and (4.104). Let Ω be unbounded. Suppose that n, l, m, p, q are fixed ($p \leq q$). Then in (4.101) $c_{26}(G) = c_{26}^{*}(d, D, M)$. Hence, $\forall k \in \mathbb{N}$

$$\|D_w^{\beta} f\|_{L_q(\Omega_k)} \leq c_{26}^{*}(d, D, M)\, \|f\|_{W_p^l(\Omega_k)}$$

and, by Corollary 13,

$$\|D_w^{\beta} f\|_{L_q(\Omega_k)} \leq M_3^{\frac{1}{q}}\, c_{26}^{*}(d, D, M)\, \|f\|_{W_p^l(\Omega)} = c_{26}^{**}(r, h)\, \|f\|_{W_p^l(\Omega)}. \quad \square$$

4.4 Estimates for intermediate derivatives

Theorem 6 *Let $l \in \mathbb{N}$, $\beta \in \mathbb{N}_0^n$ satisfy $|\beta| < l$ and let $1 \leq p \leq \infty$.*

1. If $\Omega \subset \mathbb{R}^n$ is an open set having a quasi-resolved boundary, then $\forall f \in W_p^l(\Omega)$

$$\|D_w^{\beta} f\|_{L_p(\Omega)} \leq c_{28}\, \|f\|_{W_p^l(\Omega)}, \tag{4.105}$$

where $c_{28} > 0$ is independent of f.

2. If $\Omega \subset \mathbb{R}^n$ is a bounded domain having a quasi-resolved boundary and the ball $B \subset \Omega$, then $\forall f \in w_p^l(\Omega)$

$$\|D_w^{\beta} f\|_{L_p(\Omega)} \leq c_{29}\left(\|f\|_{L_1(B)} + \|f\|_{w_p^l(\Omega)}\right), \tag{4.106}$$

where $c_{29} > 0$ is independent of f.

3. If $\Omega \subset \mathbb{R}^n$ is a bounded open set having a quasi-continuous boundary, then $\forall \varepsilon > 0$ there exists $c_{30}(\varepsilon) > 0$ such that $\forall f \in W_p^l(\Omega)$

$$\|D_w^{\beta} f\|_{L_p(\Omega)} \leq c_{30}(\varepsilon)\, \|f\|_{L_p(\Omega)} + \varepsilon\, \|f\|_{w_p^l(\Omega)}. \tag{4.107}$$

Idea of the proof. Apply successively the one-dimensional Theorem 2 to prove (4.105) and (4.106) for an elementary bounded domain Ω with a resolved boundary. In the general case apply Lemma 9 and the proof of Lemma 7. Deduce inequality (4.107) from Theorem 8 and Lemma 13 below. $\square$

Proof. 1. Suppose that Ω is a bounded elementary domain (4.87) with the parameters d, D. By inequality (4.35) it follows that $\forall \beta \in \mathbb{N}_0^n$ satisfying $|\beta| < l$ and $\forall \bar{x} \in W$

$$\|(D_w^\beta)(\bar{x}, \cdot)\|_{L_p(a_n, \varphi(\bar{x}))} = \left\| \left(\frac{\partial}{\partial x_n} \right)_w^{\beta_n} \left(D_w^{\bar{\beta}} f \right)(\bar{x}, \cdot) \right\|_{L_p(a_n, \varphi(\bar{x}))}$$

$$\leq M_1 \left(\|(D_w^{\bar{\beta}} f)(\bar{x}, \cdot)\|_{L_1(a_n + \frac{d}{2} - \delta, a_n + \frac{d}{2} + \delta)} + \left\| \left(\frac{\partial}{\partial x_n} \right)_w^{l - |\bar{\beta}|} \left(D_w^{\bar{\beta}} f \right)(\bar{x}, \cdot) \right\|_{L_p(a_n, \varphi(\bar{x}))} \right),$$

where $\bar{\beta} = (\beta_1, ..., \beta_{n-1})$, $0 < \delta \leq \frac{d}{2}$ and M_1 depends only on l, p, δ and D. (We recall that $\varphi(\bar{x}) - a_n \leq D$.)

By the theorem on the measurability of integrals depending on a parameter[17] both sides of this inequality are functions measurable on W. Therefore, taking L_p-norms with respect to $\bar{x}$ over W and applying Minkowski's inequality for sums and integrals, we have

$$\|D_w^\beta f\|_{L_p(\Omega)} \leq M_1 \left(\| \, \|(D_w^\beta f)(\bar{x}, x_n)\|_{L_{1, x_n}(a_n + \frac{d}{2} - \delta, a_n + \frac{d}{2} + \delta)} \|_{L_{p, \bar{x}}(W)} \right.$$

$$\left. + \left\| \left(\frac{\partial}{\partial x_n} \right)_w^{l - |\bar{\beta}|} D_w^{\bar{\beta}} f \right\|_{L_p(\Omega)} \right)$$

$$\leq M_1 \left(\| \, \|(D_w^{\bar{\beta}} f)(\bar{x}, x_n)\|_{L_{p, \bar{x}}(W)} \|_{L_{1, x_n}(a_n + \frac{d}{2} - \delta, a_n + \frac{d}{2} + \delta)} + \|f\|_{w_p^l(\Omega)} \right).$$

Let $\sigma = (\bar{\sigma}, \sigma_n)$, where $\sigma_n = a_n + \frac{d}{2}$, $\bar{\sigma} \in W_\delta$, and $\bar{\beta} = (\bar{\bar{\beta}}, \beta_{n-1})$, $\bar{\bar{\beta}} = (\beta_1, ..., \beta_{n-2})$. We consider the cube $Q(\sigma, \delta) = \{ x \in \mathbb{R}^n : |x_j - \sigma_j| < \delta, j = 1, ..., n \}$ and set $U = \{ \bar{\bar{x}} = (x_1, ..., x_{n-2}) \in \mathbb{R}^{n-2} : a_j < x_j < b_j, j = 1, ..., n-2 \}$. Applying the same procedure as above, we have

$$\| \, \|(D_w^{\bar{\beta}} f)(\bar{x}, x_n)\|_{L_{p, \bar{x}}(W)} \|_{L_1(a_n + \frac{d}{2} - \delta, a_n + \frac{d}{2} + \delta)}$$

$$\leq M_1 \left(\| \, \| \, \|(D_w^{\bar{\bar{\beta}}} f)(\bar{\bar{x}}, x_{n-1}, x_n)\|_{L_{p, \bar{\bar{x}}}(U)} \|_{L_{1, x_{n-1}}(\sigma_{n-1} - \delta, \sigma_{n-1} + \delta)} \|_{L_{1, x_n}(\sigma_n - \delta, \sigma_n + \delta)} \right.$$

$$\left. + \| \, \|f\|_{w_{p, \bar{x}}^l(W)} \|_{L_{1, x_n}(\sigma_n - \delta, \sigma_n + \delta)} \right).$$

Substituting from this inequalty into the previous one and applying Hölder's inequality, we get

$$\|D_w^\beta f\|_{L_p(\Omega)}$$

---

[17] We mean the following statement. Let $E \subset \mathbb{R}^m$ and $F \subset \mathbb{R}^n$ be measurable sets. Suppose that the function f is measurable on $E \times F$ and for almost all $y \in F$ the function $f(\cdot, y)$ is integrable on E. Then the function $\int_E f(x, \cdot) \, dx$ is measurable on F.

$$\leq M_2 \left(\| \, \| \, \| (D_w^{\bar{\bar{\beta}}} f)(\bar{\bar{x}}, x_{n-1}, x_n) \|_{L_{p,\bar{\bar{x}}}(U)} \|_{L_{1,x_{n-1}}(\sigma_{n-1}-\delta,\sigma_{n-1}+\delta)} \|_{L_{1,x_n}(\sigma_n-\delta,\sigma_n+\delta)} \right.$$

$$\left. + \|f\|_{w_p^l(\Omega)} \right),$$

where M_2 depends only on l, p, δ and D.

Repeating the procedure $n - 2$ times, we establish that

$$\|D_w^\beta f\|_{L_p(\Omega)} \leq M_3 \left(\|f\|_{L_1(Q(\sigma,\delta))} + \|f\|_{w_p^l(\Omega)} \right), \tag{4.108}$$

where M_3 depends only on n, l, p, δ and D.

2. Taking $\delta = \frac{d}{2}$ and applying Hölder's inequality, we establish that inequality (4.105) holds for all $\beta \in \mathbb{N}_0^n$ satisfying $|\beta| < l$ for each bounded elementary domain with a resolved boundary and c_{28} depends only on n, l, p, d and D. Hence, by Lemma 9, the first statement of Theorem 6 follows.

3. Suppose that $B \equiv B(x_0, r) \subset \Omega$, where Ω is an elementary domain considered in step 1. Without loss of generality we assume that $x_{0,n} \geq \sigma_n$ and set $\sigma = (x_{0,1}, ..., x_{0,n-1}, \sigma_n)$, $\delta = \min\{\frac{r}{\sqrt{n}}, \frac{d}{2}\}$. Then $Q(x_0, \delta) \subset B(x_0, r)$ and the parallelepiped $G = \{x \in \mathbb{R}^n : |x_j - x_{0,j}| < \delta, \ \sigma_n - \delta < x_n < x_{0,n} + \delta\} \subset \Omega$. Applying inequality (4.108) in the case, in which $\beta = 0, p = 1$ and Ω, $Q(\sigma, \delta)$ are replaced by G, $Q(x_0, \delta)$ respectively, we get

$$\|f\|_{L_1(Q(\sigma,\delta))} \leq \|f\|_{L_1(G)} \leq M_4 \left(\|f\|_{L_1(Q(x_0,\delta))} + \|f\|_{w_1^l(G)} \right)$$

$$\leq M_5 \left(\|f\|_{L_1(B)} + \|f\|_{w_p^l(G)} \right),$$

where M_4 and M_5 are independent of f. Hence, from (4.108) it follows that (4.105) holds for each $\beta \in \mathbb{N}_0^n$ satisfying $|\beta| < l$.

4. From step 3 and the proofs of Lemmas 7 and 9 it follows that for each bounded domain Ω having a resolved boundary

$$\|D_w^\beta f\|_{L_p(\Omega)} \leq M_6 \left(\sum_{j=1}^s \|f\|_{L_1(B_j)} + \|f\|_{w_p^l(\Omega)} \right), \tag{4.109}$$

where $s \in \mathbb{N}$ and B_j are arbitary balls in $\Omega \cap V_j$. (V_j and s are as in the definition of Ω having a resolved bondary.)

Let the ball $B \subset \Omega$. We choose $m \in \mathbb{N}$ and the ball B_0 in such a way that $B_0 \subset B \cap \Omega \cap V_m$. By step 3 and Hölder's inequality it follows that

$$\|f\|_{L_1(B_m)} \leq M_7 \left(\|f\|_{L_1(B_0)} + \|f\|_{w_1^l(\Omega \cap V_m)} \right) \leq M_8 \left(\|f\|_{L_1(B)} + \|f\|_{w_p^l(\Omega)} \right).$$

Let $j \neq m$. We choose a chain of parallelepipeds $U_1, ..., U_\sigma$ of the covering $\{V_j\}_{j=1}^s$, which are such that $U_1 = V_j$, $U_k \cap U_{k+1} \neq \varnothing$, $k = 1, ..., \sigma - 1$, and $U_\sigma = V_m$. Next we consider balls $\widetilde{B}_k \subset U_k \cap U_{k+1}$, $k = 1, ..., \sigma - 1$, and set $\widetilde{B}_0 := B_j$, $\widetilde{B}_\sigma := B_m$. Then by step 3

$$\|f\|_{L_1(\widetilde{B}_k)} \le M_9 \left(\|f\|_{L_1(\widetilde{B}_{k+1})} + \|f\|_{w_1^l(\Omega \cap U_k)} \right), \quad k = 0, ..., \sigma - 1.$$

Consequently, for each $j = 1, ..., s$

$$\|f\|_{L_1(B_j)} \le M_{10} \left(\|f\|_{L_1(B)} + \|f\|_{w_p^l(\Omega)} \right),$$

and inequality (4.106) for bounded domains having a resolved boundary follows from (4.109). (We note that $M_6, ..., M_{10}$ are independent of f.)

The argument for a bounded domain Ω having a quasi-resolved boundary is similar.

5. By Theorem 8 embedding (4.118) is compact. hence, by Lemma 13 inequality (4.121) holds where $q = p$ and inequality (4.117) follows. $\square$

Next we give some examples showing that assumptions on Ω in Theorem 6 are essential. The first two examples show that for open sets Ω, which do not have a resolved boundary inequality (4.105) does not always hold.

Example 8 Let $l \in \mathbb{N}$, $\beta \in \mathbb{N}_0^n$, $0 < |\beta| < l$, $1 \le p \le \infty$ and $\Omega = \bigcup_{k=1}^\infty B(x_k, r_k)$, where $r_k > 0$ and $B(x_k, r_k)$ are disjoint balls. Suppose that $r_k \to 0$ as $k \to \infty$ and $\sum_{k=1}^\infty r_k^{(|\beta| - \frac{1}{2})p} < \infty$ if $p < \infty$. We set $f(x) := r_k^{-\frac{n}{p} - \frac{1}{2}} (x - x_k)^\beta$ on $Q_k, k \in \mathbb{N}$. Then $f \in W_p^l(\Omega)$ but $D^\beta f \notin L_p(\Omega)$.

Example 9 Let $l \in \mathbb{N}$, $(\beta_1, \beta_2) \in \mathbb{N}_0^2$, $\beta_1 \neq 0$, $\beta_1 + \beta_2 < l$, $1 \le p \le \infty$ and let Ω be the domain considered in Example 3. We set $f(x_1, x_2) := x_1^{\beta_1} x_2^{\alpha_2}$ where $\alpha_2 \notin \mathbb{N}_0$. Then, for $1 \le p < \infty$,

$$\|f\|_{W_p^l(\Omega)} \le M_1 \left\| \frac{\partial^l f}{\partial x_2^l} \right\|_{L_p(\Omega)} \le M_2 \left(\int_0^1 x_1^{\beta_1 p + \gamma((\alpha_2 - l)p + 1)} \, dx_1 \right)^{\frac{1}{p}}$$

and

$$\|D^\beta f\|_{L_p(\Omega)} \ge M_3 \left(\int_0^1 x_1^{\gamma((\alpha_2 - \beta_2)p + 1)} \, dx_1 \right)^{\frac{1}{p}},$$

where M_1, M_2, $M_3 > 0$ are constants. Suppose that $0 < \gamma < 1$, $l < \frac{\beta_1}{\gamma} + \beta_2$ and $l - \frac{1}{p}(1 + \frac{1}{\gamma}) - \frac{\beta_1}{\gamma} < \alpha_2 \leq \beta_2 - \frac{1}{p}(1 + \frac{1}{\gamma})$. Then $f \in W_p^l(\Omega)$ but $D^\beta f \notin L_p(\Omega)$. (The case $p = \infty$ is similar: if $l - \frac{\beta_1}{\gamma} \leq \alpha_2 < \beta_2$, then $f \in W_\infty^l(\Omega)$ but $D^\beta f \notin L_\infty(\Omega)$.) If $\gamma > 1$ an analogous counter-example may be constructed by setting $f(x_1, x_2) := x_1^{\alpha_1} x_2^{\beta_2}$ where $\alpha_1 \notin \mathbb{N}_0$.

Example 10 For any open set $\Omega \subset \mathbb{R}^n$, which has infinite measure or is disconnected, inequality (4.106) does not hold. In the first case we arrive at a contradiction by setting $f(x) = x^\beta$. In the second case let G be any connected component of Ω containing the ball B. Inequality (4.106) does not hold if $f(x) = 0$ on G and $f(x) = x^\beta$ on $\Omega \setminus G$. We note that if Ω has a resolved boundary and is unbounded, then meas $\Omega = \infty$ because in this case $s = \infty$, meas $(\Omega \cap V_j) \geq \delta^n$, $j \in \mathbb{N}$ and $\sum\limits_{j=1}^{\infty}$ meas $(\Omega \cap V_j) \leq \varkappa$ meas Ω.

The last example shows that for bounded open sets having a quasi-resolved boundary inequality (4.107) does not necessarily hold.

Example 11 Let $1 \leq p \leq \infty$, $\omega = \bigcup\limits_{s=0}^{\infty} (2^{-(2s+1)}, 2^{-2s})$ and $\Omega = \{(x_1, x_2) \in \mathbb{R}^2 : x_1 \in \omega$ if $0 \leq x_2 < 1$ and $0 < x_1 < 1$ if $-1 < x_2 < 0\}$. Suppose that $\forall \varepsilon > 0$ there exists $M_1(\varepsilon)$ such that $\forall f \in W_p^2(\Omega)$

$$\left\| \left(\frac{\partial f}{\partial x_1} \right)_w \right\|_{L_p(\Omega)} \leq M_1(\varepsilon) \|f\|_{L_p(\Omega)} + \varepsilon \|f\|_{w_p^2(\Omega)}.$$

Let $f(x_1, x_2) = g(x_1)h(x_2)$, where g, g', $g'' \in L_p(\omega)$ and $h(x_2) = x_2^2$ if $0 \leq x_2 < 1$, $h(x_2) = 0$ if $-1 < x_2 < 0$. Then

$$\|g'\|_{L_p(\omega)} \|h\|_{L_p(0,1)} \leq M_1(\varepsilon) \|g\|_{L_p(\omega)} \|h\|_{L_p(0,1)}$$

$$+\varepsilon \left(\|g''\|_{L_p(\omega)} \|h\|_{L_p(0,1)} + \|g'\|_{L_p(\omega)} \|h'\|_{L_p(0,1)} + \|g\|_{L_p(\omega)} \|h''\|_{L_p(0,1)} \right).$$

Choosing sufficiently small ε, we establish that there exists $M_2 > 0$ such that

$$\|g'\|_{L_p(\omega)} \leq M_2 \left(\|g\|_{L_p(\omega)} + \|g''\|_{L_p(\omega)} \right)$$

for all the functions g. In fact, we have arrived at a contradiction. To verify this we set $g_k(x_1) = x_1 - 2^{-(2k+1)}$ if $x_1 \in (2^{-(2k+1)}, 2^{-2k})$ and $g_k(x) = 0$ for all other $x_1 \in \omega$ and pass to the limit as $k \to \infty$.

Corollary 14 *Let $l \in N$ and $1 \le p \le \infty$.*

1. If $\Omega \subset \mathbb{R}^n$ is an open set having a quasi-resolved boundary, then the norm

$$\|f\|_{\widetilde{W}_p^l(\Omega)} = \sum_{|\alpha| \le l} \|D_w^\alpha f\|_{L_p(\Omega)} \tag{4.110}$$

is equivalent to $\|f\|_{W_p^l(\Omega)}$.

2. If $\Omega \subset \mathbb{R}^n$ is a bounded domain having a quasi-resolved boundary and the ball $B \subset \Omega$, then the norm

$$\|f\|_{L_1(B)} + \sum_{|\alpha| = l} \|D_w^\alpha f\|_{L_p(\Omega)} \tag{4.111}$$

is equivalent to $\|f\|_{W_p^l(\Omega)}$.

Idea of the proof. Apply inequality (4.105) and inequality (4.106). $\square$

Corollary 15 *Let $l \in \mathbb{N}, 1 \le p \le \infty$ and let $\Omega \subset \mathbb{R}^n$ be an open set having a quasi-resolved boundary. If Ω is bounded, then $w_p^l(\Omega) = W_p^l(\Omega)$. If Ω is unbounded, then the inclusion $W_p^l(\Omega) \subset w_p^l(\Omega)$ is strict.*

Idea of the proof. If Ω is bounded, apply inequality (4.106) to each connected component of Ω. If Ω is unbounded, apply Example 10. $\square$

Remark 21 Since $\theta_{w_p^l(\Omega)} \ne \theta_{W_p^l(\Omega)}$, the semi-norms $\|\cdot\|_{w_p^l(\Omega)}$ and $\|\cdot\|_{W_p^l(\Omega)}$, are not equivalent. (See Corollary 3 and Remark 9.)

Generalizations of the one-dimensional inequalities (4.39), (4.40) and (4.41) hold under stronger assumptions on Ω than in Theorem 6.

Theorem 7 *Let $l \in \mathbb{N}$, $\beta \in \mathbb{N}_0^n$ satisfy $0 < |\beta| < l$ and let $1 \le p, q \le \infty$.*

1. *If $\Omega \subset \mathbb{R}^n$ is a domain star-shaped with respect to the ball B, then $\forall f \in W_p^l(\Omega)$*

$$\frac{\|D_w^\beta f\|_{L_p(\Omega)}}{(\operatorname{meas}\Omega)^{\frac{1}{p}}} \le c_{31}\left(\left(\frac{D}{d}\right)^{l-1} D^{-|\beta|}\frac{\|f\|_{L_q(B)}}{(\operatorname{meas} B)^{\frac{1}{q}}} + \left(\frac{D}{d}\right)^{n-1} D^{l-|\beta|}\frac{\|f\|_{w_p^l(\Omega)}}{(\operatorname{meas}\Omega)^{\frac{1}{p}}}\right), \tag{4.112}$$

where $d = \operatorname{diam} B, D = \operatorname{diam}\Omega$ and $c_{31} > 0$ depends only on n and l.

2. *If $\Omega \subset \mathbb{R}^n$ is an open set satisfying the cone condition and $\varepsilon_0 > 0$, then $\forall f \in W_p^l(\Omega)$ and $\forall \varepsilon \in (0, \varepsilon_0]$*

$$\|D_w^\beta f\|_{L_p(\Omega)} \le c_{32}\, \varepsilon^{-\frac{|\beta|}{l-|\beta|}} \|f\|_{L_p(\Omega)} + \varepsilon\, \|f\|_{w_p^l(\Omega)}, \tag{4.113}$$

where $c_{32} > 0$ is independent of f and ε.

Moreover, $\forall f \in W_p^l(\Omega)$

$$\|D_w^\beta f\|_{L_p(\Omega)} \leq c_{33} \|f\|_{L_p(\Omega)}^{1-\frac{|\beta|}{l}} \|f\|_{W_p^l(\Omega)}^{\frac{|\beta|}{l}}, \tag{4.114}$$

where $c_{33} > 0$ is independent of f.

Idea of the proof. Starting with inequality (3.57) apply Young's inequality for convolutions [18] to prove (4.112). If Ω satisfies the cone condition with the parameters $r, h > 0$ apply, in addition, Lemma 3 and Remark 7 of Chapter 3 and Corollary 13. Replacing r and h by $r\delta$ and $h\delta$, where $0 < \delta \leq 1$, deduce (4.113). Verify that (4.113) implies (4.114) as in the one-dimensional case considered in Section 4.2. $\square$

Proof. 1. Let Ω be a domain star-shaped with respect to the ball $B \subset \Omega$. By Corollary 10 of Chapter 3 and inequality (3.57), in particular by (4.115), we have

$$\|D_w^\beta f\|_{L_p(\Omega)} \leq M_1\left((\operatorname{meas}\Omega)^{\frac{1}{p}}\left(\frac{D}{d}\right)^{l-1} D^{-|\beta|} d^{-n} \int\limits_B |f|\, dy\right.$$

$$+\left(\frac{D}{d}\right)^{n-1} \sum_{|\alpha|=l} \left\|\int\limits_\Omega \frac{|(D_w^\alpha f)(y)|}{|x-y|^{n-l+|\beta|}}\, dy\right\|_{L_p(\Omega)}\Bigg)$$

$$\leq M_2\left((\operatorname{meas}\Omega)^{\frac{1}{p}}\left(\frac{D}{d}\right)^{l-1} D^{-|\beta|} (\operatorname{meas} B)^{-1} \int\limits_B |f|\, dy\right.$$

$$+\left(\frac{D}{d}\right)^{n-1} \big\| \,|z|^{-n+l-|\beta|}\, \big\|_{L_1(\Omega-\Omega)} \|f\|_{w_p^l(\Omega)}\Bigg),$$

---

[18] We mean its following variant: if $1 \leq p \leq \infty$, $G, \Omega \subset \mathbb{R}^n$ are measurable sets, $g \in L_p(G), f \in L_1(\Omega - G)$, where $\Omega - G$ is the vector difference of Ω and G, then

$$\left\|\int\limits_G f(x-y)g(y)\, dy\right\|_{L_p(\Omega)} \leq \|f\|_{L_1(\Omega-G)}\|g\|_{L_p(G)}.$$

In the sequel we shall also need the general case:

$$\left\|\int\limits_G f(x-y)g(y)\, dy\right\|_{L_q(\Omega)} \leq \|f\|_{L_r(\Omega-G)}\|g\|_{L_p(G)}, \tag{4.115}$$

where $1 \leq p, r \leq q \leq \infty$, $\frac{1}{r} = 1 - \frac{1}{p} + \frac{1}{q}$.

where M_1 and M_2 depend only on n and l. Since [19] $\Omega - \Omega \subset B(0, 2D)$,

$$\| \, |z|^{-n+l-|\beta|} \|_{L_1(\Omega-\Omega)} \leq \sigma_n \int\limits_0^{2d} \varrho^{l-|\beta|-1} \, d\varrho \leq 2^l \, \sigma_n \, D^{l-|\beta|}$$

and, by Hölder's inequality, (4.112) follows.

2. Next let Ω be an open set satisfying the cone condition with the parameters $r, h > 0$. By Lemma 3 and Remark 7 of Chapter 3 $\Omega = \bigcup_k \Omega_k$, where each Ω_k is a domain star-shaped with respect to a ball of radius r_1 whose diameter does not exceed $2h$, and the multiplicity of the covering does not exceed $6^n(1 + \frac{h}{r})^n$. By (4.112) and Hölder's inequality it follows that for all k

$$\|D_w^\beta f\|_{L_p(\Omega_k)} \leq M_1 \left(\left(\frac{h}{r_1}\right)^{l-1+\frac{n}{p}} |h|^{-|\beta|} \|f\|_{L_p(\Omega_k)} + \left(\frac{h}{r_1}\right)^{n-1} h^{l-|\beta|} \|f\|_{w_p^l(\Omega_k)} \right),$$

where M_1 depends only on n and l. By Corollary 13

$$\|D_w^\beta f\|_{L_p(\Omega)} \leq M_2 \left(1 + \frac{h}{r}\right)^{\frac{n}{p}} \left(\left(\frac{h}{r_1}\right)^{l-1+\frac{n}{p}} |h|^{-|\beta|} \|f\|_{L_p(\Omega)} \right.$$

$$\left. + \left(\frac{h}{r_1}\right)^{n-1} h^{l-|\beta|} \|f\|_{w_p^l(\Omega)} \right),$$

where M_2 depends only on n and l. We note that Ω satisfies the cone condition also with the parameters $r\delta$ and $h\delta$ where $0 < \delta \leq 1$ and replace r and h by $r\delta$ and $h\delta$ in this inequality. Setting $\varepsilon = M_2(1 + \frac{h}{r})^{\frac{n}{p}} (\frac{h}{r_1})^{n-1} h^{l-|\beta|} \delta^{l-|\beta|}$, we obtain inequality (4.113) for $0 < \varepsilon \leq \varepsilon_0^* = M_2(1 + \frac{h}{r})^{\frac{n}{p}} (\frac{h}{r_1})^{n-1} h^{l-|\beta|}$.

Suppose that $\varepsilon > \varepsilon_0^*$ and $\varepsilon_0^* < \varepsilon \leq \varepsilon_0$. Let $c_{32}^* = c_{32}(\varepsilon^*)$. Then

$$\|D_w^\beta f\|_{L_p(\Omega)} \leq c_{32}^* \, (\varepsilon_0^*)^{-\frac{|\beta|}{l-|\beta|}} \|f\|_{L_p(\Omega)} + \varepsilon_0^* \|f\|_{w_p^l(\Omega)},$$

$$\leq c_{32}^* \left(\frac{\varepsilon_0^*}{\varepsilon_0}\right)^{-\frac{|\beta|}{l-|\beta|}} \varepsilon^{-\frac{|\beta|}{l-|\beta|}} \|f\|_{L_p(\Omega)} + \varepsilon \|f\|_{w_p^l(\Omega)},$$

and (4.113) again follows.

Finally, inequality (4.114) follows from (4.113) in the same way as inequality (4.41) follows from (4.40).

---

[19] We apply the formula

$$\int\limits_{B(0,r)} g(|x|) \, dx = \sigma_n \int\limits_0^r g(\varrho)\varrho^{n-1} \, d\varrho, \tag{4.116}$$

where σ_n is the surface area of the unit sphere in $\mathbb{R}^n$.

Corollary 16 *Let $l \in \mathbb{N}$, $\beta \in \mathbb{N}_0^n$ satisfy $0 < |\beta| < l$ and let $1 \leq p \leq \infty$. Then $\forall r > 0$ and $\forall f \in W_p^l(B_r)$*

$$\|D_w^\beta f\|_{L_p(B_r)} \leq c_{31}(r^{-|\beta|}\|f\|_{L_p(B_r)} + r^{l-|\beta|}\|f\|_{w_p^l(B_r)}), \qquad (4.117)$$

where $c_{31} > 0$ is independent of f and r.

Idea of the proof. Apply (4.112) where $B = \Omega = B_r$. $\square$

Remark 22 The statement about equivalence of this inequality, the relevant inequality with a parameter and the multiplicative inequality, analogous to the one-dimensional Corollary 7, also holds.

Remark 23 Let $l \in \mathbb{N}$, $m \in \mathbb{N}_0$, $m < l$. By Section 4.1 inequality (4.105) for all $\beta \in \mathbb{N}_0$ satisfying $|\beta| = m$ is equivalent to the embedding

$$W_p^l(\Omega) \hookrightarrow W_p^m(\Omega) . \qquad (4.118)$$

Next we pass to the problem of compactness of this embedding and start by recalling the well-known criterion of the precompactness of a set in $L_p(\Omega)$, where $\Omega \subset \mathbb{R}^n$ is a measurable set and $1 \leq p < \infty$. We shall write f_0 for the extension by 0 of the function f to $\mathbb{R}^n$: $f_0(x) = f(x)$ if $x \in \Omega$ and $f_0(x) = 0$ if $x \notin \Omega$. The set S is precompact if, and only if,

i) S is bounded in $L_p(\Omega)$,

ii) S is equicontinuous with respect to translation in $L_p(\Omega)$, i.e.,

$$\lim_{h \to 0} \sup_{f \in S} \|f_0(x + h) - f(x)\|_{L_p(\Omega)} = 0$$

and

iii) in the case of unbounded Ω, in addition,

$$\lim_{r \to \infty} \sup_{f \in S} \|f\|_{L_p(\Omega \setminus B_r)} = 0.$$

Lemma 12 *Let $l \in \mathbb{N}_0$, $1 \leq p < \infty$. Moreover, let $\Omega \subset \mathbb{R}^n$ be an open set and $S \subset W_p^l(\Omega)$. Suppose that*

1) *S is bounded in $W_p^l(\Omega)$,*

2) $\displaystyle\lim_{\delta \to 0+} \sup_{f \in S} \|f\|_{W_p^l(\Omega \setminus \Omega_\delta)} = 0,$

3) $\displaystyle\lim_{h \to 0} \sup_{f \in S} \|f(x + h) - f(x)\|_{W_p^l(\Omega_{|h|})} = 0$

and

4) *in the case of unbounded Ω, in addition,* $\displaystyle\lim_{r \to \infty} \sup_{f \in S} \|f\|_{W_p^l(\Omega \setminus B_r)} = 0.$

Then the set S is precompact in $W_p^l(\Omega)$.

Idea of the proof. Apply the inequality

$$\|f_0(x+h) - f(x)\|_{L_p(\Omega)} \leq 2\|f\|_{L_p(\Omega\setminus\Omega_{2|h|})} + \|f(x+h) - f(x)\|_{L_p(\Omega_{|h|})} \quad (4.119)$$

and the closedness of weak differentiation. $\square$

Proof. Inequality (4.119) clearly follows from the inequality

$$\|f_0(x+h) - f(x)\|_{L_p(\Omega)} \leq \|f_0(x+h)\|_{L_p(\Omega\setminus\Omega_{|h|})}$$

$$+ \|f(x)\|_{L_p(\Omega\setminus\Omega_{|h|})} + \|f(x+h) - f(x)\|_{L_p(\Omega_{|h|})}.$$

If $l = 0$ then condition ii) follows from (4.119) and conditions 2), 3). Hence S is precompact in $L_p(\Omega)$.

Next let $l \geq 1$. From 1)–4) it follows that the set S and the sets $S_\alpha = \{D_w^\alpha f, f \in S\}$ where $\alpha \in \mathbb{N}_0^n$ and $|\alpha| = l$ are precompact in $L_p(\Omega)$. Consequently, each infinite subset of S contains a sequence $f_k, k \in \mathbb{N}$, such that $f_k \to f$ and $D_w^\alpha f \to g_\alpha$ in $L_p(\Omega)$. Since the weak differentiation operator D_w^α is closed in $L_p(\Omega)$ (see Section 1.2), $g_\alpha = D_w^\alpha f$ on Ω, $f \in W_p^l(\Omega)$ and $f_k \to f$ in $W_p^l(\Omega)$. $\square$

Theorem 8 *Let $l \in \mathbb{N}, m \in \mathbb{N}_0, m < l, 1 \leq p < \infty$ and let $\Omega \subset \mathbb{R}^n$ be a bounded open set having a quasi-continuous boundary. Then embedding (4.118) is compact.*

Idea of the proof. If Ω is a bounded elementary domain with a continuous boundary, given a set S bounded in $W_p^l(\Omega)$, apply Corollary 12 of the one-dimensional embedding Theorem 4 and Theorem 6 to prove property 2). Furthermore, apply Corollary 7 of Chapter 3 and Theorem 6 to prove property 3) with m replacing l. In the general case apply Lemma 8. $\square$

Proof. By Lemma 8 it is enough to consider the case of a bounded elementary domain with a continuous boundary Ω defined by (4.87). Let $M_1 > 0$ and $S = \{f \in W_p^l(\Omega) : \|f\|_{W_p^l(\Omega)} \leq M_1\}$. By inequality (4.79) for almost all $\bar{x} \in W$ and $0 < \gamma \leq d$

$$\|f(\bar{x}, \cdot)\|_{L_p(\varphi(\bar{x})-\gamma, \varphi(\bar{x}))} \leq \gamma^{\frac{1}{p}}\|f(\bar{x}, \cdot)\|_{L_\infty(\varphi(\bar{x})-d, \varphi(\bar{x}))}$$

$$\leq M_2 \gamma^{\frac{1}{p}}\left(\|f(\bar{x}, \cdot)\|_{L_p(\varphi(\bar{x})-d, \varphi(\bar{x}))} + \left\|\left(\frac{\partial^l f}{\partial x_n^l}\right)_w (\bar{x}, \cdot)\right\|_{L_p(\varphi(\bar{x})-d, \varphi(\bar{x}))}\right),$$

where M_2 is independent of f and γ.

By the theorem on the measurability of integrals depending on a parameter (see footnote 17) both sides of this inequality are functions measurable on W.

Therefore, taking L_p-norms with respect to $\bar{x}$ over W and applying Minkowski's inequality, we have $\forall f \in S$

$$\|f\|_{L_p(\Omega \setminus (\Omega - \gamma e_n))} \leq M_2 \, \gamma^{\frac{1}{p}} \, \|f\|_{W_p^l(\Omega)} \leq M_3 \, \gamma^{\frac{1}{p}} \, ,$$

where $M_3 = M_1 M_2$ and $e_n = (0, ..., 0, 1)$.

If f is replaced by $D_w^\beta f$, where $\beta \in \mathbb{N}_0^n$ satisfies $|\beta| = m$, then by Theorem 6 we get

$$\|D_w^\beta f\|_{L_p(\Omega \setminus (\Omega - \gamma e_n))} \leq M_4 \, \gamma^{\frac{1}{p}} \left(\|D_w^\beta f\|_{L_p(\Omega)} + \left\| \left(\frac{\partial}{\partial x_n} \right)_w^{l-m} D_w^\beta f \right\|_{L_p(\Omega)} \right)$$

$$\leq M_5 \, \gamma^{\frac{1}{p}} \, \|f\|_{W_p^l(\Omega)} \leq M_6 \, \gamma^{\frac{1}{p}} \, ,$$

where M_4, M_5 and M_6 are independent of $f \in S$ and γ.

2. Since φ is continuous on $\overline{W}$, the sets $\Gamma = \{(\bar{x}, \varphi(\bar{x})), \bar{x} \in \overline{W}\}$ and $\Gamma - \gamma e_n$ are compact and disjoint. Consequently, $\varrho(\gamma) := \mathrm{dist}\,(\Gamma, \Gamma - \gamma e_n) > 0$ and $\Gamma^{\varrho(\gamma)} \cap \Omega \subset \Omega \setminus (\Omega - \gamma \varepsilon_n)$. Hence, given $\varepsilon > 0$, there exists ϱ_0 such that $\forall f \in S \quad \|f\|_{W_p^m(G_{n1})} < \varepsilon \, 2^{-n}$, where $G_{n1} = \Gamma^{\varrho_0} \cap \Omega$.

Next let $\Gamma_{i0} = \{x \in \mathbb{R}^n : x_i = a_i; a_k \leq x_k \leq b_k, k = 1, ..., n-1, k \neq i; a_n \leq x_n \leq \varphi(\bar{x}) - \frac{\varrho_0}{2}\}$, $i = 1, ..., n-1$, and let Γ_{i1} be defined similarly with $x_i = b_i$ replacing $x_i = a_i$. Moreover, let $\Gamma_{n0} = \{x \in \mathbb{R}^n : x_n = a_n; a_k \leq x_k \leq b_k, k = 1, ..., n-1\}$. Since these sets are compact and do not intersect Γ, for sufficiently small $\varrho \in (0, \varrho_0]$ we have $G_{ij} \subset \Omega$ and $\Omega \setminus \Omega_\varrho \subset \bigcup_{i=1}^{n} \left(\bigcup_{j=0}^{1} G_{ij} \right)$.

Here, for $i = 1, ..., n-1$, $G_{i0} = \{x \in \mathbb{R}^n : a_i < x_i < a_i + \varrho; a_k < x_k < b_k, k = 1, ..., n-1, k \neq i; a_n < x_n < \varphi(\bar{x}) - \frac{\varrho}{2}\}$, G_{i1} is defined similarly with $b_i - \varrho < x_i < b_i$ replacing $a_i < x_i < a_i + \varrho$. Finally, $G_{n0} = \{x \in \mathbb{R}^n : a_n < x_n < a_n + \varrho; a_k < x_k < b_k, k = 1, ..., n-1\}$.

The same argument as above shows that for sufficiently small ϱ

$$\|f\|_{W_p^m(G_{ij})} < \varepsilon \, 2^{-n}, \quad i = 1, ..., n, \; j = 0, 1.$$

Hence

$$\|f\|_{W_p^m(\Omega \setminus \Omega_\varrho)} \leq \sum_{i=1}^{n} \sum_{j=0}^{1} \|f\|_{W_p^m(G_{ij})} < \varepsilon \, ,$$

and property 2) follows with m replacing l.

3. By Corollary 7 of Chapter 3 and Theorem 6

$$\|f(x+h) - f(x)\|_{W_p^m(\Omega_{|h|})} = \|f(x+h) - f(x)\|_{L_p(\Omega_{|h|})}$$

$$+ \sum_{|\beta|=m} \|(D_w^\beta f)(x+h) - (D_w^\beta f)(x)\|_{L_p(\Omega_{|h|})}$$

$$\leq M_7\,|h|\left(\|f\|_{w_p^1(\Omega)} + \sum_{|\beta|=m} \|D_w^\beta f\|_{w_p^1(\Omega)}\right) \leq M_8\,|h|\,\|f\|_{W_p^l(\Omega)} \leq M_9\,|h|\,,$$

where M_7, M_8 and M_9 are independent of $f \in S$ and h, and the property 3) follows.

By Lemma 12 the set S is precompact in $W_p^m(\Omega)$ and, hence, embedding (4.118) is compact. $\square$

If $m = 0$, then embedding (4.118) always holds, but it can be non-compact as the following simple examples show.

Example 12 If the unbounded set Ω contains a denumerable set of disjoint balls $B(x_k, r)$ of the same radius, then embedding (4.118) for each $m = 0, 1, ..., l-1$ is not compact. To verify this it is enough to set $f_k(x) = \varphi(x - x_k)$, where $\varphi \in C_0^\infty(B(0, r))$, $\varphi \not\equiv 0$, $k \in \mathbb{N}$. In that case $\|f_k\|_{W_p^l(\Omega)} = \|\varphi\|_{W_p^l(B(0,r))}$ and $\|f_k - f_s\|_{W_p^m(\Omega)} = 2^{\frac{1}{p}}\|\varphi\|_{W_p^m(B(0,r))}$, $k \neq s$. Hence, any subsequence of $\{f_k\}_{k\in\mathbb{N}}$ is divergent in $W_p^m(\Omega)$.

Example 13 If Ω is a bounded or unbounded open set, which is such that $\Omega = \bigcup_{k=1}^\infty \Omega_k$, where Ω_k are disjoint domains, then embedding (4.118) for each $m = 0, 1..., l-1$ is not compact. To verify this it is enough to consider functions f_k, which are such that $f_k = 0$ on $\Omega \setminus \Omega_k$, $\|f_k\|_{L_p(\Omega_k)} \geq 1$ and $\|f_k\|_{W_p^l(\Omega_k)} \leq M$, where M is independent of $k \in \mathbb{N}$. The sequence $\{f_k\}_{k\in\mathbb{N}}$ is bounded in $W_p^l(\Omega)$, but $\|f_k - f_s\|_{W_p^m(\Omega)} \geq \|f_k - f_s\|_{L_p(\Omega)} \geq 2^{\frac{1}{p}}$. Hence, again every subsequence of $\{f_k\}_{k\in\mathbb{N}}$ is divergent. If $\operatorname{meas}\Omega_k < \infty$, one may just set $f_k = (\operatorname{meas}\Omega_k)^{-1}$ on Ω_k. If $\operatorname{meas}\Omega_k = \infty$, let $f_k(x) = \eta(\frac{x}{r_k})$ on Ω_k, where $\eta \in C_0^\infty(\mathbb{R}^n)$, $\eta(x) = 1$ if $|x| \leq 1$, and $r_k \geq 1$ are chosen in such a way that $\operatorname{meas}(\Omega_k \cap B(0, r_k)) \geq 1$.

A more sophisticated example shows that embedding (4.118) for bounded domains having a guasi-resolved boundary can also be non-compact.

Example 14 Let $1 \leq p \leq \infty$ and Ω be the domain in Example 11. Then the embedding $W_p^1(\Omega) \hookrightarrow L_p(\Omega)$ is not compact. For, let $f_k(x_1, x_2) = 2^{\frac{2k+1}{p}} x_2$ if $2^{-(2k+1)} < x_1 < 2^{-2k}$ and $f_k(x_1, x_2) = 0$ for all other $(x_1, x_2) \in \Omega$. Then the sequence $\{f_k\}_{k\in\mathbb{N}}$ is bounded in $W_p^1(\Omega)$: $\|f_k\|_{W_p^1(\Omega)} = 1 + (p+1)^{-\frac{1}{p}}$. However, it does not contain a subsequence convergent in $L_p(\Omega)$ since $\|f_k - f_m\|_{L_p(\Omega)} = 2^{\frac{1}{p}}(p+1)^{-\frac{1}{p}}$ if $k \neq m$.

Lemma 13 *Let $l \in \mathbb{N}$, $m \in \mathbb{N}_0$, $m < l$, $1 \leq p, q \leq \infty$ and let $\Omega \subset \mathbb{R}^n$ be an open set.*

1. *If the embedding*

$$W_p^l(\Omega) \hookrightarrow W_q^m(\Omega) \tag{4.120}$$

is compact , then $\forall \varepsilon > 0$ there exists $c_{34}(\varepsilon) > 0$ such that $\forall f \in W_p^l(\Omega)$

$$\|f\|_{W_q^m(\Omega)} \leq c_{34}(\varepsilon) \|f\|_{L_p(\Omega)} + \varepsilon \|f\|_{w_p^l(\Omega)} . \tag{4.121}$$

2. *If $\varepsilon > 0$ (4.121) holds and the embedding $W_p^l(\Omega) \hookrightarrow L_p(\Omega)$ is compact, then embedding (4.120) is also compact.*

Idea of the proof. 1. Suppose that inequality (4.121) does not hold for all $\varepsilon > 0$, i.e., there exist $\varepsilon_0 > 0$ and functions $f_k \in W_p^l(\Omega)$, $k \in \mathbb{N}$, such that $\|f_k\|_{W_p^l(\Omega)} = 1$ and

$$\|f_k\|_{W_q^m(\Omega)} > k \, \|f_k\|_{L_p(\Omega)} + \varepsilon_0 \, \|f_k\|_{w_l^l(\Omega)} . \tag{4.122}$$

Obtain a contradiction by proving that $\lim\limits_{k\to\infty} \|f_k\|_{L_p(\Omega)} = 0$ and, consequently, $\liminf\limits_{k\to\infty} \|f_k\|_{W_q^m(\Omega)} \geq \varepsilon_0$.

2. Given a bounded set in $W_p^l(\Omega)$, it follows that it contains a sequence $\{f_k\}_{k\in\mathbb{N}}$ convergent in $L_p(\Omega)$. Applying inequality (4.121) to $f_k - f_s$, prove that $\lim\limits_{k,s\to\infty} \|f_k - f_s\|_{W_q^m(\Omega)} = 0$. $\square$

Proof. 1. Since $\|f_k\|_{W_p^l(\Omega)} = 1$, by (4.120) it follows that $\|f_k\|_{W_q^m(\Omega)} \leq M_1$, where M_1 is independent of k. Consequently, by (4.122) we have $\|f_k\|_{L_p(\Omega)} < M_1 \, k^{-1}$. Thus, $\lim\limits_{k\to\infty} \|f_k\|_{L_p(\Omega)} = 0$ and $\lim\limits_{k\to\infty} \|f_k\|_{w_p^l(\Omega)} = 1$. Hence by (4.122)

$$\liminf_{k\to\infty} \|f_k\|_{W_q^m(\Omega)} \geq \varepsilon_0 \liminf_{k\to\infty} \|f_k\|_{w_p^l(\Omega)} = \varepsilon_0.$$

Since embedding (4.120) is compact, there exists a subsequence f_{k_s} converging to a function f in $W_q^m(\Omega)$. The function f is equivalent to 0 since $f_{k_s} \to 0$ in $L_p(\Omega)$. [20] This contradicts the inequality

$$\|f\|_{W_q^m(\Omega)} = \lim_{s\to\infty} \|f_k\|_{W_q^m(\Omega)} \geq \varepsilon_0 .$$

2. Let $M_2 > 0$ and $S = \{f \in W_p^l(\Omega) : \|f\|_{W_p^l(\Omega)} \leq M_2\}$. Since the embedding $W_p^l(\Omega) \hookrightarrow L_p(\Omega)$ is compact, there exists a sequence $f_k \in S$, $k \in \mathbb{N}$, which is Cauchy in $L_p(\Omega)$. Furthermore, by (4.121)

$$\|f_k - f_s\|_{W_q^m(\Omega)} \leq c_{34}(\varepsilon) \|f_k - f_s\|_{L_p(\Omega)} + \varepsilon \|f_k - f_s\|_{w_p^l(\Omega)}$$

---

[20] If $\varphi_s \to \psi_1$ in $L_p(\Omega)$ as $s \to \infty$, then there exists a subsequence φ_{s_σ} converging to ψ_1 almost everywhere on Ω. Hence, if also $\varphi_s \to \psi_2$ in $L_q(\Omega)$, then ψ_1 is equivalent to ψ_2 on Ω.

$$\leq c_{34}(\varepsilon)\, \|f_k - f_s\|_{L_p(\Omega)} + 2\varepsilon\, M_2.$$

Given $\delta > 0$, take $\varepsilon = \frac{\delta}{4M_2}$. Since f_k is Cauchy there exists $N \in \mathbb{N}$ such that $\forall k, s > N$ we have $\|f_k - f_s\|_{L_p(\Omega)} < \frac{\delta}{2}\left(c_{34}\left(\frac{\delta}{4M_2}\right)\right)^{-1}$. Thus, $\forall k, s > N$ $\|f_k - f_s\|_{W_q^m(\Omega)} < \delta$, i.e., the sequence f_k is Cauchy in $W_q^m(\Omega)$.

By the completeness of $W_q^m(\Omega)$ there exists a function $f \in W_q^m(\Omega)$ such that $f_k \to f$ in $W_q^m(\Omega)$ as $k \to \infty$. $\square$

Corollary 17 *Let $l \in \mathbb{N}$, $m \in \mathbb{N}_0$, $m < l$, $1 \leq p, q \leq \infty$ and let $\Omega \subset \mathbb{R}^n$ be a bounded open set having a quasi-continuous boundary. Then the compactness of embedding (4.120) is equivalent to the validity of inequality (4.121) for all $\varepsilon > 0$.*

Idea of the proof. Apply Lemma 13 and Theorem 8. $\square$

Lemma 14 *Let $1 \leq q < p \leq \infty$ and let $\Omega \subset \mathbb{R}^n$ be an open set such that $\operatorname{meas}\Omega < \infty$. Then the embedding*

$$W_p^1(\Omega) \hookrightarrow L_q(\Omega) \tag{4.123}$$

is compact.

Idea of the proof. Given a bounded set $S \subset W_p^1(\Omega)$ apply Hölder's inequality and Corollary 7 of Chapter 3 to prove that conditions 1) – 4) of Lemma 12 are satisfied with $L_q(\Omega)$ replacing $W_p^l(\Omega)$. $\square$

Proof. Let $M > 0$ and $S = \{f \in W_p^1(\Omega) : \|f\|_{W_p^l(\Omega)} \leq M\}$. By Hölder's inequality $\forall f \in S$

$$\|f\|_{L_q(\Omega)} \leq (\operatorname{meas}\Omega)^{\frac{1}{q}-\frac{1}{p}}\, \|f\|_{L_p(\Omega)} \leq M\,(\operatorname{meas}\Omega)^{\frac{1}{q}-\frac{1}{p}}$$

and

$$\|f\|_{L_q(\Omega\setminus\Omega_\delta)} \leq (\operatorname{meas}(\Omega\setminus\Omega_\delta))^{\frac{1}{q}-\frac{1}{p}}\, \|f\|_{L_p(\Omega\setminus\Omega_\delta)} \leq M(\operatorname{meas}(\Omega\setminus\Omega_\delta))^{\frac{1}{q}-\frac{1}{p}}.$$

Since $\operatorname{meas}\Omega < \infty$, we have $\lim\limits_{\delta\to 0+} \operatorname{meas}(\Omega\setminus\Omega_\delta) = 0$ and $\lim\limits_{\delta\to 0+} \sup\limits_{f\in S} \|f\|_{L_q(\Omega\setminus\Omega_\delta)} = 0$. Thus, properties 1) and 2) are satisfied. Moreover, by Corollary 7 of Chapter 3 it follows that $\forall f \in S$

$$\|f(x+h) - f(x)\|_{L_q(\Omega_{|h|})} \leq |h|\, \|f\|_{W_q^l(\Omega)}$$

$$\leq |h|\,(\operatorname{meas}\Omega)^{\frac{1}{q}-\frac{1}{p}}\, \|f\|_{W_p^l(\Omega)} \leq M\,(\operatorname{meas}\Omega)^{\frac{1}{q}-\frac{1}{p}}\, |h|.$$

Hence, property 3) is satisfied (with $L_q(\Omega)$ replacing $W_p^l(\Omega)$). If Ω is unbounded, then again by Hölder's inequality

$$\|f\|_{L_q(\Omega \setminus B_r)} \leq M \text{ meas}\,(\Omega \setminus B_r)^{\frac{1}{q}-\frac{1}{p}}$$

and property 4) follows. $\square$

We conclude this section with several statements, which are based essentially on the estimates for intermediate derivatives given in Theorems $6-7$.

Lemma 15 *Let $l \in \mathbb{N}$ and let $1 \leq p, p_1, p_2 \leq \infty$ satisfy $\frac{1}{p} = \frac{1}{p_1} + \frac{1}{p_2}$. Suppose that $\Omega \subset \mathbb{R}^n$ is an open set having a quasi-resolved boundary. Then $\forall f_1 \in W_{p_1}^l(\Omega)$ and $\forall f_2 \in W_{p_2}^l(\Omega)$*

$$\|f_1\,f_2\|_{W_p^l(\Omega)} \leq c_{35}\,\|f_1\|_{W_{p_1}^l(\Omega)}\,\|f_2\|_{W_{p_2}^l(\Omega)}, \tag{4.124}$$

where $c_{35} > 0$ depends only on n and l.

Idea of the proof. Apply the Leibnitz formula, Hölder's inequality and Theorem 6. $\square$

Proof. If $f_k \in C^\infty(\Omega) \cap W_{p_k}^l(\Omega), k = 1, 2$, then starting from

$$D^\alpha(f_1\,f_2) = \sum_{0 \leq \beta \leq \alpha} \binom{\alpha}{\beta} D^\beta f_1\,D^{\alpha-\beta}f_2,$$

we have

$$\|f_1\,f_2\|_{W_p^l(\Omega)} \leq \|f_1\,f_2\|_{L_p(\Omega)} + n^l \sum_{|\alpha|=l}\sum_{0 \leq \beta \leq \alpha} \|D^\beta f_1\,D^{\alpha-\beta}f_2,\|_{L_p(\Omega)}$$

$$\leq \|f_1\|_{L_{p_1}(\Omega)}\|f_2\|_{L_{p_2}(\Omega)} + n^l \sum_{|\alpha|=l}\sum_{0 \leq \beta \leq \alpha} \|D^\beta f_1\|_{L_{p_1}(\Omega)}\|D^{\alpha-\beta}f_2\|_{L_{p_2}(\Omega)}$$

$$\leq M_1 \left(\sum_{|\beta| \leq l} \|D^\beta f_1\|_{L_{p_1}(\Omega)} \right) \left(\sum_{|\beta| \leq l} \|D^\beta f_2\|_{L_{p_2}(\Omega)} \right)$$

$$\leq M_2\,\|f_1\|_{W_{p_1}^l(\Omega)}\|f_2\|_{W_{p_2}^l(\Omega)},$$

where M_1 and M_2 depend only on n and l.

If $f_k \in W_{p_k}^l(\Omega), k = 1, 2$, then (4.124) follows by applying, in addition, Theorem 1 of Chapter 2. $\square$

Corollary 18 *Let $l \in \mathbb{N}$, $1 \le p \le \infty$ and let $\Omega \subset \mathbb{R}^n$ be an open set having a quasi-resolved boundary. Suppose that $\varphi \in C_b^l(\Omega)$. Then $\forall f \in W_p^l(\Omega)$*

$$\|f\,\varphi\|_{W_p^l(\Omega)} \le c_{36}\,\|f\|_{W_p^l(\operatorname{supp}\,\varphi\cap\Omega)}\,, \tag{4.125}$$

where $c_{36} > 0$ is independent of f.

Idea of the proof. Direct application of the proof of Lemma 15. $\square$

Lemma 16 *Let $l \in \mathbb{N}$, $1 \le p \le \infty$ and let $\Omega \subset \mathbb{R}^n$ be an open set with a quasi-resolved boundary. Moreover, let $g = (g_1, ..., g_n) : \Omega \to \mathbb{R}^n$, $g_k \in C^l(\Omega)$, $k = 1, ..., n$. Suppose that $\forall \alpha \in \mathbb{N}_0^n$ satisfying $1 \le |\alpha| \le l$ the derivatives $D^\alpha g_k$ are bounded on Ω and the Jacobian $\frac{Dg}{Dx}$ is such that $\inf\limits_{x \in \Omega} |\frac{Dg}{Dx}(x)| > 0$. Furthermore, let $g(\Omega)$ be also an open set with a quasi-resolved boundary.*
Then $\forall f \in W_p^l(\Omega)$

$$c_{37}\,\|f\|_{W_p^l(g(\Omega))} \le \|f(g)\|_{W_p^l(\Omega)} \le c_{38}\,\|f\|_{W_p^l(g(\Omega))}, \tag{4.126}$$

where $c_{37}, c_{38} > 0$ are independent of f and p.

Remark 24 By the assumptions of the lemma on g it follows that there exists the unique inverse transform $g^{(-1)} = (g_1^{(-1)}, ..., g_n^{(-1)}) : g(\Omega) \to \Omega$ such that $g_k^{(-1)} \in C^l(g(\Omega))$, $k = 1, ..., n$. Moreover, $\forall \alpha \in \mathbb{N}_0^n$ satisfying $1 \le |\alpha| \le l$ the derivatives $D^\alpha g_k^{(-1)}$ are bounded on $g(\Omega)$ and $\inf\limits_{y \in g(\Omega)} \left| \frac{Dg^{(-1)}}{Dy}(y) \right| > 0$.

Idea of the proof. Apply the formula for derivatives of $f(g)$, keeping in mind that for weak derivatives, under the assumptions of Lemma 16 on g, it has the same form as for ordinary derivatives, i.e.,

$$D_w^\alpha(f(g)) = \sum_{\beta \le \alpha, |\beta| \ge 1} (D_w^\beta f)(g) \sum_{\gamma_1 + ... + \gamma_{|\alpha|} = \alpha} c_{\beta, \gamma_1, ..., \gamma_{|\alpha|}} D^{\gamma_1} g \cdots D^{\gamma_{|\alpha|}} g\,, \tag{4.127}$$

where $\gamma_k \in \mathbb{N}_0^n$ and $c_{\beta, \gamma_1, ..., \gamma_{|\alpha|}}$ are some nonnegative integers. Apply also Theorem 6 of Chapter 4.

Proof. Let $\alpha \in \mathbb{N}_0^n$ and $|\alpha| = l$. By (4.127), Minkowski's inequality and Theorem 6 it follows that

$$\|D_w^\alpha(f(g))\|_{L_p(\Omega)} \le M_1 \sum_{\beta \le \alpha, |\beta| \ge 1} \|(D_w^\beta f)(g)\|_{L_p(\Omega)} = (\,y = g(x)\,)$$

$$= M_1 \sum_{\beta \leq \alpha, |\beta| \geq 1} \left\| (D_w^\beta f)(y) \left| \frac{Dg}{Dx}(g^{-1}(y)) \right|^{-\frac{1}{p}} \right\|_{L_p(g(\Omega))}$$

$$\leq M_2 \sum_{\beta \leq \alpha, |\beta| \geq 1} \| D_w^\beta f \|_{L_p(g(\Omega))} \leq M_3 \| f \|_{W_p^l(g(\Omega))} \,,$$

where $M_1, M_2, M_3 > 0$ are independent of f and p. Hence, the second inequality (4.126) is proved in a similar way. $\square$

Remark 25 From the above proof it follows that

$$c_{38} \leq c_{39} \left(\inf_{x \in \Omega} \left| \frac{Dg}{Dx}(x) \right| \right)^{-\frac{1}{p}} \max_{1 \leq |\alpha| \leq l} \| D^\alpha g \|_{C(\Omega)} \,, \tag{4.128}$$

where c_{39} depends only on n and l.

Theorem 9 *Let $l \in \mathbb{N}$, $l > 1$, $1 < p < \infty$ and let $\Omega \subset \mathbb{R}^n$ be an open set satisfying the cone condition. Suppose that $f \in L_p(\Omega)$, the weak derivatives $\left(\frac{\partial^l f}{\partial x_j^l} \right)_w$, $j = 1, ..., n$, exist on Ω and are in $L_p(\Omega)$. Then $\forall \beta \in \mathbb{N}_0^n$ satisfying $|\beta| = l$ the weak derivatives $D_w^\beta f$ also exist on Ω and*

$$\| D_w^\beta f \|_{L_p(\Omega)} \leq c_{40} \left(\| f \|_{L_p(\Omega)} + \sum_{j=1}^n \left\| \left(\frac{\partial^l f}{\partial x_j^l} \right)_w \right\|_{L_p(\Omega)} \right) , \tag{4.129}$$

where $c_{40} > 0$ is independent of f.

Remark 26 For an open set $\Omega \subset \mathbb{R}^n$ consider the space $W_p^{l,\dots,l}(\Omega)$ of all functions $f \in L_p(\Omega)$ whose weak derivatives $\left(\frac{\partial^l f}{\partial x_j^l} \right)_w$ exist on Ω and

$$\| f \|_{W_p^{l,\dots,l}(\Omega)} = \| f \|_{L_p(\Omega)} + \sum_{j=1}^n \left\| \left(\frac{\partial^l f}{\partial x_j^l} \right)_w \right\|_{L_p(\Omega)} < \infty.$$

By Theorem 9 $W_p^{l,\dots,l}(\Omega) = W_p^l(\Omega)$ if $1 < p < \infty$ and Ω satisfies the cone condition and the norms $\| \cdot \|_{W_p^{l,\dots,l}(\Omega)}$ and $\| \cdot \|_{W_p^l(\Omega)}$ are equivalent.

Idea of the proof. By Lemma 11 it is enough to consider the case of open sets Ω with a Lipschitz boundary. Applying the extension theorem for the spaces $W_p^{l,\dots,l}(\Omega)$ (see Remark 18 of Chapter 6) and the density of $C_0^\infty(\mathbb{R}^n)$ in $W_p^{l,\dots,l}(\mathbb{R}^n)$, which is proved as in Lemma 2 of Chapter 2, it is enough to prove (4.129) for $\Omega = \mathbb{R}^n$ and $f \in C_0^\infty(\mathbb{R}^n)$. For $p = 2$ (4.129) easily follows by taking Fourier transforms. If $1 < p < \infty$ one may apply the Marcinkiewicz multiplicator theorem: [21]

$$\|D^\beta f\|_{L_p(\mathbb{R}^n)} = \Big\| \sum_{j=1}^n F^{-1}\Big\{ (i\xi)^\beta (-i\,\mathrm{sgn}\,\xi_j)^l \Big(\sum_{k=1}^n |\xi_k|^l\Big)^{-1} F\Big(\frac{\partial^l f}{\partial x_j^l}\Big) \Big\} \Big\|_{L_p(\mathbb{R}^n)}$$

$$\leq M_1 \sum_{j=1}^n \Big\| \frac{\partial^l f}{\partial x_j^l} \Big\|_{L_p(\mathbb{R}^n)} = M_1 \|f\|_{w_p^{l,\dots,l}(\mathbb{R}^n)},$$

where M_1 depends only on n, l and p. $\square$

Example 15 Let $p = \infty$, $l \in \mathbb{N}$, $\beta \in \mathbb{N}_0^n$, $|\beta| = l$ and let f be a function defined by $f(x) = x^\beta \ln |x|\, \eta(x)$ if $x \neq 0$ $(f(0) = 0)$, where $\eta \in C_0^\infty(\mathbb{R}^n)$ and $\eta = 1$ in a neighbourhood of the origin. Then $f \in W_\infty^{l,\dots,l}(\mathbb{R}^n)$, but $D_w^\beta f \notin L_\infty(\mathbb{R}^n)$.

Thus Theorem 9 does not hold for $p = \infty$. One can also prove that it does not hold for $p = 1$.

4.5 Hardy-Littlewood-Sobolev inequality for integral of potential type

Let $f \in L_1^{loc}(\mathbb{R}^n)$. The convolution

$$|x|^{-\lambda} * f = \int\limits_{\mathbb{R}^n} \frac{f(y)}{|x-y|^\lambda}\, dy, \quad \lambda < n, \tag{4.131}$$

is called an integral of potential type.

Remark 27 One may verify that

1) if $\lambda \geq n$, f is measurable on $\mathbb{R}^n$ and f is not equivalent to 0, then $|x|^{-\lambda} * f$ does not exist on a set of positive measure,

2) if $\lambda < n$, f is measurable on $\mathbb{R}^n$ and $f \notin L_1^{loc}(\mathbb{R}^n)$, then the convolution $|x|^{-\lambda} * f$ does not exist for almost all $x \in \mathbb{R}^n$.

3) if $\lambda < n$, $f \in L_1^{loc}(\mathbb{R}^n)$ and $|x|^{-\lambda} * f$ exists for almost all $x \in \mathbb{R}^n$, then the function $|x|^{-\lambda} * f$ is measurable on $\mathbb{R}^n$.

---

[21] Let $1 < p < \infty$. Suppose that, for $\forall \alpha \in \mathbb{N}_0^n$ satisfying $0 \leq \alpha \leq 1$ (i.e., $\alpha_j = 0$ or $1, j = 1, \dots, n$), the function $\mu \in L_\infty$ has the derivatives $D^\alpha \mu$ on the set $\mathbb{R}_*^n = \{x \in \mathbb{R}^n : x_1 \cdot \ldots \cdot x_n \neq 0\}$. If $|x^\alpha (D^\alpha \mu)(x)| \leq K$, $x \in \mathbb{R}_*^n$, then

$$\|F^{-1}(\mu F f)\|_{L_p(\mathbb{R}^n)} \leq M_2 K \|f\|_{L_p(\mathbb{R}^n)}, \quad f \in C_0^\infty(\mathbb{R}^n), \tag{4.130}$$

where M_2 depends only on n and p.

Integrals of potential type are contained in the inequalities deduced from the integral representations of Chapter 3, namely (3.54), (3.58), (3.65), (3.66), (3.69) and (3.76). For this reason we are interested in conditions on f implying that $|x|^{-\lambda} * f \in L_q(\mathbb{R}^n)$.

Theorem 10 (*the Hardy-Littlewood-Sobolev inequality*). *Let* $n \in \mathbb{N}$,

$$1 < p < q < \infty \tag{4.132}$$

and

$$\lambda = n(\tfrac{1}{p'} + \tfrac{1}{q}). \tag{4.133}$$

Then $\forall f \in L_p(\mathbb{R}^n)$ *the convolution* $|x|^{-\lambda} * f$ *exists for almost all* $x \in \mathbb{R}^n$ *and*

$$\| \, |x|^{-\lambda} * f \, \|_{L_q(\mathbb{R}^n)} \le c_{41} \|f\|_{L_p(\mathbb{R}^n)}, \tag{4.134}$$

where $c_{41} > 0$ *depends only on* n, p *and* q.

Remark 28 By applying inequality (4.134) to $f(\varepsilon x)$, where $f \in L_p(\mathbb{R}^n)$, $f \not\approx 0$, is fixed and $0 < \varepsilon < \infty$, one may verify that if $\lambda \neq n(\tfrac{1}{p'} + \tfrac{1}{q})$, then inequality (4.134) does not hold for any choice of c_{41}.

We give a sketch of the proof of Theorem 10 based on the properties of maximal functions. [22]

Lemma 17 *Let* $n \in \mathbb{N}$, $\mu < n$. *Then for all functions* f *measurable on* $\mathbb{R}^n$ $\forall x \in \mathbb{R}^n$ *and* $\forall r > 0$

$$\int\limits_{B(x,r)} |x - y|^{-\mu} |f(y)| \, dy \le c_{43} \, r^{n-\mu} (Mf)(x), \tag{4.136}$$

where $c_{43} > 0$ *depends only on* n *and* μ.

---

[22] For $f \in L_1^{loc}(\mathbb{R}^n)$ the *maximal function* Mf is defined by

$$(Mf)(x) = \sup_{r>0} \frac{1}{\operatorname{meas} B(x,r)} \int_{B(x,r)} |f| \, dy, \quad x \in \mathbb{R}^n.$$

For almost all $x \in \mathbb{R}^n$ $|f(x)| \le (Mf)(x) < \infty$. Moreover, Mf is measurable on $\mathbb{R}^n$ and for $1 < p \le \infty$ there exists $c_{42} > 0$ such that $\forall f \in L_p(\mathbb{R}^n)$

$$\|Mf\|_{L_p(\mathbb{R}^n)} \le c_{42} \|f\|_{L_p(\mathbb{R}^n)}. \tag{4.135}$$

(If $p = 1$, this inequality does not hold.)

Idea of the proof. Split the ball $B(x,r)$ into a union of spherical layers $S(x, r2^{-k}) = B(x, r2^{-k}) \setminus B(x, r2^{-k-1})$, $k \in \mathbb{N}_0$, and estimate $\int_{S(x,r2^{-k})} |f|\, dy$ via the maximal function Mf. $\square$

Lemma 18 *Let $n \in \mathbb{N}$, $1 < p < \infty$, $\frac{n}{p'} < \mu < n$. Then $\forall f \in L_p(\mathbb{R}^n)$*

$$\left| |x|^{-\mu} * f \right| \le c_{44} \, \|f\|_{L_p(\mathbb{R}^n)}^{\frac{p}{n}(n-\mu)} \left((Mf)(x) \right)^{\frac{p}{n}\left(\mu - \frac{n}{p'}\right)}, \tag{4.137}$$

where $c_{44} > 0$ depends only on n, p and μ.

Idea of the proof. Split the integrals defining $|x|^{-\mu} * f$ into an integral over $B(x,r)$ and an integral over $^c B(x,r)$. Applying inequality (4.136) to the first integral and using Hölder's inequality to estimate the second one via $\|f\|_{L_p(\mathbb{R}^n)}$, establish that

$$\left| |x|^{-\mu} * f \right| \le M_1 \, r^{n-\mu}(Mf)(x) + M_2 \, r^{-\left(\mu - \frac{n}{p'}\right)} \|f\|_{L_p(\mathbb{R}^n)},$$

where M_1, M_2 depend only on n, p and μ. Finally, minimize with respect to r. $\square$

Idea of the proof of Theorem 10. Apply inequalities (4.137) and (4.135). $\square$

Proof. Since $(Mf)(x) < \infty$ for almost all $x \in \mathbb{R}^n$, the convolution $|x|^{-\lambda} * f$ exists, by (4.137), almost everywhere on $\mathbb{R}^n$ and, by Remark 27, is measurable on $\mathbb{R}^n$. Since Mf is also measurable on $\mathbb{R}^n$, taking L_q-norms in inequality (4.137) and taking into account (4.135), we get

$$\| |x|^{-\lambda} * f \|_{L_q(\mathbb{R}^n)} \le c_{44} \|f\|_{L_p(\mathbb{R}^n)}^{1 - \frac{p}{q}} \|Mf\|_{L_p(\mathbb{R}^n)}^{\frac{1}{q}} \le c_{44} \, c_{42}^{\frac{1}{q}} \|f\|_{L_p(\mathbb{R}^n)}. \quad \square$$

Remark 29 One may verify that from the above proof it follows that

$$c_{41} = (1 + o(1)) \left(\frac{v_n q}{p'} \right)^{\frac{1}{p'}} \quad \text{as} \quad q \to \infty. \tag{4.138}$$

Remark 30 Let $(Pf)(x) = |x|^{-\lambda} * f$. Theorem 10 states that for $1 < p < q < \infty$ the operator P is a bounded operator mapping the space $L_p(\mathbb{R}^n)$ into the space $L_q(\mathbb{R}^n)$. There is one more, trivial, case in which the operator P is bounded: $p = 1$ and $q = \infty$. In all other admissible cases the operator P is unbounded, thus, inequality (4.134) does not hold for any $c_{41} > 0$. If $p = 1$ or $q = \infty$, it follows from the explicit formulae for the norms of integral operators, [23] by which $\|P\|_{L_1(\mathbb{R}^n)\to L_q(\mathbb{R}^n)} = \|\,|x|^{-\frac{n}{q}}\,\|_{L_q(\mathbb{R}^n)} = \infty$ for $1 \le q < \infty$ and $\|P\|_{L_p(\mathbb{R}^n)\to L_\infty(\mathbb{R}^n)} = \|\,|x|^{-\frac{n}{p'}}\,\|_{L_{p'}(\mathbb{R}^n)} = \infty$ for $1 < p \le \infty$. If, finally, $1 < p = q < \infty$, it follows by Remark 27.

Next we discuss the case $q = \infty$ in Theorem 10, i.e., behaviour of the convolution $|x|^{-\frac{n}{p'}} * f$ for $f \in L_p(\mathbb{R}^n)$. The cases $p = 1$ and $p = \infty$ are trivial. (If $p = 1$, then this convolution is just a constant; if $p = \infty$, see Remark 27.) If $1 < p < \infty$, then in general $|x|^{-\frac{n}{p'}} * f$ does not exist on a set of positive measure.

Example 16 Let $1 < p < \infty$, $f(x) = 0$ if $|x| \le e$ and $f(x) = |x|^{-\frac{n}{p}}(\ln|x|)^{-1}$ if $|x| > e$. Then $f \in L_p(\mathbb{R}^n)$, but $|x|^{-\frac{n}{p'}} * f = \infty$ for each $x \in \mathbb{R}^n$.

For this reason we consider the case in which $1 < p < \infty$ for functions in $L_p(\mathbb{R}^n)$ with compact supports.

Theorem 11 *Let $1 < p < \infty$, $f \in L_p(\mathbb{R}^n)$, $f \not\approx 0$. If $\beta \le \frac{1}{v_n}$, then for each compact $\Omega \subset \mathbb{R}^n$*

$$\int_\Omega \exp\left(\beta\left|\frac{|x|^{-\frac{n}{p'}} * f}{\|f\|_{L_p(\mathbb{R}^n)}}\right|^{p'}\right) dx < \infty. \tag{4.141}$$

Idea of the proof (in the case $\beta < \frac{1}{v_n}$). Suppose that $\|f\|_{L_p(\mathbb{R}^n)} = 1$, the case in which $\|f\|_{L_p(\mathbb{R}^n)} \ne 1$ being similar. Following the proof of Lemma 18, establish the inequality

$$\big|\,|x|^{-\frac{n}{p'}} * f\,\big| \le M_1\, r^{\frac{n}{p}}(Mf)(x) + (\sigma_n|\ln r|)^{\frac{1}{p'}} + M_2,$$

where $0 < r \le 1$ and M_1, M_2 depend only on n, p. Take $r = (1+((Mf)(x))^p)^{-\frac{1}{n}}$ and apply inequality (4.47). $\square$

---

[23] Let $E, F \subset \mathbb{R}^n$ be measurable sets, k be a function measurable on $E \times F$ and $(Kf)(y) = \int_E k(x,y)f(y)\,dy$. Then for $1 \le q \le \infty$

$$\|K\|_{L_1(F)\to L_q(E)} = \|\,\|k(x,y)\|_{L_{q,x}(E)}\,\|_{L_{\infty,y}(F)} \tag{4.139}$$

and for $1 \le p \le \infty$

$$\|K\|_{L_p(F)\to L_\infty(E)} = \|\,\|k(x,y)\|_{L_{p',y}(E)}\,\|_{L_{\infty,x}(F)}. \tag{4.140}$$

Corollary 19 *If* $0 < \mu < p'$, *then* $\forall \beta > 0$

$$\int_\Omega \exp\left(\beta \left| |x|^{-\frac{n}{p'}} * f \right|^\mu\right) dx < \infty.$$

Idea of the proof. Apply the elementary inequality $a^\mu \le \delta^{-\frac{\mu}{p'-\mu}} + \delta a^{p'}$, where $a \ge 0, \delta > 0$, which follows from (4.47). $\square$

Remark 31 If $\beta < \frac{1}{ev_n}$, there is a simpler and more straightforward way of proving inequality (4.141), based on expanding the exponent and application of Young's inequality for convolutions (4.116).

Example 17 Let $1 < p < \infty$, $0 < \gamma < \frac{1}{p'}$ and $f(x) = |x|^{-\frac{n}{p}} |\ln|x||^{\gamma-1}$ if $0 < |x| \le \frac{1}{2}$ and $f(x) = 0$ if $|x| > \frac{1}{2}$. Then $f \in L_p(\mathbb{R}^n)$ and

$$M_1 |\ln|x||^\gamma \le |x|^{-\frac{n}{p'}} * f \le M_2 |\ln|x||^\gamma, \quad 0 < |x| \le \frac{1}{2},$$

where $M_1, M_2 > 0$ are independent of x.
Idea of the proof. To obtain the lower estimate it is convenient to estimate $|x|^{-\frac{n}{p'}} * f$ from below via the integral over $B(0, \frac{1}{2}) \setminus B(0, \frac{|x|}{2})$. To get the upper estimate one needs to split the integral defining $|x|^{-\frac{n}{p'}} * f$ into integrals over $B(0, \frac{1}{2}) \setminus B(0, 2|x|)$, $B(0, 2|x|) \setminus B(x, \frac{|x|}{2})$ and $B(x, \frac{|x|}{2})$ and to estimate them separately. $\square$

Remark 32 This example shows that the exponent p' in inequality (4.141) is sharp. Indeed, if $\mu > p'$, then for $\frac{1}{\mu} < \gamma < \frac{1}{p'}$ we have $f \in L_p(\mathbb{R}^n)$ but $\int_\Omega \exp\left(\beta| |x|^{-\frac{n}{p'}} * f|^\mu\right) dx = \infty$ for each $\beta > 0$ and for each compact $\Omega \subset \mathbb{R}^n$. A more sophisticated example can be constructed showing that for $\beta > \frac{1}{v_n}$ Theorem 11 does not hold.

4.6 Embeddings into the space of continuous functions

Theorem 12 *Let* $l \in \mathbb{N}$, $1 \le p < \infty$ *and let* $\Omega \subset \mathbb{R}^n$ *be an open set satisfying the cone condition. If*

$$l > \frac{n}{p} \text{ for } 1 < p < \infty, \quad l \ge n \text{ for } p = 1, \tag{4.142}$$

then each function $f \in W_p^l(\Omega)$ is equivalent to a function $g \in C_b(\Omega)$ and

$$\|g\|_{C(\Omega)} \leq c_{45} \|f\|_{W_p^l(\Omega)}, \tag{4.143}$$

where $c_{45} > 0$ is independent of f, i.e., $W_p^l(\Omega) \hookrightarrow C_b(\Omega)$.
If Ω is unbounded, then

$$\lim_{x \to \infty, \, x \in \Omega} g(x) = 0. \tag{4.144}$$

Idea of the proof. By Theorem 1 (4.143) is equivalent to the inequality

$$\|f\|_{L_\infty(\Omega)} = \|g\|_{C(\Omega)} \leq c_{45} \|f\|_{W_p^l(\Omega)}$$

for all $f \in W_p^l(\Omega) \cap C^\infty(\Omega)$. Since $\|f\|_{C(\Omega)} = \sup_{x \in \Omega} \|f\|_{C(K_x)}$, where K_x are the cones of the cone contition, which are congruent to the cone K defined by (3.34), it is enough to prove that

$$\|f\|_{C(K)} \leq c_{45} \|f\|_{W_p^l(K)}. \tag{4.145}$$

To prove (4.145) apply inequality (3.76). In the case of unbounded open sets Ω apply inequality (3.77) to prove (4.144). $\square$

Proof. By (3.76) where $\beta = 0$ for $\forall f \in W_p^l(\Omega) \cap C^\infty(\Omega)$ and $\forall x \in K$

$$|f(x)| \leq M_1 \left(\int_K |f|\, dx + \sum_{|\alpha|=l} \int_K \frac{|(D^\alpha f)(y)|}{|x-y|^{n-l}}\, dy \right),$$

where M_1 is independent of f and x. Hence, by Hölder's inequality,

$$|f(x)| \leq M_1 \left(\left(\operatorname{meas} K \right)^{\frac{1}{p'}} \|f\|_{L_p(K)} + \sum_{|\alpha|=l} \| \, |x-y|^{l-n} \|_{L_{p',y}(K)} \|D^\alpha f\|_{L_p(K)} \right).$$

Let D be the diameter of K $(D = \sqrt{h^2 + r^2})$. If $1 < p < \infty$ and $l > \frac{n}{p}$, then applying (4.116), we have $\forall x \in K$

$$\| \, |x-y|^{l-n} \|_{L_{p',y}(K)} \leq \| \, |z|^{l-n} \|_{L_{p'}(B(0,D))}$$

$$= \left(\sigma_n \int_0^D \varrho^{(l-n)p'+n-1}\, d\varrho \right)^{\frac{1}{p'}} = \left(\frac{\sigma_n}{p'(l-\frac{n}{p})} \right)^{\frac{1}{p'}} D^{l-\frac{n}{p}}.$$

If $p = 1$ and $l \geq n$, then

$$\| \, |x - y|^{l-n} \|_{L_{\infty,y}(K)} \leq D^{l-n}.$$

Consequently,

$$|f(x)| \leq M_2 \|f\|_{W_p^l(K)},$$

where M_2 is independent of f and x, hence, (4.145) and (4.143) follow.

If Ω is unbounded and $f \in W_p^l(\Omega)$, then, applying inequality (3.77) where $\beta = 0$ to the function g in (4.143), we get that $\forall x \in \Omega$

$$|g(x)| \leq M_2 \left(\int\limits_{K_x} |f| \, dy + \sum_{|\alpha|=l} \int\limits_{K_x} \frac{|(D_w^\alpha f)(y)|}{|x-y|^{n-l}} \, dy \right)$$

$$\leq M_3 \|f\|_{W_p^l(K_x)} \leq M_3 \|f\|_{W_p^l(\Omega \setminus B(0,|x|-D))}$$

if $|x| > D$, where M_2, M_3 are independent of f and x . Therefore, $\lim\limits_{x \to \infty} g(x) = 0$. $\square$

Corollary 20 *If Ω satisfies a Lipschitz condition, then the function $g \in \overline{C}(\Omega)$.*

Idea of the proof. For $\Omega = \mathbb{R}^n$ apply (4.143) to $f - f_k$, where the functions $f_k \in C_0^\infty(\mathbb{R}^n)$ converge to the function f in $W_p^l(\mathbb{R}^n)$. If Ω satisfies a Lipschitz condition, apply the extension Theorem 3 of Chapter 6. $\square$

Proof. If $\Omega = \mathbb{R}^n$, then from (4.143) it follows that $\|f_k - g\|_{C(\mathbb{R}^n)} \to 0$ as $k \to \infty$. Hence, $g \in \overline{C}(\mathbb{R}^n)$. Let Ω satisfy a Lipschitz condition and T be an extension operator in Theorem 3 of Chapter 6. For $f \in W_p^l(\Omega)$ consider a sequence of functions $h_k \in C_0^\infty(\mathbb{R}^n)$ converging to $Tf \in W_p^l(\mathbb{R}^n)$. Then $h_k \to f$ in $W_p^l(\Omega)$ and by (4.143) $\|g - h_k\|_{C(\Omega)} \to 0$ as $k \to \infty$. Hence, again $g \in \overline{C}(\Omega)$. $\square$

Corollary 21 *Let $l, m \in \mathbb{N}$, $1 \leq p < \infty$ and let $\Omega \in \mathbb{R}^n$ be an open set satisfying the cone condition. If*

$$l > m + \tfrac{n}{p} \quad \text{for} \quad 1 < p < \infty, \quad l \geq m + n \quad \text{for} \quad p = 1, \tag{4.146}$$

then each function $f \in W_p^l(\Omega)$ is equivalent to a function $g \in C_b^m(\Omega)$ and for $\beta \in \mathbb{N}_0^n$ satisfying $|\beta| \leq m$

$$\|D^\beta g\|_{C(\Omega)} \leq c_{46} \|f\|_{W_p^l(\Omega)}, \tag{4.147}$$

where $c_{46} > 0$ is independent of f, i.e., $W_p^l(\Omega) \hookrightarrow C_b^m(\Omega)$.

If Ω is unbounded, then

$$\lim_{x \to \infty, \, x \in \Omega} (D^\beta g)(x) = 0, \quad \beta \in \mathbb{N}_0^n, \quad |\beta| \leq m.$$

Idea of the proof. It is enough to apply Theorem 12 to $D_w^\beta f$, where $|\beta| \le m$, since by Theorem 6 $D_w^\beta f \in W_p^{l-|\beta|}(\Omega)$ and inequality (4.105) holds.

We note that conditions (4.142) and (4.146) are also necessary for the validity of (4.143) , (4.147) respectively. (See the proof of Theorem 14 below.)

Corollary 22 *Let* $1 \le p \le \infty$ *and let* $\Omega \subset \mathbb{R}^n$ *be an arbitrary open set. If* $f \in \bigcap_{l=1}^{\infty} (W_p^l)^{loc}(\Omega)$, *then* f *is equivalent to a function* $g \in C^\infty(\Omega)$.

Idea of the proof. Apply Corollary 20. $\square$

Remark 33 There exists $d_0 > 0$ depending only on n, l and p such that for convex domains Ω satisfying $D = \operatorname{diam}\Omega \le d_0$

$$\|g\|_{C(\Omega)} \le (\operatorname{meas}\Omega)^{-\frac{1}{p}} \|f\|_{\widetilde{W}_p^l(\Omega)},$$

where $\|f\|_{\widetilde{W}_p^l(\Omega)}$ is the norm defined by (4.110), equivalent to $\|f\|_{W_p^l(\Omega)}$ (coinciding if $l = 1$). The constant $(\operatorname{meas}\Omega)^{-\frac{1}{p}}$ is sharp since for $f \equiv 1$ equality holds.

This inequality follows from the proof of Theorem 12 if to start from the integral representation (3.65). Let, for $\sigma \in S^{n-1}$, where S^{n-1} is the unit sphere in $\mathbb{R}^n$, $r(x,\sigma)$ be the length of the segment of the ray $\{z \in \mathbb{R}^n : z = x + \varrho\sigma, 0 < \varrho < \infty\}$ contained in $\overline{\Omega}$. Then, for $d(x,y)$ defined in Corollary 13 of Chapter 3, we have $d(x, x + \varrho\sigma) = r(x,\sigma)$. Hence

$$\left\| \frac{d^n(x,y)}{|x-y|^{n-l}} \right\|_{L_{p',y}(\Omega)} = \left(\int\limits_{S^{n-1}} \left(\int\limits_0^{r(x,\sigma)} \left(\frac{d(x,x+\varrho\sigma)}{\varrho^{n-l}} \right)^{p'} \varrho^{n-1}\, d\varrho \right) d\sigma \right)^{\frac{1}{p'}}$$

$$= \left(p'(l - \tfrac{n}{p}) \right)^{-\frac{1}{p'}} \left(\int\limits_{S^{n-1}} (r(x,\sigma))^{lp'+n}\, d\sigma \right)^{\frac{1}{p'}} \le \left(p'(l - \tfrac{n}{p}) \right)^{-\frac{1}{p'}} D^l (\operatorname{meas}\Omega)^{\frac{1}{p'}}$$

since

$$\frac{1}{n} \int\limits_{S^{n-1}} (r(x,\sigma))^n\, d\sigma = \operatorname{meas}\Omega.$$

Thus, by (3.65) and Hölder's inequality,

$$\|g\|_{C(\Omega)} \le (\operatorname{meas}\Omega)^{-\frac{1}{p}} \left(\sum_{|\alpha|<l} \frac{D^{|\alpha|}}{\alpha!} \|D_w^\alpha f\|_{L_p(\Omega)} \right)$$

$$+\frac{l}{n}\left(p'\left(l-\frac{n}{p}\right)\right)^{-\frac{1}{p'}}\sum_{|\alpha|=l}\frac{D^{|\alpha|}}{\alpha!}\|D_w^{\alpha}f\|_{L_p(\Omega)}\Bigg),$$

and the desired statement follows. In the simplest case $p=1$, $l=n$ this inequality takes the form (3.66) and, hence, one can take $d_0=1$.

Remark 34 There is one more case, in which the sharp constant in the inequality of the type (4.143) can be computed explicitly, namely $\Omega=\mathbb{R}^n$, $p=2$, $l>\frac{n}{2}$. In this case

$$\|g\|_{C(\mathbb{R}^n)}\leq\left((2\pi)^{-n}v_n\frac{\frac{\pi n}{2l}}{\sin\frac{\pi n}{2l}}\right)^{\frac{1}{2}}\|f\|_{W_2^l(\mathbb{R}^n)}^{(2)}.$$

(See Remark 8 of Chapter 1.) Equality holds if, and only if, for some $A\in\mathbb{C}$

$$f(x)=A(F^{-1}((1+|\xi|^{2l})^{-1}))(x)$$

(if $l=n=1$, then $f(x)=B\exp(-|x|)$) for almost all $x\in\mathbb{R}^n$. If $l=n=1$, this inequality coincides with (4.75) where $p=2$. This follows since $\forall f\in W_2^l(\mathbb{R}^n)$

$$\|f\|_{L_\infty(\mathbb{R}^n)}=\|F^{-1}Ff\|_{L_\infty(\mathbb{R}^n)}=(2\pi)^{-\frac{n}{2}}\left\|\int_{\mathbb{R}^n}e^{ix\cdot\xi}(Ff)(\xi)\,d\xi\right\|_{L_\infty(\mathbb{R}^n)}$$

$$\leq(2\pi)^{-\frac{n}{2}}\|Ff\|_{L_1(\mathbb{R}^n)}=(2\pi)^{-\frac{n}{2}}\left\|(1+|\xi|^{2l})^{-\frac{1}{2}}(1+|\xi|^{2l})^{\frac{1}{2}}(Ff)(\xi)\right\|_{L_1(\mathbb{R}^n)}$$

$$\leq(2\pi)^{-\frac{n}{2}}\left\|(1+|\xi|^{2l})^{-\frac{1}{2}}\right\|_{L_2(\mathbb{R}^n)}\left\|(1+|\xi|^{2l})^{\frac{1}{2}}(Ff)(\xi)\right\|_{L_2(\mathbb{R}^n)}$$

$$=(2\pi)^{-\frac{n}{2}}\left(\int_{\mathbb{R}^n}(1+|\xi|^{2l})^{-1}\,d\xi\right)^{\frac{1}{2}}\left(\int_{\mathbb{R}^n}\left(|f|^2+|\nabla^l f|^2\right)\,dx\right)^{\frac{1}{2}}.$$

The desired inequality follows since by (4.116)

$$\int_{\mathbb{R}^n}(1+|\xi|^{2l})^{-1}\,d\xi=\sigma_n\int_0^\infty(1+\varrho^{2l})^{-1}\varrho^{n-1}\,d\varrho\,\Bigg|_{(1+\varrho^{2l})^{-1}=t}$$

$$=\frac{\sigma_n}{2l}\int_0^1 t^{-\frac{n}{2l}}(1-t)^{\frac{n}{2l}-1}\,dt=\frac{\sigma_n}{2l}B\left(1-\frac{n}{2l},\frac{n}{2l}\right)$$

$$=v_n\frac{n}{2l}\Gamma\left(\frac{n}{2l}\right)\Gamma\left(1-\frac{n}{2l}\right)=v_n\frac{\pi n}{2l}\left(\sin\frac{\pi n}{2l}\right)^{-1}.$$

In the second inequality equality holds, if, and only if, for some $A\in\mathbb{C}$ we have $(Ff)(\xi)=A(1+|\xi|^{2l})^{-1}$ for almost all $\xi\in\mathbb{R}^n$. (See footnote 11.) Since $|\int_{\mathbb{R}^n}(Ff)(\xi)\,d\xi|=\|Ff\|_{L_1(\mathbb{R}^n)}$, equality holds also in the first inequality.

4.7   Embeddings into the space L_q

Theorem 13 *Let $l \in \mathbb{N}$, $1 \le p < \infty$, $l < \frac{n}{p}$ and let $\Omega \subset \mathbb{R}^n$ be an open set satisfying the cone condition. Moreover,* [24] *let q_* be defined by*

$$l = n(\tfrac{1}{p} - \tfrac{1}{q_*}). \tag{4.148}$$

Then for each function $f \in W_p^l(\Omega)$

$$\|f\|_{L_{q_*}(\Omega)} \le c_{47} \|f\|_{W_p^l(\Omega)} , \tag{4.149}$$

where $c_{47} > 0$ is independent of f, i.e, $W_p^l(\Omega) \hookrightarrow L_{q_*}(\Omega)$.

Idea of the proof. By Lemma 10 it is enough to prove (4.149) for bounded domains Ω with star-shaped with respect to a ball. Apply inequality (3.54) and Theorem 10 to prove (4.149) for such Ω. $\square$

First proof ($p > 1$). Let Ω be a bounded domain star-shaped with respect to the ball $B \equiv B(x_0, \frac{d}{2})$ and let $\operatorname{diam}\Omega = D$. By (3.54) where $\beta = 0$

$$|f(x)| \le M_1\left(\int_B |f|\, dy + \sum_{|\alpha|=l} \int_{V_x} \frac{|(D_w^\alpha f)(y)|}{|x-y|^{n-l}}\, dy \right) \tag{4.150}$$

for almost all $x \in \Omega$, where M_1 depends only on n, l, d and D. By Hölder's inequality $\int_B |f|\, dy \le (\operatorname{meas} B)^{\frac{1}{p'}} \|f\|_{L_p(\Omega)}$. Hence

$$\|f\|_{L_{q_*}(\Omega)} \le M_2\left(\|f\|_{L_p(\Omega)} + \sum_{|\alpha|=l} \left\| \int_{\mathbb{R}^n} \frac{\Phi_\alpha(y)}{|x-y|^{n(\frac{1}{p'}+\frac{1}{q_*})}}\, dy \right\|_{L_{q_*}(\mathbb{R}^n)} \right),$$

where $\Phi_\alpha(y) = |(D_w^\alpha f)(y)|$ if $y \in \Omega$ and $\Phi_\alpha(y) = 0$ if $y \notin \Omega$ and M_2 depends only on n, l, p, d and D. By Theorem 10

$$\left\| \int_{\mathbb{R}^n} \frac{\Phi_\alpha(y)}{|x-y|^{n(\frac{1}{p'}+\frac{1}{q_*})}}\, dy \right\|_{L_{q_*}(\mathbb{R}^n)} \le M_3 \|\Phi_\alpha\|_{L_p(\mathbb{R}^n)} = M_3 \|D_w^\alpha f\|_{L_p(\Omega)},$$

where M_3 depends only on n, l and p. Thus (4.149) follows, where c_{47} depends only on n, l, p, d and D. Hence, by Lemma 10, the statement of Theorem 13 follows. $\square$

---

[24] Often q_* is called "the limiting exponent."

Remark 35 It is also possible to start, without using Lemma 10, from inequality (3.76) and argue in a similar way. (If Ω is unbounded, one should take into account that, by Remark 21 of Chapter 3, in (3.76) $\|f\|_{L_q(\mathbb{R}^n)}$ can be replaced by $\|f\|_{L_p(\Omega)}$.)

Remark 36 By Hölder's inequality and by the interpolation inequality

$$\|f\|_{L_q(\Omega)} \leq \|f\|_{L_p(\Omega)}^{\theta}\|f\|_{L_{q_*}(\Omega)}^{1-\theta}, \tag{4.151}$$

where $p < q < q_*$ and $\frac{1}{q} = \frac{\theta}{p} + \frac{1-\theta}{q_*}$, it follows that inequality (4.149) holds for open sets Ω satisfying the cone condition if q_* is replaced by q, where $1 \leq q < q_*$ for bounded sets Ω, $p \leq q < q_*$ for unbounded sets Ω respectively.

This statement may be proved, including the case $p = 1$, by simpler means – just by applying Young's inequality. By Lemma 10 it is enough to consider the case of bounded domains star-shaped with respect to a ball. Starting starting from (4.150) and (4.155), it is sufficient to note that

$$\left\| \int_{\Omega} \frac{|(D_w^\alpha f)(y)|}{|x - y|^{n-l}} \, dy \right\|_{L_q(\Omega)} \leq \big\| |z|^{l-n} \big\|_{L_r(\Omega-\Omega)} \|D_w^\alpha f\|_{L_p(\Omega)}$$

and by (4.116)

$$\big\| |z|^{l-n} \big\|_{L_r(\Omega-\Omega)} \leq \big\| |z|^{l-n} \big\|_{L_r(B(0,2D))} = \left(\sigma_n \int_0^{2D} \varrho^{(l-n)r+n-1} \, d\varrho \right)^{\frac{1}{r}}$$

$$= \left(\frac{\sigma_n}{l-n(\frac{1}{p}-\frac{1}{q})} \right)^{1-\frac{1}{p}+\frac{1}{q}} (2D)^{l-n(\frac{1}{p}-\frac{1}{q})} < \infty.$$

Remark 37 For $p = 1$ inequality (4.149) cannot be proved by applying Theorem 12, which does not hold for $p = 1$. Moreover, in this case inequality (4.149) does not follow from (4.150). More than that, inequality (4.149) can not be proved by estimating separately the L_{q_*}-norms of each summand in the remainder of Sobolev's integral representation (3.51) and not taking into account that $D_w^\alpha f$ are not arbitrary functions in $L_p(\Omega)$, but are the weak derivatives of a function $f \in L_p(\Omega)$. For, let

$$(K_\alpha \varphi)(y) = \int_{V_x} \frac{w_\alpha(x, y)}{|x - y|^{n-l}} \varphi(y) \, dy.$$

Then by (4.141)

$$\|K_\alpha\|_{L_1(\Omega)\to L_{q_*}(\Omega)} = \left\|\left\|\frac{w_\alpha(x,y)\chi_{V_x}(y)}{|x-y|^{\frac{n}{q_*}}}\right\|_{L_{q_*,x}(\Omega)}\right\|_{L_{\infty,y}(\Omega)}$$

and one can prove that $\|K_\alpha\|_{L_1(\Omega)\to L_{q_*}(\Omega)} = \infty$.

Idea of the second proof of Theorem 13. 1. Verify that by Lemma 11 and Theorem 3 of Chapter 6, it is enough to prove inequality (4.149) for $\Omega = \mathbb{R}^n$.

2. Let $\Omega = \mathbb{R}^n$, $n > 1$. First suppose that $l = 1$ and $p = 1$. Then $q_* = \frac{n}{n-1}$. Starting from the inequality

$$|f|^{\frac{n}{n-1}} = \prod_{m=1}^{n} |f|^{\frac{1}{n-1}} \leq \prod_{m=1}^{n} \|f\|_{L_{\infty,x_i}(\mathbb{R})}^{\frac{1}{n-1}}, \tag{4.152}$$

which holds almost everywhere on $\mathbb{R}^n$, apply the one-dimensional embedding inequality (4.64) and the following variant of Hölder's inequality for the product of functions $g_m \equiv g(x_1, ..., x_{m-1}, x_{m+1}, ..., x_n)$, which are independent of m-th variable: [25]

$$\left\|\prod_{m=1}^{n} g_m\right\|_{L_1(\mathbb{R}^n)} \leq \prod_{m=1}^{n} \|g_m\|_{L_{n-1}(\mathbb{R}_m^{n-1})}. \tag{4.153}$$

Here $\mathbb{R}_m^{n-1}$ is a space of $(x_1, ..., x_{m-1}, x_{m+1}, ..., x_n)$ where $x_j \in \mathbb{R}$. Obtain for $f \in W_p^1(\mathbb{R}^n)$ the inequality

$$\|f\|_{L_{\frac{n}{n-1}}(\mathbb{R}^n)} \leq \tfrac{1}{2} \prod_{m=1}^{n} \left\|\left(\frac{\partial f}{\partial x_m}\right)_w\right\|_{L_1(\mathbb{R}^n)}^{\frac{1}{n}}. \tag{4.154}$$

---

[25] This inequality can be easily proved by induction. On the other hand, it is a particular case of Hölder's inequality for mixed $L_{\bar{p}}$-norms, where $\bar{p} = (p_1, ..., p_n)$ and

$$\|f\|_{L_{\bar{p}}(\mathbb{R}^n)} = \|\cdots\|f\|_{L_{p_1,x_1}(\mathbb{R})}\cdots\|_{L_{p_n,x_n}(\mathbb{R})},$$

which has the form

$$\left\|\prod_{m=1}^{k} f_m\right\|_{L_1(\mathbb{R}^n)} \leq \prod_{m=1}^{k} \|f_m\|_{L_{\bar{p}_m}(\mathbb{R}^n)},$$

where $\bar{p}_m = (p_{1m}, ..., p_{nm})$ are such that $1 \leq p_{jm} \leq \infty$ and $\sum_{j=1}^{n} \frac{1}{p_{jm}} = 1, m = 1, ..., k$. It is proved by successive application of the one-dimensional Hölder's inequality. If $p_{jm} = n - 1$ for $j \neq m$ and $p_{mm} = \infty$, we obtain (4.153).

3. If $l = 1$ and $p > 1$, apply (4.154) to $|f|^\xi$ with appropriate $\xi > 0$ and prove that for $f \in C_0^\infty(\mathbb{R}^n)$

$$\|f\|_{L_{\frac{n}{n-1}}(\mathbb{R}^n)} \leq \frac{(n-1)p}{2(n-p)} \prod_{m=1}^{n} \left\| \frac{\partial f}{\partial x_m} \right\|_{L_p(\mathbb{R}^n)}^{\frac{1}{n}}. \tag{4.155}$$

4. If $l > 1$, apply induction and, finally, Lemma 2 of Chapter 2. $\square$

Second proof $(p \geq 1)$. 1. First let Ω be an open elementary domain with a Lipschitz boundary with the parameters d, D and M. (See Section 4.3.) Moreover, let T be an extension operator constructed in the proof in Theorem 3 of Chapter 6. By inequality (4.149) where $\Omega = \mathbb{R}^n$ we get

$$\|f\|_{L_{q_*}(\Omega)} \leq \|Tf\|_{L_{q_*}(\mathbb{R}^n)} \leq M_1 \|Tf\|_{W_p^l(\mathbb{R}^n)} \leq M_1 \|T\| \|f\|_{W_p^l(\Omega)} \leq M_2 \|f\|_{W_p^l(\Omega)}.$$

Here M_1 depends only on n, l, p and M_2 depends only on n, l, p, d, D and M. Since $q_* > p$, by Lemma 11 inequality (4.149) holds for each open set Ω satisfying the cone condition.

2. Now let $\Omega = \mathbb{R}^n$. First suppose that $l = 1$, $p = 1$ and let $f \in W_1^1(\mathbb{R}^n)$. By (4.152), (4.64) and (4.153) we have

$$\|f\|_{L_{\frac{n}{n-1}}(\mathbb{R}^n)} = \left\| |f|^{\frac{n}{n-1}} \right\|_{L_1(\mathbb{R}^n)}^{\frac{n-1}{n}} \leq \left\| \prod_{m=1}^{n} \|f\|_{L_{\infty,x_m}(\mathbb{R})}^{\frac{1}{n-1}} \right\|_{L_1(\mathbb{R}^n)}^{\frac{n-1}{n}}$$

$$\leq \frac{1}{2} \left\| \prod_{m=1}^{n} \left\| \left(\frac{\partial f}{\partial x_m}\right)_w \right\|_{L_{1,x_m}(\mathbb{R})}^{\frac{1}{n-1}} \right\|_{L_1(\mathbb{R}^n)}^{\frac{n-1}{n}}$$

$$\leq \frac{1}{2} \prod_{m=1}^{n} \left\| \left\| \left(\frac{\partial f}{\partial x_m}\right)_w \right\|_{L_{1,x_m}(\mathbb{R})} \right\|_{L_1(\mathbb{R}_m^{n-1})}^{\frac{1}{n}} = \frac{1}{2} \prod_{m=1}^{n} \left\| \left(\frac{\partial f}{\partial x_m}\right)_w \right\|_{L_1(\mathbb{R}^n)}^{\frac{1}{n}}.$$

3. Let $l = 1$ and $1 < p < n$, then $q_* = \frac{np}{n-p}$. Suppose that $f \in C_0^\infty(\mathbb{R}^n)$ and $f \not\equiv 0$. Since for $\xi > 0$ $\left|\left(\frac{\partial(|f|^\xi)}{\partial x_j}\right)_w\right| = \xi |f|^{\xi-1} \left|\frac{\partial f}{\partial x_j}\right|$, applying (4.154) to $|f|^\xi$, where $\xi = \frac{n-1}{n} q_*$, we have

$$\|f\|_{L_{q_*}(\mathbb{R}^n)} = \left\| |f|^\xi \right\|_{L_{\frac{q_*}{\xi}}(\mathbb{R}^n)}^{\frac{1}{\xi}} = \left\| |f|^\xi \right\|_{L_{\frac{n}{n-1}}(\mathbb{R}^n)}^{\frac{1}{\xi}}$$

$$\leq \left(\frac{1}{2}\right)^{\frac{1}{\xi}} \prod_{m=1}^{n} \left\| \left(\frac{\partial |f|^\xi}{\partial x_m}\right)_w \right\|_{L_1(\mathbb{R}^n)}^{\frac{1}{n\xi}} = \left(\frac{\xi}{2}\right)^{\frac{1}{\xi}} \prod_{m=1}^{n} \left\| |f|^{\xi-1} \frac{\partial f}{\partial x_m} \right\|_{L_1(\mathbb{R}^n)}^{\frac{1}{n\xi}}.$$

By Hölder's inequality

$$\left\| |f|^{\xi-1} \frac{\partial f}{\partial x_m} \right\|_{L_1(\mathbb{R}^n)} \leq \left\| |f|^{\frac{q_*}{p'}} \frac{\partial f}{\partial x_m} \right\|_{L_1(\mathbb{R}^n)}$$

$$\leq \|f\|_{L_{q_*}(\mathbb{R}^n)}^{\frac{q_*}{p'}} \left\| \frac{\partial f}{\partial x_m} \right\|_{L_p(\mathbb{R}^n)}.$$

Hence .

$$\|f\|_{L_{q_*}(\mathbb{R}^n)} \leq \left(\frac{\xi}{2}\right)^{\frac{1}{\xi}} \|f\|_{L_{q_*}(\mathbb{R}^n)}^{\frac{q_*}{p'\xi}} \prod_{m=1}^{n} \left\| \frac{\partial f}{\partial x_m} \right\|_{L_p(\mathbb{R}^n)}^{\frac{1}{n\xi}}.$$

Since $0 < \|f\|_{L_{q_*}(\mathbb{R}^n)} < \infty$ and $1 - \frac{q_*}{p'\xi} = \frac{1}{\xi}$, we obtain (4.155).

4. Next suppose that $p \geq 1, 1 < l < \frac{n}{p}$ and $f \in C_0^\infty(\mathbb{R}^n)$. We define the exponents $q_1, ..., q_{l-1}$ by $1 = n(\frac{1}{q_1} - \frac{1}{q_*})$, $1 = n(\frac{1}{q_2} - \frac{1}{q_1}), ..., 1 = n(\frac{1}{p} - \frac{1}{q_{l-1}})$. Applying (4.155) successively, we get

$$\|f\|_{L_{q_*}(\mathbb{R}^n)} \leq M_1 \sum_{m_1=1}^{n} \left\| \frac{\partial f}{\partial x_{m_1}} \right\|_{L_{q_1}(\mathbb{R}^n)} \leq \cdots$$

$$\leq M_l \sum_{m_1=1}^{n} \cdots \sum_{m_l=1}^{n} \left\| \frac{\partial^l f}{\partial x_{m_l} \cdots \partial x_{m_1}} \right\|_{L_p(\mathbb{R}^n)}$$

$$\leq M_{l+1} \sum_{|\alpha|=l} \|D^\alpha f\|_{L_p(\mathbb{R}^n)} = M_{l+1} \|f\|_{w_p^l(\mathbb{R}^n)},$$

where $M_1, ..., M_{l+1}$ depend only on n, l and p. Finally, taking into consideration Lemma 2 of Chapter 2 and passing to the limit, it follows that this inequality holds $\forall f \in W_p^l(\mathbb{R}^n)$. $\square$

Remark 38 Inequality (4.149) for $\Omega = \mathbb{R}^n$, $n > 1$, $l = 1$, $1 \leq p < n$ (steps 2 and 3 in the second proof of Theorem 13) can also be proved with the help of the *spherically symmetric rearrangements* f^* of functions $|f|$ defined by $f^*(x) = \sup\{t : \mu(t) > v_n |x|^n\}$, where $\mu(t) = \text{meas}\{x \in \mathbb{R}^n : |f(x)| > t\}$. Clearly $f^*(x) = g(|x|)$. The following properties of f^* are essential:

$$\|f^*\|_{L_p(\mathbb{R}^n)} = \|f\|_{L_p(\mathbb{R}^n)}, \quad 1 \leq p \leq \infty \tag{4.156}$$

and

$$\|\nabla_w f^*\|_{L_p(\mathbb{R}^n)} \leq \|\nabla_w f\|_{L_p(\mathbb{R}^n)}, \quad 1 \leq p \leq \infty. \tag{4.157}$$

Another tool is Hardy's inequality of the form

$$\|x^{\alpha + \frac{1}{p} - \frac{1}{q} - 1} f(x)\|_{L_q(0,\infty)} \leq c_{48} \|x^\alpha f'(x)\|_{L_p(0,\infty)}, \tag{4.158}$$

where $1 \le p \le q \le \infty$, $\alpha > \frac{1}{p'}$, f is locally absolutely continuous on $(0, \infty)$ and $\lim_{x \to \infty} f(x) = 0$. We note that for $q = q_* = \frac{np}{n-p}$, $1 \le p < n$ and $\alpha = \frac{n-1}{p}$ inequality (4.158) takes the form

$$\left(\int_0^\infty |f(x)|^{q_*} \, x^{n-1} \, dx \right)^{\frac{1}{q_*}} \le c_{49} \left(\int_0^\infty |f'(x)|^p \, x^{n-1} \, dx \right)^{\frac{1}{p}}. \tag{4.159}$$

Applying (4.156), (4.116) and (4.159), we have

$$\|f\|_{L_{q_*}(\mathbb{R}^n)} = \|f^*\|_{L_{q_*}(\mathbb{R}^n)} = \left(\sigma_n \int_0^\infty |g(\varrho)|^{q_*} \, \varrho^{n-1} \, d\varrho \right)^{\frac{1}{q_*}}$$

$$\le \sigma_n^{\frac{1}{q_*} - \frac{1}{p}} c_{49} \left(\sigma_n \int_0^\infty |g'(\varrho)|^p \, \varrho^{n-1} \, d\varrho \right)^{\frac{1}{p}} = \sigma_n^{-\frac{1}{n}} c_{49} \, \|\nabla_w f^*\|_{L_p(\mathbb{R}^n)}.$$

Hence by (4.157)

$$\|f\|_{L_{q_*}(\mathbb{R}^n)} \le c_{50} \, \|\nabla_w f\|_{L_p(\mathbb{R}^n)}, \tag{4.160}$$

where $c_{50} = \sigma_n^{-\frac{1}{n}} c_{49}$ and (4.149) follows.

Moreover, it is also possible to prove that the minimal value of c_{50} in (4.160) is equal to

$$\pi^{-\frac{1}{2}} n^{-\frac{1}{p}} \left(\frac{p-1}{n-p} \right)^{1-\frac{1}{p}} \left\{ \frac{\Gamma(\frac{n}{2} + 1)\Gamma(n)}{\Gamma(\frac{n}{p})\Gamma(1 + \frac{n}{p'})} \right\}^{\frac{1}{n}}.$$

(If $p = 1$, one must pass to the limit as $p \to 1+$.) In the case $p > 1$ equality in (4.160) holds if, and only if, for some a, $b > 0$ $|f(x)| = (a + b|x|^{\frac{p}{p-1}})^{1-\frac{n}{p}}$ almost everywhere on $\mathbb{R}^n$.

Remark 39 As in Remark 33 it can be proved that there exists d_0 depending only on n, l, p and q satisfying $1 \le q < q_*$ such that for convex domains Ω satisfying $D \equiv \text{diam}\,\Omega \le d_0$

$$\|f\|_{L_q(\Omega)} \le (\text{meas}\,\Omega)^{\frac{1}{q} - \frac{1}{p}} \|f\|_{\widetilde{W}_p^l(\Omega)},$$

where the constant $(\text{meas}\,\Omega)^{\frac{1}{q} - \frac{1}{p}}$ is sharp.

To obtain this inequality one should apply the inequality

$$\left(\int_G \left| \int_\Omega k(x, y) f(y) \, dy \right|^q dx \right)^{\frac{1}{q}} \le A \, \|f\|_{L_p(\Omega)},$$

where

$$A = \sup_{x \in G} \|k(x,\cdot)\|_{L_r(\Omega)}^{1-\frac{r}{q}} \sup_{y \in G} \|k(\cdot,y)\|_{L_r(G)}^{\frac{r}{q}},$$

$\Omega \subset \mathbb{R}^n$ and $G \subset \mathbb{R}^m$ are measurable sets, k is a measurable function on $G \times \Omega$, $1 \le p \le q \le \infty$ and $\frac{1}{r} = \frac{1}{p'} + \frac{1}{q}$. (The proof is similar to the standard proof of Young's inequality (4.115).)

As in Remark 33 one can prove that for convex domains Ω and $k(x,y) = \frac{d(x,y)^n}{|x-y|^{n-l}}$

$$A \le \left(r\left(l - n\left(\tfrac{1}{p} - \tfrac{1}{q}\right)\right)\right)^{-\frac{1}{r}} D^l (\operatorname{meas}\Omega)^{\frac{1}{p'}+\frac{1}{q}}.$$

Hence by (3.65)

$$\|f\|_{L_q(\Omega)} \le (\operatorname{meas}\Omega)^{\frac{1}{q}-\frac{1}{p}} \left(\sum_{|\alpha|<l} \frac{D^{|\alpha|}}{\alpha!} \|D_w^\alpha f\|_{L_p(\Omega)} \right.$$

$$\left. + \frac{l}{n}\left(r\left(l - n\left(\tfrac{1}{p} - \tfrac{1}{q}\right)\right)\right)^{-\frac{1}{r}} \sum_{|\alpha|=l}^{l} \frac{D^\alpha}{\alpha!} \|D_w^\alpha f\|_{L_p(\Omega)} \right),$$

and the desired statement follows.

Corollary 23 *Let $l \in \mathbb{N}$, $m \in \mathbb{N}_0$, $1 \le p,q < \infty$, $m < l < m + \frac{n}{p}$ and let $\Omega \subset \mathbb{R}^n$ be an open set satisfying the cone condition. Moreover, let q_* be defined by*

$$l = m + n\left(\tfrac{1}{p} - \tfrac{1}{q_*}\right), \tag{4.161}$$

$1 \le q \le q_*$ *if Ω is bounded and $p \le q \le q_*$ if Ω is unbounded.*
Then $\forall f \in W_p^l(\Omega)$ for $\beta \in \mathbb{N}_0^n$ satisfying $|\beta| = m$

$$\|D_w^\beta f\|_{L_q(\Omega)} \le c_{51} \|f\|_{W_p^l(\Omega)}, \tag{4.162}$$

where $c_{51} > 0$ is independent of f, i.e., $W_p^l(\Omega) \hookrightarrow W_q^m(\Omega)$.

Idea of the proof. Apply Theorem 6 and 13 to $D_w^\beta f$, Hölder's inequality if Ω is bounded, and the interpolation inequality (4.151) if Ω is unbounded. $\square$

Corollary 24 *Let $l \in \mathbb{N}$, $1 \le p \le q \le \infty$, $l > m + n\left(\tfrac{1}{p} - \tfrac{1}{q}\right)$, $\varepsilon_0 > 0$ and let $\Omega \subset \mathbb{R}^n$ be an open set satisfying the cone condition. Then $\forall f \in W_p^l(\Omega)$ for $\beta \in \mathbb{N}_0^n$ satisfying $|\beta| = m$*

$$\|D_w^\beta f\|_{L_q(\Omega)} \le c_{52}\, \varepsilon^{-\frac{m+n\left(\frac{1}{p}-\frac{1}{q}\right)}{l-m-n\left(\frac{1}{p}-\frac{1}{q}\right)}} \|f\|_{L_p(\Omega)} + \varepsilon \|f\|_{w_p^l(\Omega)}, \tag{4.163}$$

where $0 < \varepsilon \leq \varepsilon_0$ and $c_{52} > 0$ is independent of f and ε. Furthermore,

$$\|D_w^\beta f\|_{L_q(\Omega)} \leq c_{53} \|f\|_{L_p(\Omega)}^{\frac{1}{l}(l-m-n(\frac{1}{p}-\frac{1}{q}))} \|f\|_{W_p^l(\Omega)}^{\frac{1}{l}(m+n(\frac{1}{p}-\frac{1}{q}))}, \tag{4.164}$$

where $c_{53} > 0$ is independent of f.

If $\Omega = \mathbb{R}^n$ or, more generally, Ω is an arbitrary infinite cone defined by

$$\Omega = \left\{ x \in \mathbb{R}^n : x = \varrho\sigma, \, 0 < \varrho < \infty, \, \sigma \in S \right\}, \tag{4.165}$$

where S is an arbitrary open (with respect to S^{n-1}) subset of S^{n-1}, then inequality (4.163) holds for an arbitrary $\varepsilon > 0$ and in inequality (4.164) $\|f\|_{W_p^l(\Omega)}$ can be replaced by $\|f\|_{w_p^l(\Omega)}$.

Idea of the proof. If Ω has the the form (4.165), then inequality (4.163) may be obtained by applying (4.162) to $f(\varepsilon x)$, $\varepsilon > 0$, since $\varepsilon\Omega = \Omega$. Inequality (4.164) follows from (4.163) by minimization with respect to ε. To obtain (4.163) for an Ω having a Lipschitz boundary, apply the extension theorem of Chapter 6 (Theorem 3 and Remark 16) and (4.163) for $\Omega = \mathbb{R}^n$. If Ω satisfies the cone condition, apply, in addition, Lemma 6 and Corollary 13. Inequality (4.164) is derived from (4.163) as in the proof of the one-dimensional inequality (4.43). $\square$

Proof. If Ω has a Lipschitz boundary, then by Theorem 3 and Remark 16 of Chapter 6, for all $\gamma > 0$,

$$\|D_w^\beta f\|_{L_q(\Omega)} \leq \|D_w^\beta T f\|_{L_q(\mathbb{R}^n)} \leq M_1 \gamma^{-\delta} \|T f\|_{L_p(\mathbb{R}^n)} + \gamma \|T f\|_{w_p^l(\mathbb{R}^n)}$$

$$\leq M_1 \gamma^{-\delta} \|T\|_0 \|f\|_{L_p(\Omega)} + \gamma \|T\|_l \|f\|_{W_p^l(\Omega)}$$

$$= \left(M_1 \gamma^{-\delta} \|T\|_0 + \gamma \|T\|_l \right) \|f\|_{L_p(\Omega)} + \gamma \|T\|_l \|f\|_{w_p^l(\Omega)}.$$

Here $\delta = (m + n(\frac{1}{p} - \frac{1}{q}))(l - m - n(\frac{1}{p} - \frac{1}{q}))^{-1}$, T is the extension operator constructed in Theorem 3 of Chapter 6, $\|T\|_0$ – its norm as an operator acting from $L_p(\Omega)$ in $L_p(\mathbb{R}^n)$ and $\|T\|_l$ – its norm as an operator acting from $W_p^l(\Omega)$ to $W_p^l(\mathbb{R}^n)$. Both $\|T\|_0$ and $\|T\|_l$ depend only on n, l, p and the parameters of the Lipschitz boundary. Setting $\gamma\|T\|_l = \varepsilon$ and noticing that $M_1\gamma^{-\delta}\|T\|_0 + \gamma\|T\|_l \leq M_2\varepsilon^{-\delta}$ if $0 < \varepsilon \leq \varepsilon_0$, we get

$$\|D_w^\beta f\|_{L_q(\Omega)} \leq M_2 \varepsilon^{-\delta} \|f\|_{L_p(\Omega)} + \varepsilon \|f\|_{W_p^l(\Omega)},$$

where M_2 depends only on n, l, p, q, ε_0 and the parameters of the Lipschitz boundary.

Next suppose that Ω satisfies the cone condition. By Lemma 6 there exist elementary domains Ω_k, $k = \overline{1, s}$, such that $\Omega = \bigcup_k \Omega_k$, where $s \in \mathbb{N}$ for bounded Ω and $s = \infty$ for unbounded Ω. They have Lipschitz boundaries with the same parameters, and the multiplicity of the covering $\varkappa = \varkappa\left(\left\{\Omega_k\right\}_{k=1}^{s}\right)$ is finite if Ω is unbounded. Consequently for each $k = \overline{1, s}$

$$\|D_w^\beta f\|_{L_q(\Omega_k)} \leq M_2\,\varepsilon^{-\delta}\,\|f\|_{L_p(\Omega_k)} + \varepsilon\,\|f\|_{w_p^l(\Omega_k)}$$

and, by Corollary 13,

$$\|D_w^\beta f\|_{L_q(\Omega)} \leq \varkappa^{\frac{1}{q}}\left(M_2\,\varepsilon^{-\delta}\,\|f\|_{L_p(\Omega)} + \varepsilon\,\|f\|_{w_p^l(\Omega)}\right),$$

hence, (4.163) follows.

To prove (4.164) we set $\varepsilon_* = (\|f\|_{L_p(\Omega)}\,\|f\|_{w_p^l(\Omega)}^{-1})^\xi$, where $\xi = \frac{1}{l}\left(l - m - n\left(\frac{1}{p} - \frac{1}{q}\right)\right)$. If $\varepsilon_* \leq \varepsilon_0$, then (4.164) follows from (4.163) directly. If $\varepsilon_* > \varepsilon_0$, then $\|f\|_{w_p^l(\Omega)} \leq \varepsilon_0^{-\frac{1}{\xi}}\|f\|_{L_p(\Omega)}$ and by (4.163)

$$\|D_w^\beta f\|_{L_q(\Omega)} \leq M_4\,\|f\|_{L_p(\Omega)} \leq M_4\,\|f\|_{L_p(\Omega)}^\xi\,\|f\|_{W_p^l(\Omega)}^{1-\xi}\,,$$

where M_4 is independent of f. Hence (4.164) follows. $\square$

Corollary 25 *Let $l \in \mathbb{N}$, $1 \leq p < \infty$, $l < m + \frac{n}{p}$ and let Ω be defined by (4.165). Then $\forall f \in W_p^l(\Omega)$ for $\beta \in \mathbb{N}_0^n$ satisfying $|\beta| = m$*

$$\|D_w^\beta f\|_{L_{q_*}(\Omega)} \leq c_{51}\,\|f\|_{w_p^l(\Omega)}\,.$$

Idea of the proof. Applying (4.162) to $f(\varepsilon x)$ where $\varepsilon > 0$ work out that

$$\|D_w^\beta f\|_{L_q(\Omega)} \leq c_{51}\left(\varepsilon^{-l}\,\|f\|_{L_p(\Omega)} + \|f\|_{w_p^l(\Omega)}\right)$$

and pass to the limit as $\varepsilon \to \infty$. $\square$

Theorem 14 *Let $l \in \mathbb{N}$, $m \in \mathbb{N}_0$, $1 \leq p, q \leq \infty$ and let Ω be an open set satisfying the cone condition. Then the embedding*

$$W_p^l(\Omega) \hookrightarrow W_q^m(\Omega) \tag{4.166}$$

in the case of bounded Ω holds if, and only if,

$$l > m + \frac{n}{p} \quad \text{for} \quad q = \infty,\, 1 < p \leq \infty\,, \tag{4.167}$$

or

$$l \geq m + n(\tfrac{1}{p} - \tfrac{1}{q}) \quad \text{for} \quad q = \infty, p = 1 \text{ or } q < \infty, 1 \leq p \leq \infty. \qquad (4.168)$$

In the case of unbounded Ω if, and only if, in addition, $q \geq p$.

Moreover, embedding (4.166) is compact if, and only if, Ω is bounded and

$$l > m + n(\tfrac{1}{p} - \tfrac{1}{q}). \qquad (4.169)$$

Idea of the proof. Apply Corollaries 20 and 21, Example 8 of Chapter 1 and, for $q < p$, modify the function defined by (4.85). As for compactness, apply Corollaries 17 and 24 and modify the sequences defined by (4.86) and in Example 1. $\square$

Proof. 1. If conditions (4.167) or (4.168) are satisfied, then embedding (4.166) follows from Corollaries 20 and 21.

Let us assume without loss of generality that $0 \in \Omega$. Suppose that $l < m + n\left(\tfrac{1}{p} - \tfrac{1}{q}\right)$, then [26] there exists μ satisfying $l - \tfrac{n}{p} < \mu < n - \tfrac{n}{q}$, which is not a nonnegative integer. By Example 8 of Chapter 1 $|x|^\mu \in W_p^l(\Omega)$ but $|x|^\mu \notin W_q^m(\Omega)$, and it follows that embedding (4.166) does not hold. Next suppose that $l = m + \tfrac{n}{p}$, $q = \infty$ and $1 < p \leq \infty$. Let $0 < \nu < 1 - \tfrac{1}{p}$. By Example 8 of Chapter 1 $x_1^m(|\ln|x||)^\nu \in W_p^l(\Omega)$ but clearly this function does not belong to $W_\infty^m(\Omega)$. Hence again embedding (4.166) does not hold.

Let $q < p$ and let Ω be unbounded. Since Ω satisfies the cone condition, there exists $\varrho > 0$ and disjoint balls $B(x_k, \varrho) \subset \Omega$, $k \in \mathbb{N}$. We set $f(x) = \sum_{k=1}^{\infty} k^{-\frac{1}{q}} \eta(\tfrac{x - x_k}{\varrho})$, where $\eta \in C_0^\infty(\mathbb{R}^n)$, $\operatorname{supp}\eta \subset B(0,1)$ and $\eta \not\equiv 0$. Then, as in the proof of Theorem 5, $f \in W_p^l(\Omega)$ but $f \notin W_q^m(\Omega)$, hence embedding (4.166) does not hold.

2. If condition (4.169) is satisfied, then the compactness of embedding (4.166) follows from Corollaries 17 and 24.

If Ω is bounded and $l \leq m + n\left(\tfrac{1}{p} - \tfrac{1}{q}\right)$, consider the sequence $f_k(x) = k^{\frac{n}{p} - l}\eta(kx)$ where $k \in \mathbb{N}$. Then $\|f_k\|_{W_p^l(\Omega)} \leq \|\eta\|_{W_p^l(\mathbb{R}^n)}$. Suppose that, for some $g \in W_q^m(\Omega)$ and some subsequence f_{k_s}, $f_{k_s} \to g$ in $W_q^m(\Omega)$. Since $f_{k_s}(x) \to 0$ as $s \to \infty$ for all $x \neq 0$, it follows that $g \sim 0$. On the other hand, $\|f_{k_s}\|_{W_q^m(\Omega)} \geq k^{m+n(\frac{1}{p} - \frac{1}{q}) - l}\left\|\frac{\partial^m \eta}{\partial x_1^m}\right\|_{L_q(\mathbb{R}^n)}$. Hence $f_{k_s} \not\to 0$ in $W_q^m(\Omega)$.

---

[26] We note that the necessity of the inequality $l \geq m + n\left(\tfrac{1}{p} - \tfrac{1}{q}\right)$ also follows for $1 \leq p, q \leq \infty$ and $\Omega = \mathbb{R}^n$ by comparison of the differential dimensions of spaces $W_p^l(\mathbb{R}^n)$ and $W_q^m(\mathbb{R}^n)$. See footnote 14 of Chapter 1. With slight modifications a similar argument works for open sets $\Omega \neq \mathbb{R}^n$.

If Ω is unbounded, consider, as in step 1, the disjoint balls $B(x_k, \varrho)$ and set $f_k(x) = \eta\left(\frac{x-x_k}{\varrho}\right)$. As in Example 1, f_k does not contain a subsequence convergent in $W_q^m(\Omega)$. $\square$

Next let us consider in more detail the case $l = \frac{n}{p}$. By Theorem 14, for an open set Ω satisfying the cone condition, it follows that $W_p^{\frac{n}{p}}(\Omega) \subset L_q(\Omega)$ for each $p \le q < \infty$. However, $W_p^{\frac{n}{p}}(\Omega) \not\subset L_\infty$. This statement may be improved in the following way.

Theorem 15 *Let $1 < p < \infty$, $\frac{n}{p} \in \mathbb{N}$ and let Ω be a bounded open set satisfying the cone condition. Then there exists $c_{54} > 0$ depending on n, p and the parameters r, $h > 0$ of the cone condition such that $\forall f \in W_p^{\frac{n}{p}}(\Omega)$, $f \not\sim 0$,*

$$\int_\Omega \exp\left(c_{54}\left|\frac{f}{\|f\|_{W_p^{\frac{n}{p}}(\Omega)}}\right|^{p'}\right) dx < \infty. \tag{4.170}$$

Idea of the proof. Apply inequality (3.76), Remark 21 of Chapter 3 and Theorem 11. $\square$

Proof. By (3.76) and Remark 21 of Chapter 3 for almost every $x \in \Omega$

$$|f(x)| \le M_1\left(\|f\|_{L_p(\Omega)} + |x|^{\frac{n}{p'}} * \varphi\right),$$

where $\varphi(x) = \sum_{|\alpha|=\frac{n}{p}} |(D_w^\alpha f)(x)|$ for $x \in \Omega$ and $\varphi(x) = 0$ for $x \notin \Omega$. Since $\|\varphi\|_{L_p(\mathbb{R}^n)} \le \|f\|_{W_p^{\frac{n}{p}}(\Omega)}$, we have

$$\left(\frac{|f(x)|}{\|f\|_{W_p^{\frac{n}{p}}(\Omega)}}\right)^{p'} \le M_2\left(1 + \left(\frac{|x|^{\frac{n}{p'}} * \varphi}{\|\varphi\|_{L_p(\mathbb{R}^n)}}\right)^{p'}\right).$$

Here M_1, $M_2 > 0$ depend only on n, p and the parameters r, h of the cone condition. Hence inequality (4.170) where $c_{54} = M_2\, v_n^{-1}$ follows from (4.141). $\square$

Remark 40 The cone condition in Theorems 12–15 is not necessary but is sufficiently sharp, because for the domain considered in Example 6 of Chapter 3 these theorems do not hold for any $\gamma \in (0,1)$. See Remark 19 and Example 1 of Chapter 6.

Chapter 5

Trace theorems

5.1 Notion of the trace of a function

Let $f \in L_1^{loc}(\mathbb{R}^n)$ where $n > 1$. We would like to define the trace $\operatorname{tr} f \equiv \operatorname{tr}_{\mathbf{R}^m} f \equiv f\big|_{\mathbb{R}^m}$ of the function f on $\mathbb{R}^m$ where $1 \leq m < n$.

We shall represent each point $x \in \mathbb{R}^n$ as a pair $x = (u,v)$ where $u = (x_1, ..., x_m)$, $v = (x_{m+1}, ..., x_n)$ and suppose that $\mathbb{R}^m(v)$ is the m-dimensional subspace of points (u,v), where v is fixed and u runs through all possible values. We shall also write $\mathbb{R}^m$ for $\mathbb{R}^m(0)$ if this will not cause ambiguity.

If f is continuous, it is natural to define the trace $\operatorname{tr} f$ as a restriction of the function f: $(\operatorname{tr} f)(u) = f(u,0)$, $u \in \mathbb{R}^m$. However, this way of defining the trace does not make sense for an arbitrary function $f \in L_1^{loc}(\mathbb{R}^n)$, since actually it is defined only up to a set of n-dimensional measure zero. In fact, one can easily construct two functions $f, h \in L_1^{loc}(\mathbb{R}^n)$, which are equivalent on $\mathbb{R}^n$, but $f(u,0) \neq h(u,0)$ for all $u \in \mathbb{R}^m$. Finally, it is natural to define the traces themselves up to a set of m-dimensional measure zero.

The above is a motivation for the following requirements for the notion of the trace on $\mathbb{R}^n$ of a function $f \in L_1^{loc}(\mathbb{R}^n)$:

1) a trace $g \in L_1^{loc}(\mathbb{R}^n)$,

2) if $g \in L_1^{loc}(\mathbb{R}^m)$ is a trace of f, then $\psi \in L_1^{loc}(\mathbb{R}^m)$ is also a trace of f, if and only if, ψ is equivalent to g on $\mathbb{R}^m$,

3) if g is a trace of f and h is equivalent to f on $\mathbb{R}^n$, then g is also a trace of h,

4) if f is continious , then $f(u,0)$ is a trace of f.

Definition 1 *Let* $f \in L_1^{loc}(\mathbb{R}^n)$ *and* $g \in L_1^{loc}(R^m)$. *The function* g *is said to be a trace of the function* f *if there exists a function* h *equivalent to* f *on* $\mathbb{R}^n$, *which is such that* [1]

$$h(\cdot, v) \to g(\cdot) \quad \text{in} \quad L_1^{loc}(\mathbb{R}^m) \quad \text{as} \quad v \to 0. \tag{5.1}$$

Clearly the requirements 1)–4) are satisfied. In fact, if g is a trace of f and ψ is equivalent to g, then (5.1) implies $h(\cdot, v) \to \psi(\cdot)$ in $L_1^{loc}(\mathbb{R}^m)$ and ψ is also a trace of f. Next suppose that both g and ψ are traces of f, then we have (5.1) and also $H(\cdot, v) \to \psi(\cdot)$ in $L_1^{loc}(\mathbb{R}^m)$ as $v \to 0$ for some $H \sim f$ on $\mathbb{R}^n$. We note that for each compact $K \subset \mathbb{R}^m$

$$||g - \psi||_{L_1(K)} \leq ||h(\cdot, v) - g||_{L_1(K)}$$

$$+ ||h(\cdot, v) - H(\cdot, v)||_{L_1(K)} + ||H(\cdot, v) - \psi||_{L_1(K)}.$$

Since $h \sim H$ on $\mathbb{R}^n$, $h(\cdot, v) \sim H(\cdot, v)$ on $\mathbb{R}^m$ for almost all $v \in \mathbb{R}^{n-m}$. Hence, there exists a sequence $\{v_s\}_{s \in \mathbb{N}}$, $v_s \in R^{n-m}$, such that $v_s \to 0$ as $s \to \infty$ and

$$||g - \psi||_{L_1(K)} \leq ||h(\cdot, v_s) - g||_{L_1(K)} + ||H(\cdot, v_s) - \psi||_{L_1(K)}.$$

On letting $s \to \infty$, we establish that $g \sim \psi$ on $\mathbb{R}^m$.

Finally, if f is continuous, then

$$||f(u, v) - f(u, 0)||_{L_{1,u}(K)} \leq \operatorname{meas} K \max_{u \in K} |f(u, v) - f(u, 0)|.$$

Hence, $||f(\cdot, v) - f(\cdot, 0)||_{L_1(K)} \to 0$ as $v \to 0$ because f is uniformly continuous on $K \times \widetilde{B}_1$, where $\widetilde{B}_1$ is the unit ball in R^{n-m}. Thus, $f(\cdot, 0)$ is a trace of f.

Theorem 1 *Let $Z(\mathbb{R}^n)$ be a semi-normed space of functions defined on $\mathbb{R}^n$ such that*

   1) $Z(\mathbb{R}^n) \hookrightarrow L_1^{loc}(\mathbb{R}^n)$

and

   2) $C^\infty(\mathbb{R}^n) \bigcap Z(\mathbb{R}^n)$ *is dense in* $Z(\mathbb{R}^n)$.

Suppose that $1 \leq m < n$ and for each compact $K \subset \mathbb{R}^m$ there exists $c_1(K) > 0$ such that $\forall f \in C^\infty(\mathbb{R}^n) \cap Z(\mathbb{R}^n)$ and $\forall v \in R^{n-m}$ satisfying $|v| \leq 1$

$$||f(\cdot, v)||_{L_1(K)} \leq c_1(K) \, ||f||_{Z(\mathbb{R}^n)}. \tag{5.2}$$

Then $\forall f \in Z(\mathbb{R}^n)$ there exists a trace of f on $\mathbb{R}^m$.

---

[1] One may include the case $m = 0$, considering a number g satisfying $h(v) \to g$ as $v \to 0$.

Idea of the proof. Consider a function $f \in Z(R^n)$ and a sequence of functions $f_k \in C^\infty(\mathbb{R}^n) \cap Z(\mathbb{R}^n), k \in \mathbb{N}$ such that $f_k \to f$ in $Z(\mathbb{R}^n)$ as $k \to \infty$. Applying (5.2) to $f_k - f_s$, prove that $\forall v \in \mathbb{R}^{n-m} : |v| < 1$ there exists a function g_v defined on $\mathbb{R}^m$ such that $f_k(\cdot, v) \to g_v$ in $L_1^{loc}(\mathbb{R}^m)$ as $k \to \infty$. Define $h(u, v) = g_v(u), (u, v) \in \mathbb{R}^n$, and prove that the functions h and $g \equiv g_0$ satisfy Definition 1. $\square$

Proof. Let $B_r, \widetilde{B}_r$ be open balls in $\mathbb{R}^m, \mathbb{R}^{n-m}$ respectively, of radius r centered at the origin. By (5.2) with $f_k - f_s$ replacing f and $B_N, N \in \mathbb{N}$, replacing K, it follows that $f_k(\cdot, v) - f_s(\cdot, v) \to 0$ in $L_1(B_N)$ as $k, s \to \infty$ for all $v \in \widetilde{B}_1$ and all $N \in \mathbb{N}$. By completeness of $L_1(B_N)$ there exists a function $g_{v,N} \in L_1(B_N)$ such that $f_k(\cdot, v) \to g_{v,N}(\cdot)$ in $L_1(B_N)$. Consider any function $g_v \sim g_{v,N}$ on B_N for all $N \in \mathbb{N}$. Such a function exists because $g_{v,N} \sim g_{v,N+1}$ on B_N. This follows by passing to the limit in the inequality

$$||g_{v,N} - g_{v,N+1}||_{L_1(B_N)} \leq ||g_{v,N} - f_k(\cdot, v)||_{L_1(B_N)} + ||f_k(\cdot, v) - g_{v,N+1}||_{L_1(B_{N+1})}.$$

Clearly, $f_k(\cdot, v) \to g_v$ in $L_1^{loc}(\mathbb{R}^n)$ as $k \to \infty$ and, hence, for the function h, defined by $h(u, v) = g_v(u), \ (u, v) \in \mathbb{R}^n$, we have $f_k(\cdot, v) \to h(\cdot, v)$ in $L_1^{loc}(\mathbb{R}^m)$ for all $v \in \widetilde{B}_1$.

On the other hand, $f_k(\cdot, v) \to f(\cdot, v)$ for almost all $v \in \widetilde{B}_1$. This follows since by the Fatou and Fubini theorems and condition 1)

$$\int\limits_{\widetilde{B}_1} \left(\liminf_{k\to\infty} \int\limits_{B_N} |f_k(u, v) - f(u, v)| \, du \right) dv$$

$$\leq \liminf_{k\to\infty} \int\limits_{\widetilde{B}_1} \left(\int\limits_{B_N} |f_k(u, v) - f(u, v)| \, du \right) dv$$

$$= \lim_{k\to\infty} \int\limits_{B_N \times \widetilde{B}_1} |f_k(u, v) - f(u, v)| \, du \, dv = 0.$$

Thus $f(\cdot, v)$ is equivalent to $h(\cdot, v)$ on $\mathbb{R}^m$ for almost all $v \in \widetilde{B}_1$. Consequently, by Fubuni's theorem, [2] f is equivalent to h on $\mathbb{R}^m \times \widetilde{B}_1$.

Furthermore, by the continuity of a semi-norm, on letting $s \to \infty$ in (5.2), where f is replaced by $f_k - f_s$, we get

$$||f_k(\cdot, v) - h(\cdot, v)||_{L_1(K)} \leq c_1(K)||f_k - f||_{Z(\mathbb{R}^n)}.$$

---

[2] For, let $e_n = \{(u, v) \in \mathbb{R}^m \times \widetilde{B}_1 : f(u, v) \neq h(u, v)\}$ and $e_m(v) = \{u \in \mathbb{R}^m : f(u, v) \neq h(u, v)\}$. Then $\mathrm{meas}_n \, e_n = \int\limits_{\widetilde{B}_1} (\mathrm{meas}_m \, e_m(v)) \, dv = 0$.

Therefore

$$\|h(\cdot,v) - g\|_{L_1(K)} = \|h(\cdot,v) - h(\cdot,0)\|_{L_1(K)} \leq \|h(\cdot,v) - f_k(\cdot,v)\|_{L_1(K)}$$

$$+\|f_k(\cdot,v) - f_k(\cdot,0)\|_{L_1(K)} + \|f_k(\cdot,0) - h(\cdot,0)\|_{L_1(K)}$$

$$\leq 2\,c_1(K)\|f_k - f\|_{Z(\mathbb{R}^n)} + \operatorname{meas} K \max_{u \in K} |f_k(u,v) - f_k(u,0)|.$$

Given $\varepsilon > 0$, we choose $k_\varepsilon \in \mathbb{N}$ such that for $k = k_\varepsilon$ the first summand is less than $\frac{\varepsilon}{2}$. Since f_{k_ε} is uniformly continuous on $K \times \widetilde{B}_1$, there exists $\gamma = \gamma(\varepsilon) > 0$ such that for $|v| < \gamma$ the second summand is also less than $\frac{\varepsilon}{2}$ and $\|g(\cdot,v) - g\|_{L_1(K)} < \varepsilon$. Hence $\|h(\cdot,v) - g\|_{L_1(K)} \to 0$ as $v \to 0$ and $h(\cdot,v) \to g$ in $L_1^{loc}(\mathbb{R}^n)$.

Thus, by Definition 1, g is a trace on $\mathbb{R}^n$ of the function f. $\square$

Remark 1 On replacing f by f_k in (5.2) and letting $k \to \infty$, we establish that $\forall f \in Z(\mathbb{R}^n)$

$$\|\operatorname{tr} f\|_{L_1(K)} \leq c_1(K)\|f\|_{Z(\mathbb{R}^n)}.$$

Moreover, it follows that $\forall f_k \in C^\infty(\mathbb{R}^n) \bigcap Z(\mathbb{R}^n), k \in \mathbb{N}$, satisfying $f_k \to f$ in $Z(\mathbb{R}^n)$ as $k \to \infty$ we have $f_k(\cdot,0) \to \operatorname{tr} f$ in $L_1^{loc}(\mathbb{R}^m)$.

Corollary 1 *In addition to the assumptions of Theorem 1, let the following condition be satisfied*

3) if $f \in Z(\mathbb{R}^n)$, then $\forall v \in \mathbb{R}^{n-m}$ $f(\cdot, \cdot + v) \in Z(\mathbb{R}^n)$ and

$$\|f(\cdot, \cdot + v)\|_{Z(\mathbb{R}^n)} = \|f\|_{Z(\mathbb{R}^n)}.$$

Suppose that for each compact $K \subset \mathbb{R}^{n-m}$ there exists $c_2(K) > 0$ such that $\forall f \in C^\infty(\mathbb{R}^n) \bigcap Z(\mathbb{R}^n)$

$$\|f(\cdot,0)\|_{L_1(K)} \leq c_2(K)\,\|f\|_{Z(\mathbb{R}^n)}. \tag{5.3}$$

Then $\forall f \in Z(\mathbb{R}^n)$ there exists a trace on $\mathbb{R}^m$.

Idea of the proof. Given $f \in Z(\mathbb{R}^n)$, apply (5.3) to the function f_v, defined by $f_v(\cdot,\cdot) = f(\cdot, \cdot + v)$, which by condition 3) lies in $Z(\mathbb{R}^n)$, and verify that inequality (5.2) is satisfied for all $v \in \mathbb{R}^{n-m}$. $\square$

5.2 Existence of the traces on subspaces

Theorem 2 *Let $l, m, n \in \mathbb{N}, m < n$ and $1 \leq p \leq \infty$. Then traces on $\mathbb{R}^m$ exist for all $f \in W_p^l(\mathbb{R}^n)$ if, and only if,*

$$l > \tfrac{n-m}{p} \quad \text{for} \quad 1 < p \leq \infty, \quad l \geq n - m \quad \text{for} \quad p = 1, \tag{5.4}$$

i.e., if, and only if,

$$W_p^l(\mathbb{R}^{n-m}) \hookrightarrow C(\mathbb{R}^{n-m}). \tag{5.5}$$

Idea of the proof. If (5.4) is satisfied, write the inequality corresponding to embedding (5.5) for functions $f(u, \cdot)$ with fixed u, and take L_p-norms with respect to u. Next use Theorem 1. If (5.4) is not satisfied, starting from Example 8 of Chapter 1, construct counter-examples, considering the functions $f_\beta(u, v) = |v|^\beta \eta_1(u) \eta_2(v)$ if $l < \frac{n-m}{p}$ and $g_\gamma(u, v) = |\ln|v||^\gamma \eta_1(u) \eta_2(v)$ if $l = \frac{n-m}{p}, 1 < p < \infty$. Here $\eta_1 \in C_0^\infty(\mathbb{R}^m), \eta_2 \in C_0^\infty(\mathbb{R}^{n-m})$ are "cap-shaped" functions such that $\eta_1 = 1$ on $B_1, \eta_2 = 1$ on $\widetilde{B}_1$, where $B_1, \widetilde{B}_1$ are the unit balls in $\mathbb{R}^m, \mathbb{R}^{n-m}$ respectively. $\square$

Proof. Sufficiency. Let (5.4) be satisfied. First suppose that $1 \leq p < \infty$. Then $\forall f \in C^\infty(\mathbb{R}^n) \bigcap W_p^l(\mathbb{R}^n)$, by Theorem 12, we have that for almost all $u \in \mathbb{R}^m$

$$|f(u, 0)| \leq M_1 \Big(\|f(u, \eta)\|_{L_{p,\eta}(\mathbb{R}^{n-m})} + \sum_{|\gamma|=l} \|(D^{(0,\gamma)}f)(u, \eta)\|_{L_{p,\eta}(\mathbb{R}^{n-m})} \Big),$$

where $\gamma = (\gamma_{m+1}, ..., \gamma_n) \in \mathbb{N}_0^{n-m}$ and M_1 depends only on $n - m, p$ and l. By Fubuni's theorem both the left-hand and the right-hand sides are measurable with respect to u on $\mathbb{R}^m$. By Minkowski's inequality and Fubuni's theorem we get on taking L_p-norms

$$\|f(u, 0)\|_{L_{p,u}(\mathbb{R}^m)} \leq M_1 \Big(\| \, \|f(u, \eta)\|_{L_{p,\eta}(\mathbb{R}^{n-m})} \|_{L_{p,u}(\mathbb{R}^m)}$$

$$+ \sum_{|\gamma|=l} \| \, \|(D^{(0,\gamma)}f)(u, \eta)\|_{L_{p,\eta}(\mathbb{R}^{n-m})} \|_{L_{p,u}(\mathbb{R}^m)} \Big) \leq M_1 \|f\|_{W_p^l(\mathbb{R}^n)}.$$

Consequently, by Corollary 1, it follows that each function $f \in W_p^l(\mathbb{R}^n)$ has a trace on $\mathbb{R}^m$.

Necessity. Let $l < \frac{n-m}{p}$ and $l - \frac{n-m}{p} < \beta < 0$. Then, by Example 8 of Chapter 1, $f_\beta \in W_p^l(\mathbb{R}^n)$. On the other hand for each $g \in L_1^{loc}(\mathbb{R}^m)$ and $v \in \widetilde{B}_1$, by the triangle inequality,

$$\|f_\beta(\cdot, v) - g\|_{L_1(B_1)} \geq |v|^\beta \|\eta_1\|_{L_1(B_1)} - \|g\|_{L_1(B_1)} \to \infty$$

as $v \to 0$. Hence the trace of f_β does not exist. If $l = \frac{n-m}{p}, 1 < p < \infty$ and $0 < \gamma < 1 - \frac{1}{p}$, then, by Example 8 of Chapter 1, $g_\gamma \in W_p^l(\mathbb{R}^n)$, but a similar argument shows that the trace of g_γ on $\mathbb{R}^m$ does not exist. $\square$

Remark 2 Assume that (5.4) is satisfied. By Remark 1 it follows that for each $f \in W_p^l(\mathbb{R}^n)$ the trace $\operatorname{tr} f \in L_p(\mathbb{R}^m)$ and

$$\|\operatorname{tr} f\|_{L_p(\mathbb{R}^m)} \leq c_3 \|f\|_{W_p^l(\mathbb{R}^n)}, \tag{5.6}$$

where $c_3 > 0$ depends only on m, n, p and l. Moreover, if $f \in C^\infty(\mathbb{R}^n) \bigcap W_p^l(\mathbb{R}^n)$ are such that $f_k \to f$ in $W_p^l(\mathbb{R}^n)$, then $f(\cdot, 0) \to \operatorname{tr} f$ in $L_p(\mathbb{R}^m)$.

Thus, if we consider the *trace space*

$$\operatorname{tr}_{\mathbb{R}^m} W_p^l(\mathbb{R}^n) = \{\operatorname{tr} f, \ f \in W_p^l(\mathbb{R}^n)\}$$

$$= \{g \in L_1^{loc}(\mathbb{R}^n) : \exists f \in W_p^l(\mathbb{R}^n) : \operatorname{tr} f = g\},$$

then

$$\operatorname{tr}_{\mathbb{R}^m} W_p^l(\mathbb{R}^n) \subset L_p(\mathbb{R}^n). \tag{5.7}$$

The problem is to describe the trace space. In order to do this we need to introduce appropriate spaces with, in general, noninteger orders of smoothness.

5.3   Nikol'skiĭ-Besov spaces

It can be proved that for $l \in \mathbb{N}, 1 < p < \infty$ the definition of Sobolev spaces $W_p^l(\mathbb{R}^n)$ is equivalent to the following one: $f \in W_p^l(\mathbb{R}^n)$ if, and only if, f is measurable on $\mathbb{R}^n$ and [3]

$$\|f\|_{L_p(\mathbb{R}^n)} + \sup_{h \in \mathbb{R}^n, h \neq 0} \frac{\|\Delta_h^l f\|_{L_p(\mathbb{R}^n)}}{|h|^l} < \infty.$$

This definition can easily be extended to the case of an arbitrary positive l: one may define the space of functions f, measurable on $\mathbb{R}^n$, which are such that

$$\|f\|_{L_p(\mathbb{R}^n)} + \sup_{h \in \mathbb{R}^n, h \neq 0} \frac{\|\Delta_h^\sigma f\|_{L_p(\mathbb{R}^n)}}{|h|^l} < \infty,$$

where $\sigma \in \mathbb{N}$ and $0 < l \leq \sigma$.

This idea will be used in the forthcoming definition. However, for reasons, which will be clear later, in the case of integer l it will be supposed that $l < \sigma$

---

[3] One of the implications has been established in Corollary 8 of Chapter 3.

(as in the case of noninteger l). [4] Moreover, an additional parameter will be introduced, providing more delicate classification of the spaces with order of smoothness equal to l.

Definition 2 *Let* $l > 0, \sigma \in \mathbb{N}, \sigma > l, 1 \leq p, \theta \leq \infty$. *The function* f *belongs to the* Nikol'skiĭ-Besov space $B_{p,\theta}^l(\mathbb{R}^n)$ *if* f *is measurable on* $\mathbb{R}^n$ *and*

$$\|f\|_{B_{p,\theta}^l(\mathbb{R}^n)} = \|f\|_{L_p(\mathbb{R}^n)} + \|f\|_{b_{p,\theta}^l(\mathbb{R}^n)} < \infty,$$

where

$$\|f\|_{b_{p,\theta}^l(\mathbb{R}^n)} = \left(\int\limits_{\mathbb{R}^n} \left(\frac{\|\Delta_h^\sigma f\|_{L_p(\mathbb{R}^n)}}{|h|^l} \right)^\theta \frac{dh}{|h|^n} \right)^{\frac{1}{\theta}} \tag{5.8}$$

if $1 \leq \theta < \infty$ *and*

$$\|f\|_{b_{p,\infty}^l(\mathbb{R}^n)} = \sup_{h \in \mathbb{R}^n, h \neq 0} \frac{\|\Delta_h^\sigma f\|_{L_p(\mathbb{R}^n)}}{|h|^l}. \tag{5.9}$$

This definition is independent of $\sigma > l$ as the following lemma shows.

Lemma 1 *Let* $l > 0, 1 \leq p, \theta \leq \infty$. *Then the norms* [5] $\|\cdot\|_{B_{p,\theta}^l(\mathbb{R}^n)}$ *corresponding to different* $\sigma \in \mathbb{N}$ *satisfying* $\sigma > l$ *are equivalent.*

Idea of the proof. Denote temporarily semi-norms (5.9) and (5.10) corresponding to σ by $\|\cdot\|^{(\sigma)}$. It is enough to prove that $\|\cdot\|^{(\sigma)}$ and $\|\cdot\|^{(\sigma+1)}$ are equivalent on $L_p(\mathbb{R}^n)$ where $\sigma > l$. Since $\|\Delta_h^{\sigma+1} f\|_{L_p(\mathbb{R}^n)} \leq 2 \|\Delta_h^\sigma f\|_{L_p(\mathbb{R}^n)}$, it follows that $\|\cdot\|^{(\sigma+1)} \leq 2 \|\cdot\|^{(\sigma)}$. To prove the inverse inequality start with the case $0 < l < 1, \sigma = 1$ and apply the following identity for differences

$$\Delta_h f = \tfrac{1}{2} \Delta_{2h} f - \tfrac{1}{2} \Delta_h^2 f, \tag{5.10}$$

which is equivalent to the obvious identity [6] $x - 1 = \tfrac{1}{2}(x^2 - 1) - \tfrac{1}{2}(x - 1)^2$ for polynomials. To complete the proof deduce a similar identity involving $\Delta_h^\sigma f, \Delta_{2h}^\sigma f$ and $\Delta_h^{\sigma+1} f$. $\square$

Proof. 1. Suppose that $0 < l < 1$ and $\|f\|^{(2)} < \infty$. By (5.11) we have

$$\|\Delta_h f\|_{L_p(\mathbb{R}^n)} \leq \tfrac{1}{2} \|\Delta_{2h} f\|_{L_p(\mathbb{R}^n)} + \tfrac{1}{2} \|\Delta_h^2 f\|_{L_p(\mathbb{R}^n)}. \tag{5.11}$$

---

[4] The main reason for this is Theorem 3 below, which otherwise would not be valid.

[5] See footnote 1 on page 12.

[6] Here x replaces the translation operator E_h where $h \in \mathbb{R}^n$ $((E_h f)(y) = f(y+h), y \in \mathbb{R}^n)$.

First let $\theta = \infty$. Denote $\varphi(h) = |h|^{-l}\|\Delta_h f\|_{L_p(\mathbb{R}^n)}$. (Clearly, $\varphi(h) < \infty$ for all $h \in \mathbb{R}^n, h \neq 0$. Then it follows that

$$\varphi(h) \leq 2^{l-1}\varphi(2h) + 2^{-1}\|f\|^{(2)}.$$

Consequently, $\forall k \in \mathbb{N}$

$$\varphi(h) \leq 2^{l-1}(2^{l-1}\varphi(4h) + 2^{-1}\|f\|^{(2)}) + 2^{-1}\|f\|^{(2)} \leq \cdots$$

$$\leq 2^{(l-1)k}\varphi(2^k h) + 2^{-1}\|f\|^{(2)}(1 + 2^{l-1} + \cdots + 2^{(l-1)k})$$

$$\leq 2^{(l-1)k}\varphi(2^k h) + (2 - 2^l)^{-1}\|f\|^{(2)}.$$

Let k be such that $2^k|h| \geq 1$, then $\varphi(2^k h) \leq \|\Delta_{2^k h} f\|_{L_p(\mathbb{R}^n)} \leq 2\|f\|_{L_p(\mathbb{R}^n)}$. Hence,

$$\varphi(h) \leq 2^{(l-1)k+1}\|f\|_{L_p(\mathbb{R}^n)} + (2 - 2^l)^{-1}\|f\|^{(2)}.$$

On letting $k \to \infty$, we get $\|f\|^{(1)} \leq (2 - 2^l)^{-1}\|f\|^{(2)}$. Thus,

$$(2 - 2^l)\|f\|^{(1)} \leq \|f\|^{(2)} \leq 2\|f\|^{(1)}. \tag{5.12}$$

If $1 \leq \theta < \infty$, we set $\forall \varepsilon > 0$

$$\psi(\varepsilon) = \left(\int\limits_{|h| \geq \varepsilon} \left(\frac{\|\Delta_h f\|_{L_p(\mathbb{R}^n)}}{|h|^l} \right)^\theta \frac{dh}{|h|^n} \right)^{\frac{1}{\theta}}.$$

Since $\|\Delta_h f\|_{L_p(\mathbb{R}^n)} \leq 2\|f\|_{L_p(\mathbb{R}^n)}$, $\psi(\varepsilon) < \infty$ for all $\varepsilon > 0$. From (5.12) it follows, after substituting $2h = \eta$, that

$$\psi(\varepsilon) \leq 2^{l-1}\psi(2\varepsilon) + 2^{-1}\|f\|^{(2)},$$

and a similar argument leads to the same inequality (5.13).

2. If $\sigma \geq 2$, then

$$(x - 1)^\sigma = 2^{-\sigma}(x^2 - 1)^\sigma + (x - 1)^\sigma - 2^{-\sigma}(x^2 - 1)^\sigma$$

$$= 2^\sigma(x^2 - 1)^{-\sigma} + P_{\sigma-1}(x)(x - 1)^{\sigma+1},$$

where

$$P_{\sigma-1}(x) = -2^{-\sigma}(x - 1)^{-1}((x + 1)^\sigma - 2^\sigma) = -2^{-\sigma}\sum_{s=1}^{\sigma}\binom{\sigma}{s}(x - 1)^{s-1}.$$

Hence,

$$\Delta_h^\sigma f = 2^{-\sigma}\Delta_{2h}^\sigma f + P_{\sigma-1}(E_h)\Delta_h^{\sigma+1} f$$

and, since $\|E_h\|_{L_p(\mathbb{R}^n)\to L_p(\mathbb{R}^n)} = 1$,

$$\|\Delta_h^\sigma f\|_{L_p(\mathbb{R}^n)} \le 2^{-\sigma}\|\Delta_{2h}^\sigma f\|_{L_p(\mathbb{R}^n)} + 2^{-\sigma-1}(3^\sigma - 1)\|\Delta_h^{\sigma+1} f\|_{L_p(\mathbb{R}^n)}.$$

The rest is similar to step 1. $\square$

We shall prove next that the norm $\|\cdot\|_{B_{p,\theta}^l(\mathbb{R}^n)}$ is equivalent to a similar norm containing the modulus of continuity

$$\omega_\sigma(\delta, f)_p = \sup_{|h|\le\delta} \|\Delta_h^\sigma f\|_{L_p(\mathbb{R}^n)}.$$

To do this we need several auxilary statements.

Lemma 2 (Hardy's inequality) *Let $1 \le p \le \infty$ and $\alpha < \frac{1}{p'}$. Then for each function f measurable on $(0,\infty)$*

$$\left\| t^\alpha \frac{1}{t} \int_0^t |f|\, dx \right\|_{L_p(0,\infty)} \le (\tfrac{1}{p'} - \alpha)^{-1} \|x^\alpha f(x)\|_{L_p(0,\infty)}. \tag{5.13}$$

Idea of the proof. Substitute $x = yt$, apply Minkowski's inequality for integrals and substitute $t = \frac{x}{y}$. $\square$

Proof. We have [7]

$$\left\| t^\alpha \frac{1}{t} \int_0^t |f(x)|\, dx \right\|_{L_p(0,\infty)} = \left\| \int_0^1 t^\alpha |f(yt)|\, dy \right\|_{L_p(0,\infty)}$$

$$\le \int_0^1 \|t^\alpha f(yt)\|_{L_{p,t}(0,\infty)} = \int_0^1 y^{-\alpha+\frac{1}{p}} dy\, \|x^\alpha f(x)\|_{L_p(0,\infty)}$$

$$= (\tfrac{1}{p'} - \alpha)^{-1} \|x^\alpha f(x)\|_{L_p(0,\infty)}. \quad \square$$

Remark 3 The constant $(\frac{1}{p'} - \alpha)^{-1}$ in (5.14) is sharp. One may verify this considering the family of functions f_δ where $0 < \delta < p(\frac{1}{p'} - \alpha)$, defined by $f_\delta(x) = 0$ for $0 < x < 1$ and $f_\delta = x^{-\alpha-\frac{1+\delta}{p}}$ for $x \ge 1$.

---

[7] Since f is measurable on $(0,\infty)$, the function $F(y,t) := |f(yt)|$ is measurable on $(0,\infty) \times (0,\infty)$ and we can apply Minkowski's inequality for integrals.

Corollary 2 *Let $1 \leq p \leq \infty$ and $\alpha < n - \frac{1}{p}$. Then for each function f measurable on $\mathbb{R}^n$*

$$\left\| t^\alpha \frac{1}{v_n t^n} \int\limits_{|x| \leq t} |f| \, dx \right\|_{L_p(0,\infty)} \leq c_4 \, \| \, |x|^{\alpha - \frac{n-1}{p}} f(x) \|_{L_p(\mathbb{R}^n)}, \tag{5.14}$$

where $c_4 > 0$ is independent of f.

Idea of the proof. If $n = 1$, apply inequality (5.13) and its variant for the case, in which in the left-hand side the integral over $(0, t)$ is replaced by the integral over $(-t, 0)$ and in the right-hand side the norm $\| \cdot \|_{L_p(0,\infty)}$ is replaced by $\| \cdot \|_{L_p(-\infty,0)}$. If $n > 1$, take spherical coordinates, apply Minkowski's inequality, inequality (5.13) and Hölder's inequality. $\square$

Proof. Let $n > 1$. Then by (5.13)

$$\left\| t^\alpha \frac{1}{v_n t^n} \int\limits_{|x| \leq t} |f| \, dx \right\|_{L_p(0,\infty)}$$

$$= \left\| t^\alpha \frac{1}{v_n t^n} \int\limits_{S^{n-1}} \left(\int\limits_0^t \varrho^{n-1} |f(\varrho\xi)| \, d\varrho \right) dS^{n-1} \right\|_{L_{p,t}(0,\infty)}$$

$$\leq \frac{1}{v_n} \int\limits_{S^{n-1}} \left\| t^{\alpha-(n-1)} \frac{1}{t} \int\limits_0^t \varrho^{n-1} |f(\varrho\xi)| \, d\varrho \right\|_{L_{p,t}(0,\infty)} dS^{n-1}$$

$$\leq \left(v_n \left(n - \tfrac{1}{p} - \alpha \right) \right)^{-1} \int\limits_{S^{n-1}} \| \varrho^\alpha f(\varrho\xi) \|_{L_{p,\varrho}(0,\infty)} \, dS^{n-1}$$

$$\leq \left(v_n \left(n - \tfrac{1}{p} - \alpha \right) \right)^{-1} \sigma_n^{\frac{1}{p'}} \left(\int\limits_{S^{n-1}} \left(\int\limits_0^\infty \varrho^{\alpha p} |f(\varrho\xi)|^p \right) dS^{n-1} \right)^{\frac{1}{p}}$$

$$= c_4 \, \| \, |x|^{\alpha - \frac{n-1}{p}} f(x) \|_{L_p(\mathbb{R}^n)}. \quad \square$$

Next we generalize the trivial identity

$$(\Delta_h^\sigma f)(x) = (\Delta_\eta f)(x) + (\Delta_{h-\eta} f)(x + \eta),$$

where $x, h, \eta \in \mathbb{R}^n$ to the case of differences of order $\sigma > 1$.

Lemma 3 *Let $\sigma \in \mathbb{N}, h, \eta \in \mathbb{R}^n$ and $f \in L_1^{loc}(\mathbb{R}^n)$. Then for almost all $x \in R^n$*

$$(\Delta_h^\sigma f)(x) = \sum_{k=1}^{\sigma}(-1)^{\sigma-k}\binom{\sigma}{k}\left((\Delta_{\frac{k}{\sigma}\eta}^\sigma f)(x+(\sigma-k)h)\right.$$

$$\left. +(-1)^{\sigma+1}(\Delta_{h-\frac{k}{\sigma}\eta}^\sigma f)(x+k\eta)\right). \tag{5.15}$$

Idea of the proof. Replacing the translation operators E_h and $E_{\frac{\eta}{\sigma}}$ by y and z respectively, it is enough to prove the following identity for polynomials

$$(y-1)^\sigma = \sum_{k=1}^{\sigma}(-1)^{\sigma-k}\binom{\sigma}{k}(z^k-1)^\sigma y^{\sigma-k} - \sum_{k=1}^{\sigma}(-1)^k\binom{\sigma}{k}(y-z^k)^\sigma. \ \square \tag{5.16}$$

Proof. Identity (5.16) is equivalent to the identity

$$\sum_{k=0}^{\sigma}(-1)^{\sigma-k}\binom{\sigma}{k}(z^k-1)^\sigma y^{\sigma-k} = \sum_{k=0}^{\sigma}(-1)^{\sigma-k}\binom{\sigma}{k}(z^k-y)^\sigma,$$

which is clear since both its sides are equal to

$$\sum_{k,m=0}^{\sigma}(-1)^{k+m}\binom{\sigma}{k}\binom{\sigma}{m}z^{km}y^{\sigma-k}. \ \square$$

Corollary 3 *Let $\sigma \in \mathbb{N}, h, \eta \in \mathbb{R}^n, 1 \leq p \leq \infty$ and $f \in L_p(\mathbb{R}^n)$. Then*

$$\|\Delta_h^\sigma f\|_{L_p(\mathbb{R}^n)} \leq \sum_{k=1}^{\sigma}\binom{\sigma}{k}\left(\|\Delta_{\frac{k}{\sigma}\eta}^\sigma f\|_{L_p(\mathbb{R}^n)} + \|\Delta_{h-\frac{k}{\sigma}\eta}^\sigma f\|_{L_p(\mathbb{R}^n)}\right). \tag{5.17}$$

Idea of the proof. Apply (5.16), Minkowski's inequality for sums and the invariance of the norm $\|\cdot\|_{L_p(\mathbb{R}^n)}$ with respect to translations. $\square$

Lemma 4 *Let $\sigma \in \mathbb{N}, 1 \leq p \leq \infty$. Then for all functions measurable on $\mathbb{R}^n$ and $\forall h \in \mathbb{R}^n$*

$$\|\Delta_h^\sigma f\|_{L_p(\mathbb{R}^n)} \leq \frac{c_5}{v_n|h|^n}\int_{|\eta|\leq|h|}\|\Delta_\eta^\sigma f\|_{L_p(\mathbb{R}^n)}\,d\eta, \tag{5.18}$$

where $c_5 > 0$ is independent of f.

Idea of the proof. Integrate inequality (5.17) with respect to $\eta \in B(\frac{h}{2}, \frac{|h|}{2})$. $\square$

Proof. If $\eta \in B(\frac{h}{2}, \frac{|h|}{2})$, then $\frac{k\eta}{\sigma}, h - \frac{k\eta}{\sigma} \in B(0, |h|), k = 1, ..., \sigma$. Hence, by substituting $\frac{k\eta}{\sigma} = \xi$, $h - \frac{k\eta}{\sigma} = \xi$ respectively, we have

$$v_n \left(\frac{|h|}{2}\right)^n \|\Delta_h^\sigma f\|_{L_p(\mathbb{R}^n)} \leq 2 \sum_{k=1}^{\sigma} \binom{\sigma}{k} \left(\frac{\sigma}{k}\right)^n \int\limits_{|\xi| \leq |h|} \|\Delta_\xi^\sigma f\|_{L_p(\mathbb{R}^n)} \, d\xi.$$

Thus

$$\|\Delta_h^\sigma f\|_{L_p(\mathbb{R}^n)} \leq \frac{2(2\sigma)^n (2^\sigma - 1)}{v_n |h|^n} \int\limits_{|\eta| \leq |h|} \|\Delta_\eta^\sigma f\|_{L_p(\mathbb{R}^n)} \, d\eta. \; \square$$

Corollary 4 *Let $\sigma \in \mathbb{N}, 1 \leq p \leq \infty$ and $f \in L_p(\mathbb{R}^n)$. Then*

$$\omega_\sigma(t, f)_p \leq \frac{c_5}{v_n} \int\limits_{|\eta| \leq t} \|\Delta_\eta^\sigma f\|_{L_p(\mathbb{R}^n)} \frac{d\eta}{|\eta|^n}. \tag{5.19}$$

Idea of the proof. Direct application of inequality (5.18). $\square$

We note also two simple inequalities for modulae of continuity, which follow by Corollary 8 of Chapter 3:

$$\omega_\sigma(\delta, f)_p \leq 2^\sigma \|f\|_{L_p(\mathbb{R}^n)} \tag{5.20}$$

and

$$\omega_\sigma(\delta, f)_p \leq c_4 \delta^l \|f\|_{w_p^l(\mathbb{R}^n)},$$

where $l, \sigma \in \mathbb{N}, l \leq \sigma, 1 \leq p \leq \infty$ and $c_4 = 2^{\sigma-l} n^{l-1}$.

We shall also apply the following property:

$$\omega_\sigma(s\delta, f)_p \leq (s + 1)^\sigma \omega_\sigma(\delta, f)_p, \tag{5.21}$$

where $s > 0$. If $s \in \mathbb{N}$, it follows, with s^σ replacing $(s+1)^\sigma$, from the identity [8]

$$(\Delta_{sh}^\sigma f)(x) = \sum_{s_1=0}^{s-1} \cdots \sum_{s_\sigma=0}^{s-1} (\Delta_h^\sigma f)(x + s_1 h + \cdots + s_\sigma h)$$

and Minkowski's inequality. If $s > 0$, then

$$\omega_\sigma(s\delta, f)_p \leq \omega_\sigma(([s] + 1)\delta, f)_p \leq ([s] + 1)^\sigma \omega_\sigma(\delta, f)_p \leq (s + 1)^\sigma \omega_\sigma(\delta, f)_p.$$

---

[8] It follows, by induction, from the case $s = 1$, in which it is obvious.

Lemma 5 *Let $l > 0, \sigma \in \mathbb{N}, \sigma > l, 1 \leq p, \theta \leq \infty$. The norm*

$$\|f\|_{B^l_{p,\theta}(\mathbb{R}^n)}^{(1)} = \|f\|_{L_p(\mathbb{R}^n)} + \left(\int_0^\infty \left(\frac{\omega_\sigma(\delta, f)_p}{t^l} \right)^\theta \frac{dt}{t} \right)^{\frac{1}{\theta}} \tag{5.22}$$

is an equivalent norm on the space $B^l_{p,\theta}(\mathbb{R}^n)$.

Idea of the proof. Since, clearly, $\|\Delta_h^\sigma f\|_{L_p(\mathbb{R}^n)} \leq \omega_\sigma(|h|, f)_p$, the estimate $\|f\|_{B^l_{p,\theta}(\mathbb{R}^n)} \leq M_1 \|f\|_{B^l_{p,\theta}(\mathbb{R}^n)}^{(1)}$, where M_1 is independent of f, follows directly by taking spherical coordinates. To obtain an inverse estimate apply inequalities (5.19) and (5.14). $\square$

Proof. In fact, by (5.19)

$$\left(\int_0^\infty \left(\frac{\omega_\sigma(\delta, f)_p}{t^l} \right)^\theta \frac{dt}{t} \right)^{\frac{1}{\theta}} \leq M_2 \left\| t^{n-l-\frac{1}{\theta}} \frac{1}{v_n t^n} \int_{|\eta| \leq t} |\eta|^{-n} \|\Delta_\eta^\sigma f\|_{L_p(\mathbb{R}^n)} \, d\eta \right\|_{L_\theta(\mathbb{R}^n)},$$

where M_2 is independent of f.

Since $l > 0$, the assumptions of Corollary 2 are satisfied and by (5.14)

$$\|f\|_{B^l_{p,\theta}(\mathbb{R}^n)}^{(1)} \leq \|f\|_{L_p(\mathbb{R}^n)} + M_3 \| |\eta|^{-l-\frac{n}{\theta}} \|\Delta_\eta^\sigma f\|_{L_p(\mathbb{R}^n)} \|_{L_\theta(0,\infty)} \leq M_4 \|f\|_{B^l_{p,\theta}(\mathbb{R}^n)},$$

where M_3, M_4 are independent of f. $\square$

Since the modulus of continuity is a nondecreasing function, it is possible to define equivalent norms on the space $B^l_{p,\theta}(\mathbb{R}^n)$ in terms of series.

Lemma 6 *Let $l > 0, \sigma \in \mathbb{N}, \sigma > l, 1 \leq p, \theta \leq \infty$. The the norms*

$$\|f\|_{B^l_{p,\theta}(\mathbb{R}^n)}^{(2)} = \|f\|_{L_p(\mathbb{R}^n)} + \left(\sum_{k=1}^\infty \left(k^l \omega_\sigma \left(\frac{1}{k}, f \right)_p \right)^\theta \frac{1}{k} \right)^{\frac{1}{\theta}} \tag{5.23}$$

and

$$\|f\|_{B^l_{p,\theta}(\mathbb{R}^n)}^{(3)} = \|f\|_{L_p(\mathbb{R}^n)} + \left(\sum_{k=1}^\infty \left(2^{kl} \omega_\sigma \left(2^{-k}, f \right)_p \right)^\theta \right)^{\frac{1}{\theta}} \tag{5.24}$$

are equivalent norms on the space $B^l_{p,\theta}(\mathbb{R}^n)$.

Idea of the proof. Apply (5.21) and the following inequalities for nondecreasing nonnegative functions φ and $\alpha \in \mathbb{R}$:

$$c_5 \sum_{k=2}^{\infty} k^{\alpha-1} \varphi\left(\frac{1}{k}\right) \le \int_0^1 x^{\alpha} \varphi(x) \frac{dx}{x} \le c_6 \sum_{k=1}^{\infty} k^{\alpha-1} \varphi\left(\frac{1}{k}\right)$$

and

$$c_7 \sum_{k=2}^{\infty} 2^{-k\alpha} \varphi(2^{-k}) \le \int_0^{\frac{1}{2}} x^{\alpha} \varphi(x) \frac{dx}{x} \le c_8 \sum_{k=1}^{\infty} 2^{-k\alpha} \varphi(2^{-k}),$$

where $c_5, ..., c_8 > 0$ are independent of φ. $\square$

Remark 4 The norm $\|\cdot\|_{B_{p,\theta}^l(\mathbb{R}^n)}$ is the "weakest" of the considered equivalent norms on the space $B_{p,\theta}^l(\mathbb{R}^n)$ and the norm $\|\cdot\|_{B_{p,\theta}^l(\mathbb{R}^n)}^{(1)}$ (or any of its variants $\|\cdot\|_{B_{p,\theta}^l(\mathbb{R}^n)}^{(2)}$ or $\|\cdot\|_{B_{p,\theta}^l(\mathbb{R}^n)}^{(3)}$) is the "strongest" one, since the estimate $\|\cdot\|_{B_{p,\theta}^l(\mathbb{R}^n)} \le M_1 \|\cdot\|_{B_{p,\theta}^l(\mathbb{R}^n)}^{(1)}$ is trivial, while the inverse estimate $\|\cdot\|_{B_{p,\theta}^l(\mathbb{R}^n)}^{(1)} \le M_2 \|\cdot\|_{B_{p,\theta}^l(\mathbb{R}^n)}$ is nontrivial. For this reason, estimating $\|\cdot\|_{B_{p,\theta}^l(\mathbb{R}^n)}$ from above, it is convenient to use this norm itself, while estimating some quantities from above via $\|\cdot\|_{B_{p,\theta}^l(\mathbb{R}^n)}$, it is convenient to use the norm $\|\cdot\|_{B_{p,\theta}^l(\mathbb{R}^n)}^{(1)}$. This observation will be applied in the proof of Theorem 3 below.

Lemma 7 *Let* $l > 0, 1 \le p, \theta \le \infty$. *Then* $B_{p,\theta}^l(\mathbb{R}^n)$ *is a Banach space.* [9]

Idea of the proof. Obviously $B_{p,\theta}^l(\mathbb{R}^n)$ is a normed vector space. To prove the completeness, starting from the Cauchy sequence $\{f_k\}_{k\in\mathbb{N}}$ in $B_{p,\theta}^l(\mathbb{R}^n)$, deduce, using the completeness of $L_p(\mathbb{R}^n)$ and [10] $L_{p,\theta}(\mathbb{R}^{2n})$, that there exist functions $f \in L_p(\mathbb{R}^n)$ and $g \in L_{p,\theta}(\mathbb{R}^{2n})$ such that $f_k \to f$ in $L_p(\mathbb{R}^n)$ and $|h|^{-l-\frac{1}{\theta}} f_k(x) \to g(x, h)$ in $L_{p,\theta}(\mathbb{R}^{2n})$. Choosing an appropriate subsequence $\{f_{k_s}\}_{s\in\mathbb{N}}$, prove that $g(x, h) = |h|^{-l-\frac{1}{\theta}} f(x)$ for almost all $x, h \in \mathbb{R}^n$ and thus $f_k \to f$ in $B_{p,\theta}^l(\mathbb{R}^n)$. $\square$

---

[9] See footnote 1 on page 12.

[10] $L_{p,\theta}(\mathbb{R}^{2n})$ is the space of all functions g measurable on $\mathbb{R}^{2n}$, which are such that

$$\|g\|_{L_{p,\theta}(\mathbb{R}^{2n})} = \| \|g(x, h)\|_{L_{p,x}(\mathbb{R}^n)} \|_{L_{\theta,h}(\mathbb{R}^n)} < \infty.$$

Lemma 8 *Let $l > 0$. The norm*

$$\|f\|_{B_{2,2}^l(\mathbb{R}^n)}^{(4)}\| = \|(1 + |\xi|^{2l})^{\frac{1}{2}}(Ff)(\xi)\|_{L_2(\mathbb{R}^n)} \tag{5.25}$$

is an equivalent norm on the space $B_{2,2}^l(\mathbb{R}^n)$.

Idea of the proof. Apply Parseval's equality (1.26) and the equality

$$(F(\Delta_h^\sigma f))(\xi) = \left(e^{ih\cdot\xi} - 1\right)^\sigma (Ff)(\xi) = \left(2ie^{i\frac{h\cdot\xi}{2}}\right)^\sigma \left(\sin\frac{h\cdot\xi}{2}\right)^\sigma (Ff)(\xi)$$

for $f \in L_2(\mathbb{R}^n)$. $\square$

Proof. Since $\sqrt{a^2 + b^2} \leq \sqrt{a} + \sqrt{b} \leq 2\sqrt{a^2 + b^2}, a, b \geq 0$, the norm $\|f\|_{B_{2,2}^l(\mathbb{R}^n)}$ is equivalent to

$$\left(\|f\|_{L_2(\mathbb{R}^n)}^2 + \int\limits_{\mathbb{R}^n} |h|^{-2l} \|\Delta_h^\sigma f\|_{L_2(\mathbb{R}^n)}^2 \frac{dh}{|h|^n}\right)^{\frac{1}{2}}$$

$$= \left(\int\limits_{\mathbb{R}^n} (1 + 2^{2\sigma}\Lambda_n(\xi))|(Ff)(\xi)|^2 \, d\xi\right)^{\frac{1}{2}},$$

where

$$\Lambda_n(\xi) = \int\limits_{\mathbb{R}^n} |h|^{-2l-n} \sin^{2\sigma}\frac{h\cdot\xi}{2} \, dh.$$

If $n = 1$, then after substituting $h = \frac{t}{|\xi|}$, we have

$$\Lambda_1(\xi) = M_1|\xi|^{2l}, \quad M_1 = \int\limits_{-\infty}^{\infty} |t|^{-2l-1} \sin^{2\sigma}\frac{t}{2} \, dt < \infty,$$

since $l > 0$ and $\sigma > l$. If $n > 1$, we first substitute $h = A_\xi \eta$, where A_ξ is a rotation in $\mathbb{R}^n$ such that $h\cdot\xi = |\xi|\eta_1$, and afterwards $\eta = \frac{t}{|\xi|}$. Hence

$$\Lambda_n(\xi) = \int\limits_{\mathbb{R}^n} |\eta|^{-2l-n} \sin^{2\sigma}\frac{|\xi|\eta_1}{2} \, d\eta = M_n|\xi|^{2l}, \quad M_n = \int\limits_{\mathbb{R}^n} |t|^{-2l-n} \sin^{2\sigma}\frac{t_1}{2} \, dt.$$

If $t_k = |t_1|\tau_k, k = 2, ..., n$, then $|t| = |t_1|\sqrt{1 + |\tau|^2}$, where $|\tau| = (\sum\limits_{k=2}^{n} \tau_k^2)^{\frac{1}{2}}$. Hence, applying (4.116), we have

$$M_n = M_1 \int\limits_{\mathbb{R}^{n-1}} \sqrt{1 + |\tau|^2}^{-2l-n} \, d\tau = M_1\sigma_{n-1}\int\limits_{0}^{\infty} \sqrt{1 + \varrho^2}^{-2l-2} \, d\varrho < \infty.$$

To complete the proof it is enough to note that $K_1(1 + |\xi|^{2l}) \leq 1 + 2^{2\sigma}\Lambda_n(\xi) \leq K_2(1 + |\xi|^{2l})$, where $K_1, K_2 > 0$ are independent of ξ. $\square$

Corollary 5 *If $l \in \mathbb{N}$, then*

$$B_{2,2}^l(\mathbb{R}^n) = W_2^l(\mathbb{R}^n).$$

The corresponding norms are equivalent. Moreover, $\|f\|_{B_{2,2}^l(\mathbb{R}^n)}^{(4)} = \|f\|_{W_2^l(\mathbb{R}^n)}^{(2)}$.

Idea of the proof. Apply Lemmas 8 of this chapter and of Chapter 1. $\square$

Next we state, without proofs, several properties of the spaces $B_{p,\theta}^l(\mathbb{R}^n)$, which will not be directly used in the sequel, but provide better understanding of the trace theorems.

Remark 5 If $l > 0$, $1 \leq p \leq \infty$, $1 \leq \theta_1 < \theta_2 \leq \infty$, then

$$B_{p,\theta_1}^l(\mathbb{R}^n) \subset B_{p,\theta_2}^l(\mathbb{R}^n).$$

Moreover, if $l > 0$, $0 < \varepsilon < l$, $1 \leq p, \theta, \theta_1, \theta_2 \leq \infty$, then

$$B_{p,\theta_1}^{l+\varepsilon}(\mathbb{R}^n) \subset B_{p,\theta}^l(\mathbb{R}^n) \subset B_{p,\theta_2}^{l-\varepsilon}(\mathbb{R}^n).$$

Hence the parameter θ, which is also a parameter describing smoothness, is a weaker parameter compared with the main smoothness parameter l.

Remark 6 If $l \in \mathbb{N}$, $1 \leq p \leq \infty$ and $p \neq 2$, then for each $\theta, 1 \leq \theta \leq \infty$,

$$B_{p,\theta}^l(\mathbb{R}^n) \neq W_p^l(\mathbb{R}^n).$$

Moreover, if $l \in \mathbb{N}$, $1 \leq p < \infty$, then

$$B_{p,\theta_1}^l(\mathbb{R}^n) \subset W_p^l(\mathbb{R}^n) \subset B_{p,\theta_2}^l(\mathbb{R}^n),$$

where $\theta_1 = \min\{p, 2\}, \theta_2 = \max\{p, 2\}$. If $\theta_1 > \min\{p, 2\}, \theta_2 < \max\{p, 2\}$, the corresponding embeddings do not hold.

Remark 7 The following norms are equivalent to $\|f\|_{B_{p,\theta}^l(\mathbb{R}^n)}$:

$$\|f\|_{B_{p,\theta}^l(\mathbb{R}^n)}^{(k)} = \|f\|_{L_p(\mathbb{R}^n)} + \|f\|_{b_{p,\theta}^l(\mathbb{R}^n)}^{(k)}, \quad k = 5, 6, 7, 8,$$

where

$$\|f\|^{(5)}_{b^l_{p,\theta}(\mathbb{R}^n)} = \sum_{|\alpha|=m} \Big(\int_{|h|\leq H} \Big(\frac{\|\Delta^\sigma_h D^\alpha_w f\|_{L_p(\mathbb{R}^n)}}{|h|^{l-m}}\Big)^\theta \frac{dh}{|h|^n}\Big)^{\frac{1}{\theta}},$$

$$\|f\|^{(6)}_{b^l_{p,\theta}(\mathbb{R}^n)} = \sum_{|\alpha|=m} \Big(\int_0^H \Big(\frac{\omega_\sigma(t, D^\alpha_w f)_p\|_{L_p(\mathbb{R}^n)}}{t^{l-m}}\Big)^\theta \frac{dt}{t}\Big)^{\frac{1}{\theta}},$$

$$\|f\|^{(7)}_{b^l_{p,\theta}(\mathbb{R}^n)} = \sum_{j=1}^n \Big(\int_0^H \Big(\frac{\|\Delta^\sigma_{te_j}\big(\frac{\partial^m f}{\partial x_j^m}\big)_w\|_{L_p(\mathbb{R}^n)}}{t^{l-m}}\Big)^\theta \frac{dt}{t}\Big)^{\frac{1}{\theta}},$$

$$\|f\|^{(8)}_{b^l_{p,\theta}(\mathbb{R}^n)} = \sum_{j=1}^n \Big(\int_0^H \Big(\frac{\omega_{\sigma,j}\big(t, \big(\frac{\partial^m f}{\partial x_j^m}\big)_w\big)_p\|_{L_p(\mathbb{R}^n)}}{t^{l-m}}\Big)^\theta \frac{dt}{t}\Big)^{\frac{1}{\theta}}.$$

Here $m \in \mathbb{N}_0$, $m < l < \sigma+m$, $0 < H \leq \infty$, e_j is the unit vector in the direction of the axis Ox_j and $\omega_{\sigma,j}(\cdot, \varphi)$ is the modulus of continuity of the function φ of order σ in the direction of the axis Ox_j. If $\theta = \infty$, then, as in Definition 2, the integrals must be replaced by the appropriate suprema.

There also exist other eqivalent ways of defining the space $B^l_{p,\theta}(\mathbb{R}^n)$: with the help of Fourier transforms (not only for $p = \theta = 2$ as in Lemma 8), with the help of the best approximations by entire functions of exponential type, by means of the theory of interpolation, etc.

Remark 8 It can be proved that

$$W^l_p(\mathbb{R}^n) \hookrightarrow B^{l-\frac{n}{p}}_{\infty,p}(\mathbb{R}^n), \quad l > \tfrac{n}{p},\ 1 \leq p \leq \infty,$$

and

$$W^l_p(\mathbb{R}^n) \hookrightarrow B^{l-n(\frac{1}{p}-\frac{1}{q})}_{q,p}(\mathbb{R}^n), \quad n(\tfrac{1}{p} - \tfrac{1}{q}) < l < \tfrac{n}{p},\ 1 \leq p < q < \infty.$$

These embeddings are sharp in terms of the considered spaces: the second lower index p can not be replaced by $\theta < p$.

Remark 9 In the sequel we shall use only the spaces $B^l_p(\mathbb{R}^n) \equiv B^l_{p,p}(\mathbb{R}^n)$. One can easily verify by changing variables that in this case $\|f\|_{B^l_p(\mathbb{R}^n)}$ is equivalent to

$$\|f\|^{(9)}_{B^l_p(\mathbb{R}^n)} = \|f\|_{L_p(\mathbb{R}^n)} + \Big(\int_{\mathbb{R}^n} \int_{\mathbb{R}^n} \frac{|(\Delta^\sigma f)(x,y)|^p}{|x-y|^{n+pl}}\, dx\, dy\Big)^{\frac{1}{p}},$$

where $(\Delta^\sigma f)(x,y) = (\Delta^\sigma_{\frac{x-y}{\sigma}} f)(y)$.

5.4   Description of the traces on subspaces

We recall that by Corollary 8 of Chapter 3

$$\|\Delta_h^l f\|_{L_p(\mathbb{R}^n)} \le c_9 \, |h|^l \|f\|_{w_p^l(\mathbb{R}^n)} \tag{5.26}$$

or

$$\left\| \frac{(\Delta_h^l f)(x)}{|h|^l} \right\|_{L_{p,x}(\mathbb{R}^n)} \le c_9 \, \|f\|_{w_p^l(\mathbb{R}^n)},$$

where $l \in \mathbb{N}$, $1 \le p \le \infty$ and $c_9 = n^{l-1}$.

Lemma 9 *Let* $l \in \mathbb{N}, 1 \le p \le \infty$ *and* $f \in w_p^l(\mathbb{R}^n)$. *Then* $\forall h \in \mathbb{R}^n$ *for almost all* $x \in \mathbb{R}^n$

$$(\Delta_h^l f)(x) = \int_0^{l|h|} K_l(|h|,\tau) \left(\frac{\partial^l f}{\partial \xi^l}\right)_w (x + \xi\tau)\, d\tau, \tag{5.27}$$

where $\xi = \frac{h}{|h|}$ *and* $\left(\frac{\partial^l f}{\partial \xi^l}\right)_w$ *is the weak derivative of* f *in the direction of* ξ,

$$K_l(\varrho, \tau) = \underbrace{(\chi_\varrho * \cdots * \chi_\varrho)}_{l}(\tau), \quad 0 < \varrho < \infty, \quad -\infty < \tau < \infty,$$

and χ_ϱ *is the characteristic function of the interval* $(0, \varrho)$.

Idea of the proof. Starting from Lemma 5 of Chapter 1, prove, by induction, that for almost all $x \in \mathbb{R}^n$

$$(\Delta_h^l f)(x) = \int_0^{|h|} \cdots \int_0^{|h|} \left(\frac{\partial^l f}{\partial \xi^l}\right)_w (x + \xi(\tau_1 + \cdots \tau_l))\, d\tau_1 \cdots d\tau_l,$$

and apply the following formula

$$\int_{-\infty}^{\infty} \cdots \int_{-\infty}^{\infty} \psi(\tau_1) \cdots \psi(\tau_l) \varphi(\tau_1 + \cdots + \tau_l)\, d\tau_1 \cdots d\tau_l = \int_{-\infty}^{\infty} K_l(\tau)\varphi(\tau)\, d\tau,$$

where $K_l(\tau) = \underbrace{(\psi * \cdots * \psi)}_{l}(\tau)$, $\varphi, \psi \in L_1^{loc}(\mathbb{R})$ and ψ has a compact support. $\square$

Corollary 6 *Under the assumptions of Lemma 4 $\forall h \in \mathbb{R}^n$ and for almost all $x \in \mathbb{R}^n$*

$$\left|\left(\Delta_h^l f\right)(x)\right| \leq \frac{1}{(l-1)!} \int_0^{l|h|} \tau^{l-1} \left|\left(\frac{\partial^l f}{\partial \xi^l}\right)_w (x + \xi\tau)\right| d\tau. \qquad (5.28)$$

Idea of the proof. It is enough to notice that $K_l(\varrho, \tau) = 0$ for $\tau \leq 0$ or $\tau \geq l\varrho$ and to prove, by induction, that $K_l(\varrho, \tau) \leq \frac{\tau^{l-1}}{(l-1)!}$ for $0 \leq \tau \leq l\varrho$.

Lemma 10 *Let $l \in \mathbb{N}, 1 \leq p \leq \infty$ and $l > \frac{n}{p}$. Then $\forall f \in w_p^l(\mathbb{R}^n)$ for almost all $x \in \mathbb{R}^n$*

$$\left\|\frac{(\Delta_h^l f)(x)}{|h|^l}\right\|_{L_{p,h}(\mathbb{R}^n)} \leq c_{10} \|f\|_{w_p^l(\mathbb{R}^n)}, \qquad (5.29)$$

where [11] *$c_{10} > 0$ is independent of f.*

Idea of the proof. Take spherical coordinates and apply Corollary 6 and Lemma 2. $\square$

Proof. After setting $h = \varrho\xi$, where $\varrho = |h|$ and $\xi = \frac{h}{|h|} \in S^{n-1}$, substituting $\varrho = \frac{r}{l}$ and applying (5.28) and (5.14) we get

$$I \equiv \left\| \, |h|^{-l}(\Delta_h^l f)(x)\right\|_{L_{p,h}(\mathbb{R}^n)} = \left\| \left\| \varrho^{-l+\frac{n-1}{p}} (\Delta_{\varrho\xi}^l f)(x)\right\|_{L_{p,\varrho}(0,\infty)}\right\|_{L_{p,\xi}(S^{n-1})}$$

$$\leq \frac{1}{(l-1)!} \left\| \left\| \varrho^{-l+\frac{n-1}{p}} \int_0^{l\varrho} \tau^{l-1} \left|\left(\frac{\partial^l f}{\partial \xi^l}\right)_w (x + \xi\tau)\right| d\tau\right\|_{L_{p,\varrho}(0,\infty)}\right\|_{L_{p,\xi}(S^{n-1})}$$

$$= \frac{l^{l-\frac{n}{p}}}{(l-1)!} \left\| \left\| r^{-l+1+\frac{n-1}{p}} \frac{1}{r} \int_0^r \tau^{l-1} \left|\left(\frac{\partial^l f}{\partial \xi^l}\right)_w (x + \xi\tau)\right| d\tau\right\|_{L_{p,r}(0,\infty)}\right\|_{L_{p,\xi}(S^{n-1})}$$

$$\leq \frac{l^{l-\frac{n}{p}}}{(l-\frac{n}{p})(l-1)!} \left\| \left\| \tau^{\frac{n-1}{p}} \left(\frac{\partial^l f}{\partial \xi^l}\right)_w (x + \xi\tau)\right\|_{L_{p,\tau}(0,\infty)}\right\|_{L_{p,\xi}(S^{n-1})}.$$

Since $\frac{l!}{\alpha!} \leq n^{l-1}$, where $\alpha \in \mathbb{N}_0^n$ and $|\alpha| = l$, we have, for almost all $z \in \mathbb{R}^n$,

$$\left|\left(\frac{\partial^l f}{\partial \xi^l}\right)_w (z)\right| = \left|\sum_{j_1=1}^n \cdots \sum_{j_l=1}^n \xi_{j_1} \cdots \xi_{j_l} \left(\frac{\partial^l f}{\partial x_{j_1} \cdots \partial x_{j_l}}\right)(z)\right|$$

---

[11] If $n = l = 1, p > 1$, then $c_{10} = p'$ and (5.29) is equivalent to (5.13), where $\alpha = 0$ and f is replaced by f'_w.

$$\leq \sum_{|\alpha|=l} \frac{l!}{\alpha!} \left| (D^\alpha f)_w(z) \right| \leq n^{l-1} \sum_{|\alpha|=l} \left| (D^\alpha f)_w(z) \right|.$$

Consequently, by Minkowski's inequality,

$$I \leq \frac{n^{l-1} l^{l-\frac{n}{p}}}{(l-\frac{n}{p})(l-1)!} \sum_{|\alpha|=l} \left\| \left\| \tau^{\frac{n-1}{p}} (D^\alpha f)_w(x+\xi\tau) \right\|_{L_{p,\tau}(0,\infty)} \right\|_{L_{p,\xi}(S^{n-1})}$$

$$= \frac{n^{l-1} l^{l-\frac{n}{p}}}{(l-\frac{n}{p})(l-1)!} \sum_{|\alpha|=l} \|(D^\alpha f)_w(x+y)\|_{L_{p,y}(\mathbb{R}^n)} = \frac{n^{l-1} l^{l-\frac{n}{p}}}{(l-\frac{n}{p})(l-1)!} \|f\|_{W_p^l(\mathbb{R}^n)}. \quad \square$$

Theorem 3 *Let $l, m, n \in \mathbb{N}$, $m < n$, $1 \leq p \leq \infty$ and $l > \frac{n-m}{p}$. Then*

$$\mathrm{tr}_{\mathbb{R}^m} W_p^l(\mathbb{R}^n) = B_p^{l-\frac{n-m}{p}}(\mathbb{R}^m). \tag{5.30}$$

Remark 10 Assertion (5.30) consists of two parts: the direct trace theorem, stating that $\forall f \in W_p^l(\mathbb{R}^n)$ there exists a trace $g \in B^{l-\frac{n-m}{p}}(\mathbb{R}^m)$, and the extension theorem, or the inverse trace theorem, stating that $\forall g \in B_p^{l-\frac{n-m}{p}}(\mathbb{R}^m)$ there exists a function $f \in W_p^l(\mathbb{R}^n)$ such that $f\big|_{\mathbb{R}^m} = g$. Actually stronger assertions hold. In the first case it will be proved that the trace operator $\mathrm{tr} : W_p^l(\mathbb{R}^n) \rightarrow B_p^{l-\frac{n-m}{p}}(\mathbb{R}^m)$ (clearly linear) is bounded. In the second case it will be proved that there exists a bounded linear extension [12] operator $T : B_p^{l-\frac{n-m}{p}}(\mathbb{R}^m) \rightarrow W_p^l(\mathbb{R}^n)$.

Idea of the proof of the direct trace theorem. Start with the case $m = 1, n = 2$. If $l = 1$, apply for $f \in C^\infty(\mathbb{R}^2) \bigcap W_p^1(\mathbb{R}^2)$ the identity

$$f(u+h,0) - f(u,0) = f(u+h,0) - f(u+h,h)$$

$$+ f(u+h,h) - f(u,h) + f(u,h) - f(u,0) \tag{5.31}$$

and inequality (5.29) with $l = 1$. If $l > 1$, deduce a similar identity for differences of order l. In general case take, in addition, spherical coordinates in $\mathbb{R}^m$ and $\mathbb{R}^{n-m}$. $\square$

Proof of the direct trace theorem. 1. Let $\forall f \in C^\infty(\mathbb{R}^n) \bigcap W_p^l(\mathbb{R}^n)$. It is enough to prove that

$$\|f(\cdot,0)\|_{B_p^{l-\frac{n-m}{p}}(\mathbb{R}^m)} \leq M_1 \|f\|_{W_p^l(\mathbb{R}^n)}, \tag{5.32}$$

---

[12] I.e., $\forall g \in B_p^{l-\frac{n-m}{p}}(\mathbb{R}^m), g$ is a trace of $Tg \in W_p^l(\mathbb{R}^n)$ on $\mathbb{R}^m$.

where M_1 is independent of f. In fact, if $f \in W_p^l(\mathbb{R}^n)$, then we consider any functions $f_k \in C^\infty(\mathbb{R}^n) \bigcap W_p^l(\mathbb{R}^n)$, such that $f_k \to f$ in $W_p^l(\mathbb{R}^n)$ as $k \to \infty$. By a standard limiting procedure (see, for example, the proof of Theorem 1 of Chapter 4), it follows, since the space $B_p^{l-\frac{n-m}{p}}(\mathbb{R}^m)$ is complete, that there exists a function $g \in B_p^{l-\frac{n-m}{p}}(\mathbb{R}^m)$ such that $f_k(\cdot,0) \to g$ in $B_p^{l-\frac{n-m}{p}}(\mathbb{R}^m)$. By Remark 2 the function g is a trace of the function f on $\mathbb{R}^m$. Moreover,

$$\|\operatorname{tr} f\|_{B_p^{l-\frac{n-m}{p}}(\mathbb{R}^m)} \le M_1 \|f\|_{W_p^l(\mathbb{R}^n)}. \tag{5.33}$$

Since inequality (5.6) is already proved, it is enough to show that the inequality

$$\|f(\cdot,0)\|_{b_p^{l-\frac{n-m}{p}}(\mathbb{R}^m)} \le M_2 \|f\|_{W_p^l(\mathbb{R}^n)}$$

holds $\forall f \in C^\infty(\mathbb{R}^n) \bigcap W_p^l(\mathbb{R}^n)$, where M_2 is independent of f.

2. Let $l = 1, m = 1, n = 2, 1 < p \le \infty$ and $f \in C^\infty(\mathbb{R}^2) \bigcap W_p^l(\mathbb{R}^2)$. By (5.31) and (5.26) we get

$$\|(\Delta_{u,h}f)(u,0)\|_{L_{p,u}(\mathbb{R})} \le \|(\Delta_{v,h}f)(u+h,0)\|_{L_{p,u}(\mathbb{R})}$$

$$+\|(\Delta_{u,h}f)(u,h)\|_{L_{p,u}(\mathbb{R})} + \|(\Delta_{v,h}f)(u,0)\|_{L_{p,u}(\mathbb{R})}$$

$$\le 2\|(\Delta_{v,h}f)(u,0)\|_{L_{p,u}(\mathbb{R})} + |h| \cdot \left\|\frac{\partial f}{\partial u}(u,h)\right\|_{L_{p,u}(\mathbb{R})}.$$

Hence, applying Fubuni's theorem and inequality (5.29), we get

$$\|f(\cdot,0)\|_{b_p^{1-\frac{1}{p}}(\mathbb{R})} = \| |h|^{-1}\|(\Delta_{u,h}f)(u,0)\|_{L_{p,u}(\mathbb{R})}\|_{L_{p,h}(\mathbb{R})}$$

$$\le 2\|\,\| |h|^{-1}(\Delta_{v,h}f)(u,0)\|_{L_{p,h}(\mathbb{R})}\|_{L_{p,u}(\mathbb{R})} + \left\|\,\left\|\frac{\partial f}{\partial u}(u,h)\right\|_{L_{p,u}(\mathbb{R})}\right\|_{L_{p,h}(\mathbb{R})}$$

$$\le 2p'\left\|\,\left\|\frac{\partial f}{\partial v}(u,v)\right\|_{L_{p,v}(\mathbb{R})}\right\|_{L_{p,u}(\mathbb{R})} + \left\|\frac{\partial f}{\partial u}\right\|_{L_p(\mathbb{R}^2)} \le 2p' \|f\|_{w_p^1(\mathbb{R}^2)}.$$

3. Let next $l > 1, m = 1, n = 2$. The following identity

$$(\Delta_{u,h}^l f)(u,0) = \sum_{\lambda=0}^{l}(-1)^\lambda \binom{l}{\lambda}(\Delta_{v,h}^l f)(u+\lambda h,0)$$

$$-\sum_{\lambda=1}^{l}(-1)^\lambda \binom{l}{\lambda}(\Delta_{u,h}^l f)(u,\lambda h) \tag{5.34}$$

is an appropriate generalization of (5.31) for differences of order $l > 1$. [13] By (5.26), as in step 2,

$$\|(\Delta_{u,h}^l f)(u,0)\|_{L_{p,u}(\mathbb{R})} \leq M_3 \Big(\|(\Delta_{v,h}^l f)(u,0)\|_{L_{p,u}(\mathbb{R})}$$

$$+ |h|^l \sum_{\lambda=1}^{l} \Big\| \Big(\frac{\partial^l f}{\partial u^l}\Big)(u,\lambda h)\Big\|_{L_{p,u}(\mathbb{R})} \Big),$$

where $M_3 = 2^l$. Hence, by inequality (5.29)

$$\|f(\cdot,0)\|_{b_p^{l-\frac{1}{p}}(\mathbb{R})} = \| \, |h|^{-l} \|(\Delta_{u,h}^l f)(u,0)\|_{L_{p,u}(\mathbb{R})}\|_{L_{p,h}(\mathbb{R})}$$

$$\leq M_4 \Big(\| \, \| \, |h|^{-l}(\Delta_{u,h}^l f)(u,0)\|_{L_{p,h}(\mathbb{R})}\|_{L_{p,u}(\mathbb{R})}$$

$$+ \sum_{\lambda=1}^{l} \Big\| \Big\| \Big(\frac{\partial^l f}{\partial u^l}\Big)(u,\lambda h)\Big\|_{L_{p,u}(\mathbb{R})}\Big\|_{L_{p,h}(\mathbb{R})} \Big)$$

$$\leq M_5 \Big(\Big\|\frac{\partial^l f}{\partial v^l}\Big\|_{L_p(\mathbb{R}^2)} + \sum_{\lambda=1}^{l} \Big\| \Big(\frac{\partial^l f}{\partial u^l}\Big)(u,\lambda h)\Big\|_{L_p(\mathbb{R}^2)} \Big) \leq M_6 \|f\|_{w_p^l(\mathbb{R}^2)},$$

where M_4, M_5 and M_6 are independent of f.

4. In the general case, in which $1 \leq m < n$, $l > \frac{n-m}{p}$, we apply the identity

$$(\Delta_{u,h}^l f)(u,0) = \sum_{\lambda=0}^{l}(-1)^l \binom{l}{\lambda}(\Delta_{v,|h|\eta}^l f)(u+\lambda h,0)$$

$$- \sum_{\lambda=1}^{l}(-1)^l \binom{l}{\lambda}(\Delta_{u,h}^l f)(u,\lambda|h|\eta), \tag{5.36}$$

where $\eta \in S^{n-m-1}$, which also follows from (5.35) if we replace x by $E_{u,h}$ and y by $E_{v,|h|\eta}$. Taking spherical coordinates in $\mathbb{R}^m$ and using equality (4.116), we get

$$\|f(\cdot,0)\|_{b_p^{l-\frac{n-m}{p}}(\mathbb{R}^m)} \leq M_7 \Big(\| \, \| \, |h|^{-l+\frac{n-m}{p}-\frac{m}{p}}(\Delta_{v,|h|\eta}^l f)(u,0)\|_{L_{p,h}(\mathbb{R}^m)}\|_{L_{p,u}(\mathbb{R}^m)}$$

---

[13] This follows from the obvious identity for polynomials

$$(x-1)^l = (-1)^l(x-1)^l(y-1)^l + (x-1)^l(1-(-1)^l(y-1)^l)$$

$$= \sum_{\lambda=0}^{l}(-1)^\lambda \binom{l}{\lambda}x^\lambda(y-1)^l - \sum_{\lambda=1}^{l}(-1)^\lambda \binom{l}{\lambda}y^\lambda(x-1)^l. \tag{5.35}$$

if x is replaced by $E_{u,h}$ and y by $E_{v,h}$.

$$+ \sum_{|\gamma|=l} \sum_{\lambda=1}^{l} \| \, \| \, |h|^{\frac{n-m}{p}-\frac{m}{p}} (D^{(\gamma,0)}f)(u,\lambda|h|\eta)\|_{L_{p,h}(\mathbb{R}^m)} \|_{L_{p,u}(\mathbb{R}^m)} \Big)$$

$$= \sigma_m^{\frac{1}{p}} M_7 \Big(\| \, \| \varrho^{-l+\frac{n-m-1}{p}} (\Delta_{v,\varrho\eta}^l f)(u,0)\|_{L_{p,\varrho}(0,\infty)} \|_{L_{p,u}(\mathbb{R}^m)}$$

$$+ \sum_{|\gamma|=l} \sum_{\lambda=1}^{l} \| \, \| \varrho^{\frac{n-m-1}{p}} (D^{(\gamma,0)}f)(u,\lambda\varrho\eta)\|_{L_{p,\varrho}(0,\infty)} \|_{L_{p,u}(\mathbb{R}^m)} \Big).$$

Here M_7 is independent of f. Taking L_p-norms with respect to $\eta \in S^{n-m-1}$ and applying inequality (5.29), we get

$$\|f(\cdot,0)\|_{b_p^{l-\frac{n-m}{p}}(\mathbb{R}^m)} \leq \sigma_{n-m}^{-\frac{1}{p}} \sigma_m^{\frac{1}{p}} M_7 \Big(\| \, \| \, |h|^{-l} (\Delta_{v,h}^l)(u,0)\|_{L_{p,h}(\mathbb{R}^{n-m})} \|_{L_{p,u}(\mathbb{R}^m)}$$

$$+ \sum_{|\gamma|=l} \sum_{\lambda=1}^{l} \| \, \| (D^{(\gamma,0)}f)(u,\lambda v)\|_{L_{p,v}(\mathbb{R}^{n-m})} \|_{L_{p,u}(\mathbb{R}^m)} \Big)$$

$$\leq M_8 \Big(\| \sum_{|\beta|=l} \| (D^{(0,\beta)}f)(u,v)\|_{L_{p,v}(\mathbb{R}^{n-m})} \|_{L_{p,u}(\mathbb{R}^m)}$$

$$+ \sum_{|\gamma|=l} \sum_{\lambda=1}^{l} \| D^{(\gamma,0)}(u,\lambda v)\|_{L_p(\mathbb{R}^n)} \Big) \leq M_9 \|f\|_{w_p^l(\mathbb{R}^n)},$$

where M_8 and M_9 are independent of f. $\square$

In the proof of the second part of Theorem 3 we shall need the following statement.

Lemma 11 *Let $l \in \mathbb{N}, l > 1$. Suppose that the functions $\lambda, \nu \in L_\infty(\mathbb{R}^n)$, have compact supports and satisfy the equality* [14]

$$\lambda(z) = \sum_{k=1}^{l} (-1)^{l-k} \binom{l}{k} \frac{1}{k^n} \nu\Big(\frac{z}{k}\Big), \quad z \in \mathbb{R}^n. \tag{5.37}$$

Then $\forall f \in L_1^{loc}(\mathbb{R}^n)$ for almost all $x \in \mathbb{R}^n$

$$\int_{\mathbb{R}^n} (\Delta_h f)(x)\lambda(h)\,dh = \int_{\mathbb{R}^n} (\Delta_h^l f)(x)\nu(h)\,dh. \tag{5.38}$$

---

[14] We note that from (5.37) it follows that $\int_{\mathbb{R}^n} z^s \lambda(z)\,dz = 0$, $s = 1,...,l-1$.

Idea of the proof. Notice that from (5.38) it follows that

$$\int_{\mathbb{R}^n} \lambda(h)\,dh = (-1)^{l+1} \int_{\mathbb{R}^n} \nu(h)\,dh, \tag{5.39}$$

expand the difference $\Delta_h^l f$ in a sum and use appropriate change of variables for each term of that sum. $\square$

Proof. By (5.37) and (5.38)

$$\int_{\mathbb{R}^n} (\Delta_h^l f)(x)\nu(h)\,dh$$

$$= \sum_{k=1}^{l} (-1)^{l-k} \binom{l}{k} \int_{\mathbb{R}^n} f(x+kh)\nu(h)\,dh + (-1)^l f(x) \int_{\mathbb{R}^n} \nu(h)\,dh$$

$$= \sum_{k=1}^{l} (-1)^{l-k} \binom{l}{k} \frac{1}{k^n} \int_{\mathbb{R}^n} f(x+z)\nu\Big(\frac{z}{k}\Big)\,dz - f(x) \int_{\mathbb{R}^n} \lambda(z)\,dz$$

$$= \int_{\mathbb{R}^n} (f(x+z) - f(x))\lambda(z)\,dz = \int_{\mathbb{R}^n} (\Delta_h f)(x)\lambda(h)\,dh. \quad \square$$

Let $\omega \in C_0^\infty(\mathbb{R}^n)$ and let ω_δ where $\delta > 0$ be defined by $\omega_\delta(x) = \frac{1}{\delta^n}\omega(\frac{x}{\delta})$. We denote by $A_{\delta,\omega}$ the operator defined by $A_{\delta,\omega} f = \omega_\delta * f$ for $f \in L_1^{loc}(\mathbb{R}^n)$. (If, in addition, $\operatorname{supp}\omega \subset \overline{B(0,1)}$ and $\int_{\mathbb{R}^n} \omega\,dx = 1$, then $A_{\delta,\omega} = A_\delta$ is a standard mollifier, considered in Chapters 1 and 2).

Lemma 12 *Let* $l \in \mathbb{N}, 1 \le p \le \infty, \nu \in C^\infty(\mathbb{R}^n), \int_{\mathbb{R}^n} \nu\,dx = (-1)^{l+1}$ *and let* λ *be defined by* (5.37). *Then* [15] $\forall f \in L_p(\mathbb{R}^n)$

$$\|A_{\delta,\lambda} f - f\|_{L_p(\mathbb{R}^n)} \le c_{11}\omega_l(\delta, f)_p, \tag{5.40}$$

where $c_{11} > 0$ *is independent of* f *and* δ.

Idea of the proof. Notice that

$$(A_{\delta,\lambda} f)(x) - f(x) = \int_{\mathbb{R}^n} (f(x-z\delta) - f(x))\lambda(z)\,dz = \int_{\mathbb{R}^n} (\Delta_{-z\delta}^l f)(x)\nu(z)\,dz \tag{5.41}$$

---

[15] If $l = 1$, then $\lambda = \nu$ and (5.40) coincides with (1.8).

since by (5.39) $\int_{\mathbb{R}^n} \lambda \, dz = 1$, and apply Minkowski's inequality for integrals and (5.21). $\square$

Proof. Since the functions $\lambda(-\frac{h}{\delta})$ and $\nu(-\frac{h}{\delta})$ also satisfy (5.37), equality (5.38) still holds if we replace $\mu(h)$ and $\nu(h)$ by $\lambda(-\frac{h}{\delta})$ and $\nu(-\frac{h}{\delta})$. After substituting $h = -z\delta$ we obtain (5.41). Let $r > 0$ be such that $\operatorname{supp} \nu \subset \overline{B(0,r)}$. By Minkowski's inequality for integrals and (5.21)

$$\|A_{\delta,\lambda} f - f\|_{L_p(\mathbb{R}^n)} \le \int_{\mathbb{R}^n} \|\Delta^l_{-z\delta} f\|_{L_p(\mathbb{R}^n)} |\nu(z)| \, dz \le \sup_{|h| \le r\delta} \|\Delta^l_h f\|_{L_p(\mathbb{R}^n)} \|\nu\|_{L_1(\mathbb{R}^n)}$$

$$= \omega_l(r\delta, f)_p \|\nu\|_{L_1(\mathbb{R}^n)} \le (r+1)^l \|\nu\|_{L_1(\mathbb{R}^n)} \omega_l(\delta, f)_p. \quad \square$$

Corollary 7 *In addition to the assumptions of Lemma 12, let $\mu \in C_0^\infty(\mathbb{R}^n)$. If $\int_{\mathbb{R}^n} \mu \, dx = 1$, then $\forall f \in L_p(\mathbb{R}^n)$*

$$\|A_{\delta,\lambda*\mu} f - f\|_{L_p(\mathbb{R}^n)} \le c_{12} \, \omega_l(\delta, f)_p, \tag{5.42}$$

and if $\int_{\mathbb{R}^n} \mu \, dx = 0$, then $\forall f \in L_p(\mathbb{R}^n)$

$$\|A_{\delta,\lambda*\mu} f\|_{L_p(\mathbb{R}^n)} \le c_{13} \, \omega_l(\delta, f)_p, \tag{5.43}$$

where $c_{12}, c_{13} > 0$ are independent of f and δ.

Idea of the proof. Inequality (5.42) is a direct corollary of (5.40) because in this case $\int_{\mathbb{R}^n} (\lambda * \mu) \, dx = \int_{\mathbb{R}^n} \lambda \, dx \cdot \int_{\mathbb{R}^n} \mu \, dx = 1$. If $\int_{\mathbb{R}^n} \mu \, dx = 0$, starting from the equality

$$(A_{\delta,\lambda*\mu} f)(x) = \int_{\mathbb{R}^n} \left(\int_{\mathbb{R}^n} (f(x - z\delta - \xi\delta) - f(x))\lambda(z) \, dz \right) \mu(\xi) \, d\xi,$$

argue as in the proof of Lemma 12. $\square$

Idea of the proof of the inverse trace theorem. Define the "strips" G_k by

$$G_k = \{v \in \mathbb{R}^{n-m} : 2^{-k-1} < |v| \le 2^{-k}\}, \quad k \in \mathbb{Z}.$$

Consider an appropriate partition of unity (see Lemma 5 of Chapter 2), i.e., functions $\psi_k \in C_0^\infty(\mathbb{R}^n)$, $k \in \mathbb{Z}$, satisfying the following conditions: $0 \le \psi_k \le 1$, $\sum_{k=-\infty}^{\infty} \psi_k(v) = 1$, $v \ne 0$,

$$G_k \subset \operatorname{supp} \psi_k \subset \left\{v \in \mathbb{R}^{n-m} : \tfrac{7}{8} 2^{-k-1} \le |v| \le \tfrac{9}{8} 2^{-k}\right\}$$

$$\subset G_{k-1} \cup G_k \cup G_{k+1} \tag{5.44}$$

and

$$|(D^\gamma \psi_k)(v)| \le c_{14}\, 2^{k|\gamma|}, \quad k \in \mathbb{Z}, v \in \mathbb{R}^{n-m}, \gamma \in \mathbb{N}_0^{n-m}, \tag{5.45}$$

where $c_{14} > 0$ is independent of v and k.

Keeping in mind Definition 2 of Chapter 2, for $g \in B_p^{l-\frac{n-m}{p}}(\mathbb{R}^m)$ set

$$(Tg)(u, v) = \sum_{k=1}^{\infty} \psi_k(v)(A_{2^{-k},\omega}g)(u), \tag{5.46}$$

where

$$\omega = \lambda * \lambda \tag{5.47}$$

and the function λ is defined by equality (5.37), in which n is replaced by m and $\nu \in C_0^\infty(\mathbb{R}^m)$ is a fixed function satisfying [16] $\int_{\mathbb{R}^m} \nu\, du = (-1)^{l+1}$.

Prove that g is a trace of Tg on $\mathbb{R}^m$ by applying Definition 1 and property (1.8). To estimate $\|Tg\|_{L_p(\mathbb{R}^n)}$ apply inequality (1.7). Estimate $\|D^\alpha Tg\|_{L_p(\mathbb{R}^n)}$, where $\alpha = (\beta, \gamma)$, $\beta \in \mathbb{N}_0^m$, $\gamma \in \mathbb{N}_0^{n-m}$ and $|\alpha| = l$, via $\|g\|_{B_p^{l-\frac{n-m}{p}}(\mathbb{R}^m)}^{(3)}$. To do this differentiate (5.46) term by term, apply inequalities (2.58), (5.42) and (5.43) and the estimate

$$\|D^\gamma \psi_k\|_{L_p(\mathbb{R}^{n-m})} \le c_{15}\, 2^{k(|\gamma|-\frac{l-m}{p})}, \tag{5.48}$$

where $c_{15} > 0$ is independent of k, which follows directly from (5.45). $\square$

Proof. 1. By the properties of the functions ψ_k it follows that the sum in (5.42) is in fact finite. Moreover,

$$(Tg)(u, v) = \sum_{k=s-1}^{s+1} \psi_k(v)(A_{2^{-k},\omega}g)(u) \quad \text{on} \ \ \mathbb{R}^m \times G_s \tag{5.49}$$

and $(Tg)(u, v) = 0$ if $|v| \ge \frac{7}{16}$. Hence $Tg \in C^\infty(\mathbb{R}^n \setminus \mathbb{R}^m)$ and $\forall \alpha = (\beta, \gamma)$ where $\beta \in \mathbb{N}_0^m, \gamma \in \mathbb{N}_0^{n-m}$

$$(D^\alpha(Tg))(u, v) = \sum_{k=1}^{\infty} (D^\gamma \psi_k)(v) D^\beta((A_{2^{-k},\omega}g)(u))$$

---

[16] By (5.39) and the properties of convolutions it follows that $\int_{\mathbb{R}^m} \omega\, du = 1$. If $l = 1$, then $\lambda = \nu$. In this case one may consider an arbitrary $\omega \in C_0^\infty(\mathbb{R}^m)$ satisfying $\int_{\mathbb{R}^m} \omega\, du = 1$.

$$= \sum_{k=1}^{\infty} (D^{\gamma}\psi_k)(v) 2^{k|\beta|} (A_{2^{-k},\lambda*D^{\beta}\lambda}g)(u) \tag{5.50}$$

since, by the properties of mollifiers and convolutions,

$$D^{\beta}(A_{2^{-k},\lambda*\lambda}g) = 2^{k|\beta|}A_{2^{-k},D^{\beta}(\lambda*\lambda)}g = 2^{k|\beta|}A_{2^{-k},\lambda*D^{\beta}\lambda}g.$$

2. Let $|v| \leq \frac{7}{16}$. By (5.44) $\psi_k(v) = 0$ if $k \leq 0$. Hence

$$\sum_{k=1}^{\infty} \psi_k(v) = 1. \tag{5.51}$$

Let $s = s(v)$ be such that $2^{-s-1} < |v| \leq 2^{-s}$. Then by (5.51), (5.44), (5.42) and Minkowski's inequality

$$\|(Tg)(\cdot,v) - g(\cdot)\|_{L_p(\mathbb{R}^m)} = \left\| \sum_{k=s-1}^{s+1} \psi_k(v)(A_{2^{-k},\lambda*\lambda}g - g) \right\|_{L_p(\mathbb{R}^m)}$$

$$= \sum_{k=s-1}^{s+1} \psi_k(v) \|A_{2^{-k},\lambda*\lambda}g - g\|_{L_p(\mathbb{R}^m)} \leq M_1 \sum_{k=s-1}^{s+1} \omega_l(2^{-k},g)_p$$

$$\leq M_2\, 2^{-(s-1)(l-\frac{n-m}{p})} \sum_{k=s-1}^{s+1} 2^{k(l-\frac{n-m}{p})}\omega_l(2^{-k},g)_p$$

$$\leq M_3\, |v|^{l-\frac{n-m}{p}} \sum_{k=s-1}^{s+1} 2^{k(l-\frac{n-m}{p})}\omega_l(2^{-k},g)_p,$$

where M_1, M_2, M_3 are independent of g and v.

Since the function $g \in B_p^{l-\frac{n-m}{p}}(\mathbb{R}^m)$, by Lemma 6 it follows that the quantity $2^{k(l-\frac{n-m}{p})}\omega_l(2^{-k},g)_p \to 0$ as $k \to \infty$ if $1 \leq p < \infty$ and is bounded if $p = \infty$. Hence

$$\|(Tg)(\cdot,v) - g(\cdot)\|_{L_p(\mathbb{R}^m)} = o\left(|v|^{l-\frac{n-m}{p}}\right), \quad 1 \leq p < \infty \tag{5.52}$$

and

$$\|(Tg)(\cdot,v) - g(\cdot)\|_{L_\infty(\mathbb{R}^m)} = O\left(|v|^l\right) \tag{5.53}$$

as $v \to 0$ (hence $s \to \infty$). In particular, by Definition 1, if follows that g is a trace of Tg on R^m.

3. By (1.7)

$$\|(Tg)(\cdot,v)\|_{L_p(\mathbb{R}^m)} \le \sum_{k=1}^{\infty} \psi_k(v)\|A_{2^{-k},\omega}g\|_{L_p(\mathbb{R}^m)} \le M_4\,\|g\|_{L_p(\mathbb{R}^m)},$$

where $M_4 = \|\omega\|_{L_1(\mathbb{R}^m)}$. Since $(Tg)(u,v) = 0$ if $|v| \ge \frac{9}{16}$, we have

$$\|Tg\|_{L_p(\mathbb{R}^n)} \le M_5\,\|g\|_{L_p(\mathbb{R}^m)}, \tag{5.54}$$

where $M_5 = M_4\, v_{n-m}^{\frac{1}{p}}$.

4. Let $\alpha = (\beta,\gamma)$, where $\beta \in \mathbb{N}_0^m$, $\gamma \in \mathbb{N}_0^{n-m}$ and $|\alpha| = |\beta| + |\gamma| = l$. First suppose that $1 \le p < \infty$ and $\beta \ne 0$. Since the multiplicity of the covering $\{\psi_k\}_{k\in\mathbb{Z}}$ is equal to 2, by (2.58) we have

$$\|D^\alpha Tg\|_{L_p(\mathbb{R}^n)} \le 2^{1-\frac{1}{p}}\Big(\sum_{k=1}^{\infty}\|D^\gamma\psi_k\|_{L_p(\mathbb{R}^{n-m})}^p\, 2^{k|\beta|p}\Big\|A_{2^{-k},\lambda*D^\beta\lambda}g\Big\|_{L_p(\mathbb{R}^m)}^p\Big)^{\frac{1}{p}}.$$

Since $\int\limits_{\mathbb{R}^m} D^\beta\lambda\,du = 0$, by (5.43) and (5.48) we have

$$\|D^\alpha Tg\|_{L_p(\mathbb{R}^n)} \le M_6\Big(\sum_{k=1}^{\infty} 2^{k(l-\frac{n-m}{p})p}\omega_l(2^{-k},g)_p^p\Big)^{\frac{1}{p}} = M_6\,\|g\|_{B_p^{l-\frac{n-m}{p}}(\mathbb{R}^m)}^{(3)}, \tag{5.55}$$

where M_6 is independent of g.

If $\beta = 0$, then $\gamma \ne 0$ and by (5.51) $\sum\limits_{k=1}^{\infty}(D^\gamma\psi_k)(v) = 0$ for v satisfying $0 < |v| \le \frac{7}{16}$. Hence

$$(D^{(0,\gamma)}(Tg))(u,v) = \sum_{k=1}^{\infty}(D^\gamma\psi_k)(v)((A_{2^{-k},\lambda*\lambda}g)(u) - g(u)),\quad 0 < |v| \le \tfrac{7}{16}.$$

Furthermore, $\psi_k(v) = 0$ if $|v| \ge \frac{7}{16}$ and $k \ge 2$. Therefore

$$(D^{(0,\gamma)}(Tg))(u,v) = (D^\gamma\psi_1)(v)(A_{2^{-1},\lambda*\lambda}g)(u),\quad |v| \ge \tfrac{7}{16}.$$

Consequently, by (2.58), (5.42), (5.48)

$$\|D^{(0,\gamma)}(Tg)\|_{L_p(\mathbb{R}^m\times\tilde{B}_{\frac{7}{16}})} \le 2^{1-\frac{1}{p}}\Big(\sum_{k=1}^{\infty}\|D^\gamma\psi_k\|_{L_p(\mathbb{R}^{n-m})}^p\|A_{2^{-k},\lambda*\lambda}g - g\|_{L_p(\mathbb{R}^m)}^p\Big)^{\frac{1}{p}}$$

$$\leq M_7 \, \|g\|^{(3)}_{B_p^{l-\frac{n-m}{p}}(\mathbb{R}^m)} \tag{5.56}$$

and by (1.7)

$$\|D^{(0,\gamma)}(Tg)\|_{L_p(\mathbb{R}^m \times {}^c\tilde{B}_{\frac{7}{16}})}$$

$$= \|D^\gamma \psi_1\|_{L_p(\mathbb{R}^{n-m})} \|A_{2^{-1},\lambda*\lambda}g\|_{L_p(\mathbb{R}^m)} \leq M_8 \, \|g\|_{L_p(\mathbb{R}^m)}, \tag{5.57}$$

where M_7 and M_8 are independent of g.

If $p = \infty$, then the argument is similar. For example, if $\beta \neq 0$, then

$$\|D^\alpha Tg\|_{L_\infty(\mathbb{R}^n)} \leq 2 \sup_{k \in \mathbb{N}} \|D^\gamma \psi_k\|_{L_\infty(\mathbb{R}^{n-m})} 2^{k|\beta|} \|A_{2^{-k},\lambda*D^\beta\lambda}g\|_{L_\infty(\mathbb{R}^m)}$$

$$\leq M_9 \sup_{k \in \mathbb{N}} 2^{kl}\omega_l(2^{-k}, g)_\infty = M_9 \, \|g\|^{(3)}_{B_\infty^l(\mathbb{R}^m)},$$

where M_9 is independent of g.

From (5.54) – (5.57) it follows that

$$\|Tg\|_{W_p^l(\mathbb{R}^n)} \leq M_{10} \, \|g\|^{(3)}_{B_p^{l-\frac{n-m}{p}}(\mathbb{R}^m)}, \tag{5.58}$$

where M_{10} is independent of g. $\square$

Corollary 8 *Let* $l, m, n \in \mathbb{N}$, $m < n$, $1 \leq p \leq \infty$, $l > \frac{n-m}{p}$ *and let the operator* T *be defined by* (5.46). *Then*

$$Tg\Big|_{\mathbb{R}^m} = g; \quad D^\alpha(Tg)\Big|_{\mathbb{R}^m} = 0, \quad 0 < |\alpha| < l - \frac{n-m}{p}. \tag{5.59}$$

Idea of the proof. Establish, as in step 2 of the proof of the second part of Theorem 3, that, in addition to (5.52) and (5.53),

$$\|(D^\alpha(Tg))(\cdot, v)\|_{L_p(\mathbb{R}^m)} = o(|v|^{l-|\alpha|-\frac{n-m}{p}}), \quad 1 \leq p < \infty. \tag{5.60}$$

and

$$\|D^\alpha(Tg)(\cdot, v)\|_{L_\infty(\mathbb{R}^m)} = O(|v|^{l-|\alpha|}) \tag{5.61}$$

as $v \to 0$. $\square$

Proof. Let $\alpha = (\beta, \gamma)$, where $\beta \in \mathbb{N}_0$, $\gamma \in \mathbb{N}_0^{n-m}$, and $2^{-s-1} < |v| \leq 2^{-s}$. If $\beta \neq 0$, then by (5.49) and (5.43)

$$\|(D^\alpha(Tg))(\cdot, v)\|_{L_p(\mathbb{R}^m)} \leq \sum_{k=s-1}^{s+1} |(D^\gamma \psi_k)(v)| \, 2^{k|\beta|} \|A_{2^{-k},\lambda*D^\beta\lambda}g\|_{L_p(\mathbb{R}^m)}$$

$$\leq M_1 \sum_{k=s-1}^{s+1} 2^{k|\alpha|}\,\omega_l(2^{-k}, g)_p$$

$$\leq M_2 \,|v|^{l-|\alpha|-\frac{n-m}{p}} \sum_{k=s-1}^{s+1} 2^{k(l-\frac{n-m}{p})}\,\omega_l(2^{-k}, g)_p, \tag{5.62}$$

where M_1, M_2 are independent of g and v.

If $\beta = 0$, then by (5.49) and (5.51)

$$\left\| (D^{(0,\gamma)}(Tg))(\cdot, v) \right\|_{L_p(\mathbb{R}^m)} = \left\| \sum_{k=s-1}^{s+1} (D^\gamma \psi_k)(v)\,(A_{2^{-k}, \lambda * \lambda} g - g) \right\|_{L_p(\mathbb{R}^m)}$$

$$\leq M_3 \sum_{k=s-1}^{s+1} 2^{k|\alpha|}\,\left\| A_{2^{-k}, \lambda * \lambda} g - g \right\|_{L_p(\mathbb{R}^m)}$$

and by (5.42) we again obtain (5.62).

Relations (5.60) and (5.61) follow from (5.62) as in step 2 of the proof of the second part of Theorem 3. $\square$

The following stronger statement follows from the proof of the second part of Theorem 3.

Theorem 4 *Let* $l, m, n \in \mathbb{N}$, $m < n$, $1 \leq p \leq \infty$, $l > \frac{n-m}{p}$. *Then there exists a bounded linear extension operator*

$$T : B_p^{l-\frac{n-m}{p}}(\mathbb{R}^m) \to W_p^l(\mathbb{R}^n) \bigcap C^\infty(\mathbb{R}^n \setminus \mathbb{R}^m) \tag{5.63}$$

satisfying the inequalities

$$\| \, |v|^{|\alpha|-l} D^\alpha(Tg) \|_{L_p(\mathbb{R}^n)} \leq c_{16}\, \|g\|_{B_p^{l-\frac{n-m}{p}}(\mathbb{R}^m)}, \quad |\alpha| > 0, \tag{5.64}$$

and

$$\| \, |v|^{-l}(Tg - g) \|_{L_p(\mathbb{R}^n)} \leq c_{17}\, \|g\|_{B_p^{l-\frac{n-m}{p}}(\mathbb{R}^m)}, \tag{5.65}$$

where $c_{16}, c_{17} > 0$ *are independent of* g.

In (5.64) the exponent $|\alpha| - l$ *can not be replaced by* $|\alpha| - l - \varepsilon$ *for any* $\varepsilon > 0$ *and for any extension operator (5.63).*

Idea of the proof. Consider the extension operator (5.44) used in Theorem 3. To prove (5.64) apply, in addition, the inequality $2^{-k-2} \leq |v| \leq 2^{-k+1}$ for $v \in \operatorname{supp} \psi_k$. To prove the second statement apply Remark 11. $\square$

Proof of the first statement of Theorem 4. 1. Let $\alpha = (\beta, \gamma)$, where $\beta \in \mathbb{N}_0^m$, $\gamma \in \mathbb{N}_0^{n-m}$, $|\alpha| > 0$ and $s = |\alpha| - l = |\beta| + |\gamma| - l$. Then as for (5.55) we obtain

$$\| \, |v|^s (D^\alpha(Tg)) \|_{L_p(\mathbb{R}^n)}$$

$$\leq 2^{1-\frac{1}{p}} \Big(\sum_{k=1}^\infty \| \, |v|^s D^\gamma \psi_k \|^p_{L_p(\mathbb{R}^{n-m})} \, \| A_{2^{-k}, \lambda * D^\beta \lambda} g \|^p_{L_p(\mathbb{R}^m)} \Big)^{\frac{1}{p}}$$

$$\leq M_1 \Big(\sum_{k=1}^\infty 2^{-ksp} 2^{k(|\gamma| - \frac{n-m}{p})p} 2^{k|\beta|p} \, \omega_l(2^{-k}, g)_p^p \Big)^{\frac{1}{p}} \tag{5.66}$$

$$= M_1 \Big(\sum_{k=1}^\infty \Big(2^{k(l - \frac{n-m}{p})} \omega_l(2^{-k}, g)_p \Big)^p \Big)^{\frac{1}{p}} = M_1 \| g \|^{(3)}_{B_p^{l - \frac{n-m}{p}}(\mathbb{R}^m)},$$

where M_1 is independent of g. The proof of the appropriate analogues of (5.56) and (5.57) is similar and we arrive at (5.64). $\square$

2. Furthermore, as for (5.56) and (5.57)

$$(Tg)(u, v) - g(v) = \sum_{k=1}^\infty \psi_k(v)((A_{2^{-k}, \lambda * \lambda} g)(u) - g(u)), \quad 0 < |v| \leq \tfrac{7}{16},$$

and

$$(Tg)(u, v) - g(v) = \psi_1(v)(A_{2^{-1}, \lambda * \lambda} g)(u) - g(u), \quad |v| \geq \tfrac{7}{16}.$$

Hence by (5.42)

$$\| \, |v|^{-l}((Tg)(u, v) - g(v)) \|_{L_p \left(\mathbb{R}^m \times \tilde{B}_{\frac{7}{16}} \right)}$$

$$\leq 2^{1-\frac{1}{p}} \Big(\sum_{k=1}^\infty \| \, |v|^{-l} \psi_k \|^p_{L_p(\mathbb{R}^{n-m})} \, \| A_{2^{-k}, \lambda * \lambda} g - g \|^p_{L_p(\mathbb{R}^m)} \Big)^{\frac{1}{p}}$$

$$\leq M_2 \| g \|_{B_p^{l - \frac{n-m}{p}}(\mathbb{R}^m)}$$

and

$$\| \, |v|^{-l}((Tg)(u, v) - g(v)) \|_{L_p \left(\mathbb{R}^m \times {}^c\tilde{B}_{\frac{7}{16}} \right)}$$

$$\leq \| \, |v|^{-l} \|_{L_p \left({}^c\tilde{B}_{\frac{7}{16}} \right)} \Big(\| A_{2^{-1}, \lambda * \lambda} g \|_{L_p(\mathbb{R}^m)} + \| g \|_{L_p(\mathbb{R}^n)} \Big) \leq M_3 \| g \|_{L_p(\mathbb{R}^m)},$$

where M_2 and M_3 are independent of g, and (5.65) follows. $\square$

Remark 11 Let $m, n \in \mathbb{N}$, $m < n$, $\mathbb{R}^n_+ = \{x = (u, v) \in \mathbb{R}^n : v > 0\}$[17], $l \in \mathbb{N}$, $1 \leq p \leq \infty$, $s \geq 0$. We shall say that the function f belongs to the *weighted Sobolev space* $W^l_{p,\,|v|^s}(\mathbb{R}^n_+)$ if $f \in L_p(\mathbb{R}^n_+)$, if it has weak derivatives $D^\alpha_w f$ on $\mathbb{R}^n_+$ for all $\alpha \in \mathbb{N}^n_0$ satisfying $|\alpha| = l$ and

$$\|f\|_{W^l_{p,\,|v|^s}(\mathbb{R}^n_+)} = \|f\|_{L_p(\mathbb{R}^n_+)} + \sum_{|\alpha|=l} \big\| \, |v|^s D^\alpha_w f \big\|_{L_p(\mathbb{R}^n_+)} < \infty. \tag{5.67}$$

We note that the set $C^\infty(\mathbb{R}^n) \bigcap W^l_{p,\,|v|^s}(\mathbb{R}^n_+)$ is dense in $W^l_{p,\,|v|^s}(\mathbb{R}^n_+)$. This is proved as in Lemma 25 of Chapter 6.

Suppose that $l - s - \frac{n-m}{p} > 0$. Then

$$\mathrm{tr}_{\mathbb{R}^m} W^l_{p,\,|v|^s}(\mathbb{R}^n_+) = B_p^{l-s-\frac{n-m}{p}}(\mathbb{R}^m). \tag{5.68}$$

The idea of the proof is essentially the same as in Theorem 3.

The proof of the extension theorem is like that of Theorem 4. If in (5.65) $|\alpha| = l$, then the same argument shows that

$$\big\| \, |v|^s D^\alpha T g \big\|_{L_p(\mathbb{R}^n_+)} \leq M_1 \|g\|_{B_p^{l-s-\frac{n-m}{p}}(\mathbb{R}^m)},$$

etc.

To prove the direct trace theorem one needs to follow, step by step, the proof of the first part of Theorem 3 and apply the inequality

$$\left\| \, |h|^s \frac{(\Delta^l_h f)(0)}{|h|^l} \right\|_{L_{p,h}(\mathbb{R}^{n-m})} \leq M_2 \|f\|_{W^l_{p,\,|v|^s}(\mathbb{R}^{n-m})},$$

where $1 \leq p \leq \infty$, $l - s - \frac{n-m}{p} > 0$ and M_2 is independent of f, instead of (5.29) (with $n - m$ replacing n and $x = 0$). The last inequality, as (5.29), is also proved by applying inequality (5.15).

Proof of the second statement of Theorem 4. Suppose that (5.64) holds with $|\alpha| - l - \varepsilon$ replacing $|\alpha| - l$, where $\varepsilon > 0$. Let $g \in B_p^{l-\frac{n-m}{p}}(\mathbb{R}^m) \setminus B_p^{l+\varepsilon-\frac{n-m}{p}}(\mathbb{R}^m)$. Then $Tg \in W^{l_1}_{p,\,|v|^{l_1-l-\varepsilon}}(\mathbb{R}^n_+)$ where $l_1 \in \mathbb{N}$, $l_1 > l + \varepsilon$. Since g is a trace of Tg, by (5.67) $g \in B_p^{l+\varepsilon-\frac{n-m}{p}}(\mathbb{R}^m)$ and we have arrived at a contradiction. $\square$

We note that from (5.65) it follows, in particular, that $Tg\big|_{\mathbb{R}^m} = g$. This may be deduced as a corollary of the following more general statement.

---

[17] We recall that $v = (x_{m+1}, ..., x_n) > 0$ means that $x_{m+1} > 0, ..., x_n > 0$.

Lemma 13 *Let $l, m, n \in \mathbb{N}$, $m < n$, $1 \leq p \leq \infty$, $l > \frac{n-m}{p}$ if $p > 1$ and $l \geq n - m$ if $p = 1$. Suppose that λ is a nonnegative function measurable on $\mathbb{R}^{n-m}$, which is such that $\|\lambda\|_{L_p(\widetilde{B}_\varepsilon)} = \infty$ for each $\varepsilon > 0$. Moreover, let $f \in L_1^{loc}(\mathbb{R}^n)$, for $\gamma \in \mathbb{N}_0^{n-m}$ satisfying $|\gamma| = l$ the weak derivatives $D_w^{(0,\gamma)} f$ exist on $\mathbb{R}^n$ and*

$$\|\lambda f\|_{L_p(\mathbb{R}^n)} + \sum_{|\gamma|=l} \|D_w^{(0,\gamma)} f\|_{L_p(\mathbb{R}^n)} < \infty.$$

Then $f|_{\mathbb{R}^m} = 0$.

Idea of the proof. Using the embedding Theorem 12 and the proof of Corollary 20 of Chapter 4, establish that there exists a function G, which is equivalent to f on $\mathbb{R}^n$ and is such that the function $\|G(\cdot, v)\|_{L_p(\mathbb{R}^m)}$ is uniformly continuous on $\mathbb{R}^{n-m}$. $\square$

Proof. Let us consider the case $p < \infty$, the case $p = \infty$ being similar. By Theorem 6 of Chapter 4 $f \in L_p(\mathbb{R}^n)$, hence, $f \in W_p^l(\mathbb{R}^n)$ and for almost all $u \in \mathbb{R}^m$ we have $f(u, \cdot) \in W_p^l(\mathbb{R}^{n-m})$. By Theorem 12 of Chapter 4 there exists a function $g_u(\cdot) \in C(\mathbb{R}^{n-m})$ such that $\forall v \in \mathbb{R}^{n-m}$

$$|g_u(v)| \leq M_1 \left(\|f(u, \cdot)\|_{L_p(\mathbb{R}^{n-m})} + \sum_{|\gamma|=l} \|(D_w^{(0,\gamma)} f)(u, \cdot)\|_{L_p(\mathbb{R}^{n-m})} \right),$$

where M_1 is independent of f and u. Let $G(u, v) = g_u(v)$, $u \in \mathbb{R}^m$, $v \in \mathbb{R}^{n-m}$. Then $G \sim f$ on $\mathbb{R}^n$ and

$$\| \|G(u, v)\|_{L_{p,u}(\mathbb{R}^m)} \|_{C_v(\mathbb{R}^{n-m})} \leq M_1 \left(\|f\|_{L_p(\mathbb{R}^n)} + \sum_{|\gamma|=l} \|D_w^{(0,\gamma)} f\|_{L_p(\mathbb{R}^n)} \right).$$

As in the proof of Corollary 20 of Chapter 4, let $f_k \in C_0^\infty(\Omega)$ and [18]

$$\|f - f_k\|_{L_p(\mathbb{R}^n)} + \sum_{|\gamma|=l} \|D_w^{(0,\gamma)} f - D^{(0,\gamma)} f_k\|_{L_p(\mathbb{R}^n)} \to 0$$

as $k \to \infty$. Then, by the triangle inequality,

$$\| \|G(u, v)\|_{L_{p,u}(\mathbb{R}^m)} - \|f_k(u, v)\|_{L_{p,u}(\mathbb{R}^m)} \|_{C_v(\mathbb{R}^{n-m})}$$

$$\leq \| \|G(u, v) - f_k(u, v)\|_{L_{p,u}(\mathbb{R}^m)} \|_{C_v(\mathbb{R}^{n-m})} \to 0$$

as $k \to \infty$. Since the functions $\|f_k(u, \cdot)\|_{L_{p,u}(\mathbb{R}^m)}$ are uniformly continuous on $\mathbb{R}^{n-m}$, the function $H(\cdot) = \|G(u, \cdot)\|_{L_{p,u}(\mathbb{R}^m)}$ is also uniformly

---

[18] The existence of such f_k is establised as in Lemma 2 of Chapter 2.

continuous on $\mathbb{R}^{n-m}$. So there exists $\lim_{v \to 0} H(v) = A$. If $A > 0$, then $\frac{A}{2}\|\lambda\|_{L_p(\tilde{B}_\varepsilon)} \leq \|\lambda H\|_{L_p(\mathbb{R}^{n-m})} = \|\lambda f\|_{L_p(\mathbb{R}^n)}$ for sufficiently small $\varepsilon > 0$. This is impossible because $\|\lambda\|_{L_p(\tilde{B}_\varepsilon)} = \infty$ and $\|\lambda f\|_{L_p(\mathbb{R}^n)} < \infty$. Hence $A = 0$, i.e., $\lim_{v \to 0} \|G(\cdot, v)\|_{L_p(\mathbb{R}^m)} = 0$ and by Definition 1 $f|_{\mathbb{R}^m} = 0$. $\square$

The next theorem deals with the case $p = 1$, $l = n - m$, which was not considered in Theorem 3.

Theorem 5 *Let $m, n \in \mathbb{N}$, $m < n$. Then*

$$\mathrm{tr}_{\mathbb{R}^m} W_1^{n-m}(\mathbb{R}^m) = L_1(\mathbb{R}^m). \tag{5.69}$$

Idea of the proof. The direct trace trace theorem follows from Theorem 2 and, in particular, from inequality (5.6). To prove the inverse trace ($\equiv$ extension) theorem it is enough to construct an extension operator $T : L_1(\mathbb{R}^m) \to W_1^1(\mathbb{R}^{m+1})$ and iterate it to obtain an extension operator $T : L_1(\mathbb{R}^m) \to W_1^{n-m}(\mathbb{R}^n)$. However, it is more advantageous to give a direct construction for arbitrary $n > m$. Start from an arbitrary sequence $\{\delta_k\}_{k \in \mathbb{Z}}$ of posivite numbers δ_k satisfying $\delta_{k+1} \leq \frac{\delta_k}{2}$, $\sum_{k=0}^{\infty} \delta_k \leq 1$ and consider the sets $G_k = \{v \in \mathbb{R} : \mu_{k+1} < |v| \leq \mu_k\}$, where $\mu_k = \sum_{s=k}^{\infty} \delta_s$. (Note that $\leq 2\delta_k$.) Verify that from the proof of Lemma 4 of Chapter 2 it follows that there exist functions $\psi_k \in C_0^\infty(\mathbb{R}^{n-m})$ satisfying the following conditions: $0 \leq \psi_k \leq 1$,
$$\sum_{k=-\infty}^{\infty} \psi_k(v) = 1, v \neq 0,$$

$$G_k \subset \mathrm{supp}\, \psi_k \subset \{v \in \mathbb{R} : \mu_{k+1} - \tfrac{\delta_{k+1}}{4} \leq |v| \leq \mu_k + \tfrac{\delta_k}{4}\}$$

$$\subset G_{k-1} \cup G_k \cup G_{k+1} \tag{5.70}$$

and

$$\|D^\gamma \psi_k\|_{L_1(\mathbb{R}^{n-m})} \leq M_1 \delta_k^{n-m-|\gamma|}, \quad \gamma \in \mathbb{N}_0^{n-m}, |\gamma| \leq n - m, \tag{5.71}$$

where M_1 is independent of k.

For $g \in L_1(\mathbb{R}^m)$ set

$$(Tg)(u, v) = \sum_{k=1}^{\infty} \psi_k(v)(A_{\delta_k, \omega} g)(u), \tag{5.72}$$

where ω is the same as in (5.46). Prove as in the second part of the proof of Theorem 3 that

$$\|Tg\|_{W_1^{n-m}(\mathbb{R}^n)} \leq M_2 \left(\|g\|_{L_1(\mathbb{R}^m)} + \sum_{k=1}^{\infty} \omega_l(\delta_k, g)_1 \right), \qquad (5.73)$$

where M_2 is independent of g and δ_k. Since $\omega_l(\delta_k, g)_1 \to 0$ as $k \to \infty$, choose δ_k depending on g in such a way that, in addition, $\omega_l(\delta_k, g)_1 \leq 2^{-k} \|g\|_{L_1(\mathbb{R}^m)}$. Hence

$$\|Tg\|_{W_1^{n-m}(\mathbb{R}^n)} \leq M_3 \|g\|_{L_1(\mathbb{R}^m)}, \qquad (5.74)$$

where $M_3 = 2M_2$. $\square$

Proof. 1. Since $\frac{3}{4}\delta_k \leq \mu_k - \frac{\delta_k}{4} \leq \mu_k + \frac{\delta_k}{4} \leq \frac{9}{4}\delta_k$ and $|\gamma| \leq n - m$, inequality (5.71) follows from (2.10):

$$\|D^\gamma \psi_k\|_{L_1(\mathbb{R}^{n-m})} = \left\| D^\gamma A_{\frac{\delta_{k+1}}{4}} \chi_k \right\|_{L_1\left(\mu_{k+1} - \frac{\delta_{k+1}}{4} \leq |v| \leq \mu_{k+1} + \frac{\delta_{k+1}}{4} \right)}$$

$$+ \left\| D^\gamma A_{\frac{\delta_k}{4}} \chi_k \right\|_{L_1\left(\mu_k - \frac{\delta_k}{4} \leq |v| \leq \mu_k + \frac{\delta_k}{4} \right)}$$

$$\leq M_4 \left(\delta_{k+1}^{-|\gamma|} \operatorname{meas}_{n-m} \widetilde{B}(0, 3\delta_{k+1}) + \delta_k^{-|\gamma|} \operatorname{meas}_{n-m} \widetilde{B}(0, 3\delta_k) \right)$$

$$= M_5 \left(\delta_{k+1}^{n-m-|\gamma|} + \delta_k^{n-m-|\gamma|} \right) \leq 2 M_5 \, \delta_k^{n-m-|\gamma|},$$

where M_4 and M_5 depend only on $n - m$.

2. Let $|v| \leq \mu_1 - \frac{\delta_1}{4}$. By (5.70) $\psi_k(v) = 0$ if $k \leq 0$ and hence

$$\sum_{k=1}^{\infty} \psi_k(v) = 1. \qquad (5.75)$$

Let $s = s(v)$ be such that $v \in G_s$. Then by (5.75) and (1.9)

$$\|(Tg)(\cdot, v) - v(\cdot)\|_{L_1(\mathbb{R}^m)} = \left\| \sum_{k=s-1}^{s+1} \psi_k(v) \left(A_{2^{-k}, \omega} g - g \right) \right\|_{L_1(\mathbb{R}^m)}$$

$$\leq \sum_{k=s-1}^{s+1} \left\| A_{2^{-k}, \omega} g - g \right\|_{L_1(\mathbb{R}^m)} \to 0 \qquad (5.76)$$

as $v \to 0$ (hence $s \to \infty$). Thus by Definition 1 g is a trace of Tg on $\mathbb{R}^m$.

3. Since $(Tg)(u,v) = 0$ if $|v| \geq \mu_1 + \frac{\delta_1}{4}$ and $\mu_1 + \frac{\delta_1}{4} \leq \mu_0$, we have

$$\|Tg\|_{L_1(\mathbb{R}^n)} = \left\| \sum_{k=1}^{\infty} \psi_k \|A_{\delta_k,\omega}g\|_{L_1(\mathbb{R}^m)} \right\|_{L_1(\mathbb{R}^{n-m})}$$

$$\leq M_6 \left\| \sum_{k=1}^{\infty} \psi_k \right\|_{L_1(\mathbb{R}^{n-m})} \|g\|_{L_1(\mathbb{R}^m)} \leq M_7 \mu_0 \|g\|_{L_1(\mathbb{R}^m)} \leq M_7 \|g\|_{L_1(\mathbb{R}^m)}, \quad (5.77)$$

where M_6 and M_7 are independent of g and δ_k.

4. Let $\alpha = (\beta, \gamma)$, where $\beta \in \mathbb{N}_0^m, \gamma \in \mathbb{N}_0^{n-m}$ and $|\alpha| = |\beta| + |\gamma| = l$. If $\beta \neq 0$, as for (5.55) we obtain

$$\|D^\alpha Tg\|_{L_1(\mathbb{R}^n)} \leq \sum_{k=1}^{\infty} \|D^\gamma \psi_k\|_{L_1(\mathbb{R}^{n-m})} \delta_k^{-|\beta|} \|A_{\delta_k,\lambda * D^\beta \lambda} g\|_{L_1(\mathbb{R}^m)}$$

$$\leq M_8 \sum_{k=1}^{\infty} \delta_k^{n-m-|\gamma|-|\beta|} \omega_l(\delta_k, g)_1 = M_8 \sum_{k=1}^{\infty} \omega_l(\delta_k, g)_1, \quad (5.78)$$

where M_8 is independent of g and δ_k.

If $\beta = 0$, then starting from (5.56), where now $0 < |v| \leq \mu_1 + \frac{\delta_1}{4}$, and (5.57), where $|v| \geq \mu_1 + \frac{\delta_1}{4}$, we have as for (5.58) and (5.59)

$$\|D^{(0,\gamma)}(Tg)\|_{L_1\left(\mathbb{R}^m \times \tilde{B}_{\mu_1+\frac{\delta_1}{4}}\right)} \leq \sum_{k=1}^{\infty} \|D^\gamma \psi_k\|_{L_1(\mathbb{R}^{n-m})} \|A_{\delta_k,\lambda*\lambda}g - g\|_{L_1(\mathbb{R}^m)}$$

$$\leq M_9 \sum_{k=1}^{\infty} \omega_l(\delta_k, g)_1 \quad (5.79)$$

and

$$\|D^{(0,\gamma)}(Tg)\|_{L_1\left(\mathbb{R}^m \times^c \tilde{B}_{\mu_1+\frac{\delta_1}{4}}\right)} = \|D^\gamma \psi_1\|_{L_1(\mathbb{R}^{n-m})} \|A_{\delta_k,\lambda*\lambda}g\|_{L_1(\mathbb{R}^m)}$$

$$\leq M_{10} \|g\|_{L_1(\mathbb{R}^m)}, \quad (5.80)$$

where M_9 and M_{10} are independent of g and δ_k. So we have established (5.73). $\square$

Remark 12 If $m = n - 1$, then in fact

$$\|\psi_k\|_{L_1(\mathbb{R})} = 2\delta_k, \quad \|\psi_k'\|_{L_1(\mathbb{R})} = 4, \quad \|\psi_k'\|_{L_1(|v|\geq\mu_k-\frac{\delta_k}{4})} = 2. \quad (5.81)$$

Given $\varepsilon > 0$, this allows one to construct, choosing appropriate $\delta_k = \delta_k(g)$, an extension operator $T : L_1(\mathbb{R}^{n-1}) \to W_1^1(\mathbb{R}^n)$ satisfying $\|T\| \leq 2 + \varepsilon$.

Remark 13 The extension operator $T : L_1(\mathbb{R}^m) \to W_1^{n-m}(\mathbb{R}^n)$ defined by (5.65) is a bounded nonlinear operator since δ_k depend on g. It can be proved that a bounded linear extension operator $T : L_1(\mathbb{R}^m) \to W_1^{n-m}(\mathbb{R}^n)$ does not exist. However, there exists a bounded linear extension operator $T : L_1(\mathbb{R}^m) \to B_{1,2}^{n-m}(\mathbb{R}^n)$ acting from $L_1(\mathbb{R}^m)$ into slightly larger space $B_{1,2}^{n-m}(\mathbb{R}^n)$ than $W_1^{n-m}(\mathbb{R}^n)$ (see Remark 6).

Theorem 6 *Let $m, n \in \mathbb{N}$, $m < n$. Then there exists a bounded nonlinear extension operator*

$$T : L_1(\mathbb{R}^m) \to W_1^{n-m}(\mathbb{R}^n) \bigcap C^\infty(\mathbb{R}^n \setminus \mathbb{R}^m) \tag{5.82}$$

satisfying the inequalities

$$\| \, |v|^{|\alpha|-(n-m)} D^\alpha(Tg) \|_{L_1(\mathbb{R}^n)} \leq c_{18} \|g\|_{L_1(\mathbb{R}^m)}, \quad |\alpha| > 0, \tag{5.83}$$

and [19]

$$\| \, |v|^{-(n-m)} (Tg - g) \|_{L_1(\mathbb{R}^m \times \tilde{B}_1)} \leq c_{19} \|g\|_{L_1(\mathbb{R}^m)}, \tag{5.84}$$

where $c_{18}, c_{19} > 0$ are independent of g.

In (5.83) the exponent $|\alpha| - (n-m)$ can not be replaced by $|\alpha| - (n-m) - \varepsilon$ for any $\varepsilon > 0$ and for any extension operator (5.82).

Idea of the proof. As in Theorem 5 consider the extension operator (5.72) . To prove (5.83) and (5.84) note, in addition, that $|v| \leq \frac{9}{4}\delta_k$ on $\operatorname{supp}\psi_k$ and

$$\| \, |v|^{|\gamma|-(n-m)} (D^\gamma \psi_k)(v) \|_{L_1(\mathbb{R}^{n-m})} \leq M_1, \quad \gamma \in \mathbb{N}_0^{n-m}, \tag{5.85}$$

where M_1 is independent of k. The second statement of the theorem is proved as the second statement of Theorem 4. $\square$

Proof. 1. Since $\frac{3}{4}\delta_k \leq \mu_k - \frac{\delta_k}{4} \leq \mu_k - \frac{\delta_k}{4} \leq \frac{9}{4}\delta_k$ and $|D^\gamma(A_{\frac{\delta_k}{4}}\chi_k)(v)| \leq M_2 \delta_k^{-|\gamma|}$, where M_2 is independent of v and k, by (2.10) we have

$$\sup_{v \in \mathbb{R}^{n-m}} |v|^\gamma |(D^\gamma \psi_k)(v)| = \max \Bigg\{ \sup_{\frac{3}{4}\delta_{k+1} \leq |v| \leq \frac{9}{4}\delta_{k+1}} |v|^{|\gamma|} |(D^\gamma(A_{\frac{\delta_{k+1}}{4}}\chi_k)(v)|, $$

$$\sup_{\frac{3}{4}\delta_k \leq |v| \leq \frac{9}{4}\delta_k} |v|^{|\gamma|} |(D^\gamma(A_{\frac{\delta_k}{4}}\chi_k)(v)| \Bigg\} \leq M_3,$$

---

[19] By Lemma 13 from (5.84) it follows directly that $Tg\big|_{\mathbb{R}^m} = g$.

where M_3 is independent of k. Hence

$$\big\| |v|^{|\gamma|-(n-m)} \, (D^\gamma \psi_k)(v) \big\|_{L_1(\mathbb{R}^{n-m})} \leq M_3 \, \big\| \, |v|^{-(n-m)} \big\|_{L_1\left(\frac{3}{4}\delta_{k+1} \leq |v| \leq \frac{9}{4}\delta_k\right)}$$

$$= M_3 \, \sigma_{n-m} \int\limits_{\frac{3}{4}\delta_{k+1}}^{\frac{9}{4}\delta_k} \varrho^{-1} \, d\varrho = M_3 \, \sigma_{n-m} \ln \frac{3\delta_k}{\delta_{k+1}}$$

and (5.85) is established with $M_1 = M_3 \, \sigma_{n-m} \ln 6$.

2. If $\alpha = (\beta, \gamma)$, where $\beta \in \mathbb{N}_0^m$, $\gamma \in \mathbb{N}_0^{n-m}$ and $\beta \neq 0$, then as in step 4 of the proof of Theorem 5

$$\big\| \, |v|^{|\alpha|-(n-m)} D^\alpha T g \big\|_{L_1(\mathbb{R}^n)}$$

$$\leq M_5 \sum_{k=1}^{\infty} \big\| \, |v|^{|\beta|+|\gamma|-(n-m)} \psi_k \big\|_{L_1(\mathbb{R}^{n-m})} \, \delta_k^{-|\beta|} \, \big\| A_{\delta_k,\lambda * D^\beta \lambda} g \big\|_{L_1(\mathbb{R}^m)}$$

$$\leq M_6 \sum_{k=1}^{\infty} \big\| \, |v|^{|\gamma|-(n-m)} \psi_k \big\|_{L_1(\mathbb{R}^{n-m})} \, \omega_l(\delta_k, g)_1$$

$$\leq M_6 \, M_1 \sum_{k=1}^{\infty} \omega_l(\delta_k, g)_1 \leq M_6 \, M_1 \, \|g\|_{L_1(\mathbb{R}^m)}$$

The case $\alpha = (0, \gamma)$ where $\gamma \neq 0$ is similar.

3. As in the second step of the proof of Theorem 4

$$(Tg)(u,v) - g(u) = \sum_{k=1}^{\infty} \psi_k(v) \Big((A_{\delta_k,\lambda * \lambda} g)(u) - g(u) \Big), \quad 0 < |v| \leq \mu_1 - \tfrac{\delta_1}{4},$$

and

$$(Tg)(u,v) - g(u) = \psi_1(v) \Big(A_{\delta_1,\lambda * \lambda} g \Big)(u) - g(u), \quad \mu_1 - \tfrac{\delta_1}{4} \leq |v| \leq 1.$$

Hence by (5.42)

$$\big\| \, |v|^{-(n-m)} \big((Tg)(u,v) - g(u) \big) \big\|_{L_1(\mathbb{R}^m \times \tilde{B}_1)}$$

$$\leq \Big\| \sum_{k=1}^{\infty} |v|^{-(n-m)} \psi_k(v) \Big((A_{\delta_k,\lambda * \lambda} g)(u) - g(u) \Big) \Big\|_{L_1(\mathbb{R}^n)}$$

$$+ \big\| \, |v|^{-(n-m)} \psi_1(v) (A_{\delta_1,\lambda * \lambda} g)(u) \big\|_{L_1(\mathbb{R}^n)} + \big\| \, |v|^{-(n-m)} g(u) \big\|_{L_1(\mathbb{R}^m \times \{\frac{3}{4}\delta_1 \leq |v| \leq 1\})}$$

$$\leq \sum_{k=1}^{\infty} \| \, |v|^{-(n-m)} \psi_k(v)\|_{L_1(\mathbb{R}^m)} \, \omega_l(\delta_k, g)_1 + \| \, |v|^{-(n-m)} \psi_1(v)\|_{L_1(\mathbb{R}^m)} \, \|g\|_{L_1(\mathbb{R}^m)}$$

$$+ \| \, |v|^{-(n-m)} \|_{L_1(\frac{3}{4}\delta_1 \leq |v| \leq 1)} \|g\|_{L_1(\mathbb{R}^m)}$$

$$\leq M_7 \Big(\|g\|_{L_1(\mathbb{R}^m)} + \sum_{k=1}^{\infty} \omega_l(\delta_k, g)_1 \Big) \leq 2 \, M_7 \|g\|_{L_1(\mathbb{R}^m)},$$

where M_7 is independent of g. $\square$

Remark 14 Here we give the proof of the second part of Theorem 8 of Chapter 2. Let $\Omega = \mathbb{R}_+^n = \{ x \in \mathbb{R}^n : x_n > 0 \}$. First suppose that $l > \frac{1}{p}$ and $g \in B_p^{l-\frac{1}{p}}(\mathbb{R}^{n-1}) \setminus B_p^{l+\varepsilon-\frac{1}{p}}(\mathbb{R}^{n-1})$. By Theorem 3 there exists a function $f \in W_p^l(\mathbb{R}^n)$ such that $f\Big|_{\mathbb{R}^{n-1}} = g$. Suppose that there exist functions $\varphi_s \in C^\infty(\mathbb{R}_+^n) \cap W_p^l(\mathbb{R}_+^n)$, which satisfy property 4) and are such that $\|D^\alpha \varphi_s \, x^{|\alpha|-l-\varepsilon}\|_{L_p(\mathbb{R}_+^n)} < \infty$ for all $\alpha \in \mathbb{N}_0^0$ satisfying $|\alpha| = m > l + \varepsilon$. By Lemma 13 from (2.86) it follows that $\varphi_s\Big|_{\mathbb{R}^{n-1}} = f\Big|_{\mathbb{R}^{n-1}} = g$. Since $\varphi_s \in W_{p,x_n^{m-l-\varepsilon}}^m(\mathbb{R}_+^n)$, where $m \in \mathbb{N}$, $m > l + \varepsilon$, by the trace theorem (5.68) $g \in B_p^{l+\varepsilon-\frac{1}{p}}(\mathbb{R}^{n-1})$ and we arrive at a contradiction.

If $l = p = 1$, the argument is similar: one should consider $g \in L_1(\mathbb{R}^{n-1}) \setminus B_1^\varepsilon(\mathbb{R}^{n-1})$ and apply Theorem 5 instead of Theorem 3.

Let $l, m, n \in \mathbb{N}$, $\alpha \in \mathbb{N}_0^n$. Suppose that $|\alpha| < l - \frac{n-m}{p}$ for $1 < p \leq \infty$ and $|\alpha| \leq l - (n-m)$ for $p = 1$. By Theorem 6 of Chapter 4 and Theorem 2 it follows that $\forall f \in W_p^l(\mathbb{R}^n)$ there exist traces $\mathrm{tr}_{\mathbb{R}^m} D_w^\alpha f$. We note that these traces are not independent. In fact, let $\alpha = (\beta, \gamma)$, where $\beta \in \mathbb{N}_0^m$, $\gamma \in \mathbb{N}_0^{n-m}$. Then[20] $\mathrm{tr}_{\mathbb{R}^m} D_w^{(\beta,\gamma)} f = D_w^\beta (\mathrm{tr}_{\mathbb{R}^m} D_w^{(0,\gamma)} f)$. For this reason we consider only weak derivatives $D_w^{(0,\gamma)} f$ and introduce the *total trace* of a function $f \in W_p^l(\mathbb{R}^n)$ by setting

$$\mathrm{Tr}_{\mathbb{R}^m} f = \Big(\mathrm{tr}_{\mathbb{R}^m} D_w^{(0,\gamma)} f \Big)_{|\gamma| < l - \frac{n-m}{p}}, \quad 1 < p \leq \infty, \tag{5.86}$$

and

$$\mathrm{Tr}_{\mathbb{R}^m} f = \Big(\mathrm{tr}_{\mathbb{R}^m} D_w^{(0,\gamma)} f \Big)_{|\gamma| \leq l - (n-m)}, \quad p = 1. \tag{5.87}$$

---

[20] If $f \in W_p^l(\mathbb{R}^n) \cap C^\infty(\mathbb{R}^n)$, then this formula is clear. If $f \in W_p^l(\mathbb{R}^n)$, it can be obtained by choosing a sequence of functions $f_k \in W_p^l(\mathbb{R}^n) \cap C^\infty(\mathbb{R}^n)$, which converges to f in $(W_1^l)^{loc}(\mathbb{R}^n)$, and passing to the limit in the definition of the weak derivative.

In particular,

$$\mathrm{Tr}_{\mathbb{R}^{n-1}} f = \left(f\big|_{\mathbb{R}^m}, \left(\frac{\partial f}{\partial x_n}\right)_w\big|_{\mathbb{R}^m}, \cdots, \left(\frac{\partial^{l-1} f}{\partial x_n^{l-1}}\right)_w\big|_{\mathbb{R}^m}\right), \quad 1 \le p \le \infty.$$

We also define the *total trace space* by setting

$$\mathrm{Tr}_{\mathbb{R}^m} W_p^l(\mathbb{R}^n) = \left\{ \mathrm{Tr}_{\mathbb{R}^m} f, \ f \in W_p^l(\mathbb{R}^n) \right\}.$$

Theorem 7 *Let $l, m, n \in \mathbb{N}$, $m < n$, $1 \le p \le \infty$. Then*

$$\mathrm{Tr}_{\mathbb{R}^m} W_p^l(\mathbb{R}^n) = \prod_{|\gamma| < l - \frac{n-m}{p}} B_p^{l-|\gamma|-\frac{n-m}{p}}(\mathbb{R}^m), \quad 1 < p \le \infty, \tag{5.88}$$

and

$$\mathrm{Tr}_{\mathbb{R}^m} W_1^l(\mathbb{R}^n) = \prod_{|\gamma| < l-(n-m)} B_1^{l-|\gamma|-(n-m)}(\mathbb{R}^m) \times \prod_{|\gamma| = l-(n-m)} L_1(\mathbb{R}^m). \tag{5.89}$$

Idea of the proof. The direct trace theorem follows from the first part of Theorem 3. To prove the extension theorem (i.e., the inverse trace theorem), given the functions g_γ, where $|\gamma| < l - \frac{n-m}{p}$ for $1 < p \le \infty$ and $|\gamma| \le l - (n - m)$ for $p = 1$, lying in the appropriate spaces, set

$$(T\{g_\gamma\})(u, v) = \sum_{|\gamma| < l - \frac{n-m}{p}} \frac{v^\gamma}{\gamma!} \left(T_1 g_\gamma\right)(u, v), \quad 1 < p \le \infty, \tag{5.90}$$

where $v^\gamma = x_{m+1}^{\gamma_{m+1}} \cdots x_n^{\gamma_n}$, $\gamma! = \gamma_{m+1}! \cdots \gamma_n!$ and the operator T_1 is defined by (5.46) and

$$(T\{g_\gamma\})(u, v) = \sum_{|\gamma| < l-(n-m)} \frac{v^\gamma}{\gamma!}(T_1 g_\gamma)(u, v) + \sum_{|\gamma| = l-(n-m)} \frac{v^\gamma}{\gamma!}(T_2 g_\gamma)(u, v), \tag{5.91}$$

where T_1 is defined by (5.46) while T_2 is defined by (5.72). Apply (5.59), (5.64) and (5.83). $\square$

Proof. 1. Suppose that $1 < p \le \infty$ and let $\gamma, \mu \in \mathbb{N}_0^{n-m}$ and $|\gamma|, |\mu| < l - \frac{n-m}{p}$. If $\gamma \le \mu$, i.e., $\gamma_j \le \mu_j$, $j = m+1, ..., n$, then by Leibnitz' formula $D_w^{(0,\mu)}(v^\gamma T_1 g_\gamma)$ is equal to $\gamma! \, D_w^{(0,\mu-\gamma)}(T_1 g)$ plus a sum of the terms containing the factor v^σ where $\sigma \ne 0$. Otherwise $D_w^{(0,\mu)}(v^\gamma T_1 g_\gamma)$ is a sum, each term of which contains the factor v^σ where $\sigma \neq 0$. Hence $\operatorname{tr} D_w^{(0,\mu)}(v^\gamma T_1 g_\gamma)$ is equal to $\gamma! \operatorname{tr} D_w^{(0,\mu-\gamma)}(T_1 g)$ if $\gamma \leq \mu$ and is equal to 0 otherwise. So by (5.59)

$$\operatorname{tr} D_w^{(0,\mu)}(T\{g_\gamma\}) = \sum_{0 \leq \gamma \leq \mu} \operatorname{tr} D_w^{(0,\mu-\gamma)}(T_1 g_\gamma) = g_\mu.$$

2. Since $(T_1 g_\gamma)(u,v) = 0$ if $|v| \geq 1$, by (5.54) we have

$$\|T\{g_\gamma\}\|_{L_p(\mathbb{R}^n)} \leq \sum_{|\gamma| < l - \frac{n-m}{p}} \|T_1 g_\gamma\|_{L_p(\mathbb{R}^n)} \leq M_1 \sum_{|\gamma| < l - \frac{n-m}{p}} \|g_\gamma\|_{L_p(\mathbb{R}^m)},$$

where M_1 is independent of f.

3. Finally, let $\alpha = (\beta, \mu)$, where $\beta \in \mathbb{N}_0^m$, $\mu \in \mathbb{N}_0^{n-m}$ and $|\alpha| = |\beta| + |\mu| = l$. By Leibnitz' formula

$$D^\alpha(v^\alpha T_1 g_\gamma) = \sum_{0 \leq \nu \leq \mu,\, \nu \geq \mu - \gamma} c_{\beta,\gamma,\mu}\, v^{\gamma - \mu + \nu} D^{(\beta,\gamma)}(T_1 g_\gamma)$$

for certain $c_{\beta,\gamma,\mu} \in \mathbb{N}_0$. Hence by (5.64)

$$\|D^\alpha(v^\alpha T_1 g_\gamma)\|_{L_p(\mathbb{R}^n)} \leq M_2 \sum_{0 \leq \nu \leq \mu,\, \nu \geq \mu - \gamma} \|v^{|\beta|+|\gamma|-(l-|\gamma|)} D^{(\beta,\nu)}(T_1 g_\gamma)\|_{L_p(\mathbb{R}^n)}$$

$$\leq M_3 \|g_\gamma\|_{B_p^{l-|\gamma|-\frac{n-m}{p}}(\mathbb{R}^m)}$$

and, consequently,

$$\|T(\{g_\gamma\})\|_{W_p^l(\mathbb{R}^n)} \leq M_4 \sum_{|\gamma| < l - \frac{n-m}{p}} \|g_\gamma\|_{B_p^{l-|\gamma|-\frac{n-m}{p}}(\mathbb{R}^m)}, \qquad (5.92)$$

where M_2, M_3, M_4 are independent of g_γ.

4. If $p = 1$, then
$$\|T\{g_\gamma\}\|_{W_1^l(\mathbb{R}^n)}$$

$$\leq M_5\left(\sum_{|\gamma| < l - (n-m)} \|g_\gamma\|_{B_1^{l-|\gamma|-(n-m)}(\mathbb{R}^m)} + \sum_{|\gamma| = l - (n-m)} \|g_\gamma\|_{L_1(\mathbb{R}^m)} \right), \qquad (5.93)$$

where M_5 is independent of g_γ. $\square$

Remark 15 As in Theorems 4 and 6, in addition to (5.92) and (5.93), we have the following estimates

$$\| |v|^{|\alpha|-l} D^\alpha(T\{g_\gamma\})\|_{L_p(\mathbb{R}^n)} \leq c_{20} \sum_{|\gamma| < l - \frac{n-m}{p}} \|g_\gamma\|_{B_p^{l-|\gamma|-\frac{n-m}{p}}(\mathbb{R}^m)}, \qquad (5.94)$$

where $1 < p \leq \infty$, $|\alpha| \geq l - \frac{n-m}{p}$, and

$$\| |v|^{|\mu|-l}(D^{(0,\mu)}(T\{g_\gamma\}) - g_\mu)\|_{L_p(\mathbb{R}^n)} \leq c_{21} \sum_{|\gamma| < l - \frac{n-m}{p}} \|g_\gamma\|_{B_p^{l-|\gamma|-\frac{n-m}{p}}(\mathbb{R}^m)}, \quad (5.95)$$

where $1 < p \leq \infty$, $|\mu| < l - \frac{n-m}{p}$. In (5.94) the exponent $|a| - l$ can not be replaced by $|\alpha| - l - \varepsilon$ for any $\varepsilon > 0$ and for any extension operator T. We also note that by Lemma 13 from (5.95) it follows directly that $D^{(0,\mu)}(T\{g_\gamma\})\big|_{\mathbb{R}^m} = g_\mu$. Similar statements hold for $p = 1$.

Remark 16 From the concluding statements of Theorems 4, 6 and Remark 15 it follows that the extension operators defined by (5.44), (5.72), (5.90) and (5.91) are in a certain sense the best possible extension operators, namely, in the sense that the derivatives of higher orders of Tg, $T\{g_\gamma\}$ respectively, have the minimal possible growth on approaching $\mathbb{R}^m$.

5.5   Traces on smooth surfaces

Let $\Omega \subset \mathbb{R}^n$ be an open set with a C^1-boundary. We should like to extend Definition 1 to the case, in which $\mathbb{R}^n, \mathbb{R}^m$ are replaced by $\Omega, \partial\Omega$ respectively.

We start with the case of a bounded elementary domain $\Omega \subset \mathbb{R}^n$ with a C^1-boundary with the parameters d, D, M. Thus Ω has the form

$$\Omega = \{x \in \mathbb{R}^n : a_n < x_n < \varphi(\bar{x}), \ \bar{x} \in W\},$$

where $\bar{x} = (x_1, ..., x_{n-1})$, $W = \{\bar{x} \in \mathbb{R}^{n-1} : a_i < x_i < b_i, i = 1, ..., n-1\}$, $-\infty \leq a_i < b_i \leq \infty, i = 1, ..., n-1$, $-\infty \leq a_n < b_n < \infty$, and satisfies the definition of Section 4.3.

Suppose that $f \in L_1(\Omega)$. In the spirit of Definition 1 we say that the function $g \in L_1(\Gamma)$, where $\Gamma = \{x \in \mathbb{R}^n : x_n = \varphi(\bar{x}), \bar{x} \in W\}$, is a trace of the function f on Γ if there exists a function h equivalent to f on Ω such that

$$h(\cdot + te_n) \to g(\cdot) \ \text{in} \ L_1(\Gamma) \ \text{as} \ t \to 0-, \quad (5.96)$$

where $e_n = (0, ..., 0, 1)$. Since $\left|\left(\frac{\partial\varphi}{\partial x_i}\right)(\bar{x})\right| \leq M, \bar{x} \in W, i = 1, ..., n-1$, we have

$$(1 + (n-1)M^2)^{-\frac{1}{2}}\|F(\bar{x}, \varphi(\bar{x}))\|_{L_1(W)} \leq \|F\|_{L_1(\Gamma)}$$

$$= \int_W |F(\bar{x}, \varphi(\bar{x}))|\left(1 + \sum_{i=1}^{n}\left(\left(\frac{\partial\varphi}{\partial x_i}\right)(\bar{x})\right)^2\right)^{-\frac{1}{2}} d\bar{x} \leq \|F(\bar{x}, \varphi(\bar{x}))\|_{L_1(W)}.$$

Consequently (5.96) is equivalent to

$$\int\limits_{W} |f(\bar{x}, \varphi(\bar{x}) + t) - g(\bar{x}, \varphi(\bar{x}))| \, d\bar{x} \to 0 \quad \text{as} \quad t \to 0 - . \tag{5.97}$$

Let the transformation $y = \Phi(\bar{x})$ be defined by

$$\bar{y} = \bar{x}, \quad y_n = x_n - \varphi(\bar{y}), \quad x \in \Omega, \tag{5.98}$$

then

$$\Phi(\Omega) = \{y \in \mathbb{R}^n : a_n - \varphi(\bar{y}) < y_n < 0, \bar{y} \in W\}$$

and

$$\Phi(\Gamma) = \{y \in \mathbb{R}^n : y_n = 0, \bar{y} \in W\} \equiv W^*.$$

Relation (5.97) means that $g(\Phi^{(-1)})$ is a trace of $f(\Phi^{(-1)})$ on $\Phi(\Gamma)$.

Next suppose that an open set $\Omega \subset \mathbb{R}^n$ is such that for a certain map λ, which is a composition of rotations, reflections and translations, the set $\lambda(\Omega)$ is a bounded elementary domain with a C^1-boundary and Γ is such that $\lambda(\Gamma) = \{x \in \mathbb{R}^n : x_n = \varphi(\bar{x}), \bar{x} \in W\}$. Then we say that g is a trace of f on Γ if $g(\lambda^{(-1)})$ is a trace of $f(\lambda^{(-1)})$ on $\lambda(\Gamma)$ in the above sense.

Finally, suppose that $\Omega \subset \mathbb{R}^n$ is an arbitrary open set with a C^1-boundary with the parameters $d, D, \varkappa$ and M, and let V_j be open parallelepipeds satisfying conditions 1)–4) in the definition of Section 4.3. From the proof of Lemma 3 of Chapter 2 it follows that there exists an appropriate partition of unity, i.e., there exist functions $\psi_j \in C^\infty(\mathbb{R}^n)$ such that $0 \le \psi_j \le 1, \operatorname{supp} \psi_j \subset (V_j)_{\frac{d}{2}}, j = \overline{1, s}, \sum\limits_{j=1}^{s} \psi_j(x) = 1$ on Ω and

$$|(D^\alpha \psi_j)(x)| \le c_{22} \, d^{-|\alpha|}, \quad x \in \mathbb{R}^n, \ \alpha \in \mathbb{N}_0, \ j = \overline{1, s}, \tag{5.99}$$

where $c_{22} > 0$ is independent of x, j and d.

Definition 3 *Let $\Omega \subset \mathbb{R}^n$ be an open set with a C^1-boundary and $f \in L_1(\Omega \bigcap B)$ for each ball $B \subset \mathbb{R}^n$. Suppose that $f = \sum\limits_{j=1}^{s} f_j$, where* supp $f_j \subset V_j$ *and $f_j \in L_1(\Omega \bigcap V_j)$. If the functions g_j are traces of the functions f_j on $V_j \bigcap \partial\Omega$, $j = \overline{1, s}$, then the function $\sum\limits_{j=1}^{s} g_j$ is said to be a trace of the function f on $\partial\Omega$.*

Remark 17 One can show that Definition 3 does not depend on the covering $\{V_j\}$ and on the representation $f = \sum\limits_{j=1}^{s} f_j$ and satisfy the requirements to the notion of the trace analogous to the requirements 1)–4) at the beginning of Section 5.1.

Next we need to define the spaces $B_p^l(\partial\Omega)$ where $l > 0, 1 \leq p \leq \infty$. We follow the same scheme as in defining the notion of the trace. If Ω is a bounded elementary domain with a C^1-boundary and the function f is defined on Γ, we say that $f \in B_p^l(\Gamma)$ if $f(\bar{x}, \varphi(\bar{x})) \in B_p^l(W)$ and set

$$\|f\|_{B_p^l(\Gamma)} = \|f(\Phi^{(-1)})\|_{B_p^l(\Phi(\Gamma))} \equiv \|f(\Phi^{(-1)})\|_{B_p^l(W^\bullet)} \equiv \|f(\bar{x}, \varphi(\bar{x}))\|_{B_p^l(W)}.$$

Here the norm $\|\cdot\|_{B_p^l(W)}$ is defined as in Definition 2, where n is replaced by $n-1$, $\mathbb{R}^n$ by W and, in (5.8), (5.9), $\|\Delta_h^\sigma f\|_{L_p(\mathbb{R}^n)}$ by $\|\Delta_h^\sigma f\|_{L_p(W_{\sigma|h|})}$.

If Ω is such that for a certain map λ, which is a composition of rotations, reflections and translations, the set $\lambda(\Omega)$ is a bounded elementary domain with a C^1-boundary, then $f \in B_p^l(\Gamma)$ if $f(\lambda^{(-1)}) \in B_p^l(\lambda(\Gamma))$ and

$$\|f\|_{B_p^l(\Gamma)} = \|f(\lambda^{(-1)})\|_{B_p^l(\lambda(\Gamma))} = \|f(\Lambda^{(-1)})\|_{B_p^l(\Lambda(\Gamma))},$$

where $\Lambda = \Phi(\lambda)$.

Definition 4 *Let $l > 0, 1 \leq p \leq \infty$ and let Ω be an open set with a C^1-boundary. We say that $f \in B_p^l(\partial\Omega)$ if $f\psi_j \in B_p^l(V_j \cap \partial\Omega)$, $j = \overline{1,s}$, and*

$$\|f\|_{B_p^l(\partial\Omega)} = \left(\sum_{j=1}^{s} \|f\psi_j\|_{B_p^l(V_j \cap \partial\Omega)}^p \right)^{\frac{1}{p}}$$

$$= \left(\sum_{j=1}^{s} \|(f\psi_j)(\Lambda_j^{(-1)})\|_{B_p^l(W_j^\bullet)}^p \right)^{\frac{1}{p}}$$

$$\equiv \left(\sum_{j=1}^{s} \|(f\psi_j)(\lambda_j^{(-1)}(\bar{x}, \varphi_j(\bar{x})))\|_{B_p^l(W_j)}^p \right)^{\frac{1}{p}} < \infty. \tag{5.100}$$

Here $\Lambda_j = \Phi_j(\lambda_j)$ and Φ_j is defined by (5.98), where φ, W are replaced by φ_j, W_j respectively.

Remark 18 In the case $l = k - \frac{1}{p}$, where $k \in \mathbb{N}$ and $1 < p \leq \infty$, which will be of interest for us, from Theorem 8 below it will follow that, for open sets Ω having a C^k-boundary, this definition is independent of $\{V_j\}$ and $\{\psi_j\}$. As for the general case, see Remark 19.

For the function f defined on an open set $\Omega \subset \mathbb{R}^n$ we write f_0 for its extension by 0 to $\mathbb{R}^n$. If $f \in W_p^l(\Omega)$ and $\operatorname{supp} f \subset \Omega$, then, by the additivity of the Lebesgue integral and the properties of weak derivatives, $f_0 \in W_p^l(\mathbb{R}^n)$ and $\|f_0\|_{W_p^l(\mathbb{R}^n)} = \|f\|_{W_p^l(\Omega)}$. We shall need an analogue of this statement for the spaces $B_p^l(\Omega)$, which has the following form.

Lemma 14 *Let $l > 0$, $1 \le p \le \infty$ and $\delta > 0$. Then for each open set $\Omega \subset \mathbb{R}^n$ and $\forall f \in B_p^l(\Omega)$ satisfying* $\operatorname{supp} f \subset \overline{\Omega_\delta}$

$$\|f_0\|_{B_p^l(\mathbb{R}^n)} \le c_{23} \|f\|_{B_p^l(\Omega)}, \tag{5.101}$$

where $c_{23} > 0$ is independent of f and Ω.

Idea of the proof. Note that for $|h| < \frac{\delta}{2\sigma}$

$$\|\Delta_h^\sigma f_0\|_{L_p(\mathbb{R}^n)} \le \|\Delta_h^\sigma f\|_{L_p(\Omega_{\sigma|h|})} + \|\Delta_h^\sigma f_0\|_{L_p({}^c\Omega_{\sigma|h|})} = \|\Delta_h^\sigma f\|_{L_p(\Omega_{\sigma|h|})}. \quad \square$$

Proof. From the definition of the spaces $B_p^l(\Omega)$ we have

$$\|f_0\|_{b_p^l(\mathbb{R}^n)} \le \left(\int\limits_{\mathbb{R}^n} \left(|h|^{-l} \|\Delta_h^\sigma f\|_{L_p(\Omega_{\sigma|h|})} \right)^p \frac{dh}{|h|^n} \right)^{\frac{1}{p}}$$

$$+ \left(\int\limits_{|h| \ge \frac{\delta}{2\sigma}} \left(|h|^{-l} \|\Delta_h^\sigma f_0\|_{L_p({}^c\Omega_{\sigma|h|})} \right)^p \frac{dh}{|h|^n} \right)^{\frac{1}{p}}$$

$$\le \|f\|_{b_p^l(\Omega)} + 2^\sigma \|f_0\|_{L_p(\mathbb{R}^n)} \left(\int\limits_{|h| \ge \frac{\delta}{2\sigma}} \frac{dh}{|h|^{n+pl}} \right)^{\frac{1}{p}} \le M_1 \|f\|_{B_p^l(\Omega)},$$

where M_1 depends only on n, l, σ, p and δ, and (5.101) follows. $\square$

Theorem 8 *Let $l \in \mathbb{N}$, $1 \le p \le \infty$ and let $\Omega \subset \mathbb{R}^n$ be an open set with a C^l-boundary. Then*

$$\operatorname{tr}_{\partial\Omega} W_p^l(\Omega) = B_p^{l-\frac{1}{p}}(\partial\Omega), \quad l > \tfrac{1}{p}, \tag{5.102}$$

and

$$\operatorname{tr}_{\partial\Omega} W_1^1(\Omega) = L_1(\partial\Omega). \tag{5.103}$$

Idea of the proof. 1. To prove the direct trace theorem establish, by Theorems 3 and 5, that the trace g_j of $f\psi_j$ exists on $V_j \cap \partial\Omega$, and $g_j \in B_p^{l-\frac{1}{p}}(V_j \cap \partial\Omega)$ if $l > \frac{1}{p}$ and $g_j \in L_1(V_j \cap \partial\Omega)$ if $l = p = 1$.

2. To prove the inverse trace ($\equiv$ extension) theorem, given a function $g \in B_p^{l-\frac{1}{p}}(\partial\Omega)$, consider the functions $(g\psi_j)(\Lambda_j^{(-1)})$ on W_j^*, extend them by zero to $\mathbb{R}^{n-1}$ preserving the same notation, and set

$$Tg = \sum_{j=1}^{s} \Big(T_0((g\psi_j)(\Lambda_j^{(-1)})) \Big)(\Lambda_j), \tag{5.104}$$

where T_0 is a modification of the extension operator (5.46) for $l > \frac{1}{p}$, respectively (5.72) for $l = p = 1$. Namely, the sum $\sum\limits_{k=1}^{\infty}$ in (5.46), (5.72) must be replaced by $\sum\limits_{k=k_0}^{\infty}$, where k_0 is such that

$$\operatorname{supp} T_0 h \subset (\operatorname{supp} h)^d \times \widetilde{B}(0,d). \quad \Box \tag{5.105}$$

Proof. 1. By Corollary 18 of Chapter 4 $f\psi_j \in W_p^l(V_j \cap \Omega)$ and by Lemma 16 of Chapter 4 $(f\psi_j)(\Lambda_j^{(-1)}) \in W_p^l(\Lambda_j(V_j \cap \Omega))$. Since $\operatorname{supp}(f\psi_j)(\Lambda_j^{(-1)}) \subset \Lambda_j(V_j \cap \Omega)$, the extension [21] by 0 to $\mathbb{R}^n$ of the function $(f\psi_j)(\Lambda_j^{(-1)})$ is such that $(f\psi_j)(\Lambda_j^{(-1)}) \in W_p^l(\mathbb{R}^n)$. Hence by Theorem 2 there exists a trace h_j of this function on $\mathbb{R}^{n-1}$ and therefore on $W_j^* = \Lambda_j(V_j \cap \partial\Omega)$. This means that $g_j = h_j(\Lambda_j)$ is a trace of $f\psi_j$ on $V_j \cap \partial\Omega$. So by Definition 3 $g = \sum\limits_{j=1}^{s} g_j$ is a trace of f on $\partial\Omega$. Moreover, if $l > \frac{1}{p}$, then by Theorem 3

$$\|g_j\|_{B_p^{l-\frac{1}{p}}(V_j \cap \partial\Omega)} = \|h_j\|_{B_p^{l-\frac{1}{p}}(W_j^*)} \leq \|h_j\|_{B_p^{l-\frac{1}{p}}(\mathbb{R}^{n-1})} \leq M_1 \|f\psi_j\|_{W_p^l(\mathbb{R}^n)}.$$

Finally, since $g_j = g\psi_j$, by Definition 4, Corollary 18 of Chapter 4, (5.99), Minkowski's inequality for sums and (2.59), we have

$$\|g\|_{B_p^{l-\frac{1}{p}}(\partial\Omega)} = \Big(\sum_{j=1}^{s} \|g_j\|_{B_p^{l-\frac{1}{p}}(V_j \cap \partial\Omega)}^p \Big)^{\frac{1}{p}}$$

$$\leq M_1 \Big(\sum_{j=1}^{s} \|f\psi_j\|_{W_p^l(\mathbb{R}^n)}^p \Big)^{\frac{1}{p}} \leq M_2 \Big(\sum_{j=1}^{s} \|f\|_{W_p^l(V_j \cap \Omega)}^p \Big)^{\frac{1}{p}}$$

---

[21] We preserve the same notation for the extended function.

$$\leq M_2 \left(\left(\sum_{j=1}^{s}\|f\|_{L_p(V_j\cap\Omega)}^p\right)^{\frac{1}{p}} + \sum_{|\alpha|=l}\left(\sum_{j=1}^{s}\|D_w^\alpha\|_{L_p(V_j\cap\Omega)}^p\right)^{\frac{1}{p}}\right) \leq M_2\,\varkappa^{\frac{1}{p}}\|f\|_{W_p^l(\Omega)},$$

where M_1 and M_2 are independent of f.

If $l = p = 1$, then, by Theorem 5, in the above argument $B_p^{l-\frac{1}{p}}$ should be replaced by L_1.

2. If T_0 is defined by (5.72), then by (1.4) supp $T_0 h \subset \overline{(\text{supp }h)^{2^{-k_0}}} \times \widetilde{B}(0, 2^{-k_0+1})$.Hence (5.105) follows if $2^{-k_0+1} \leq d$. If T_0 is defined by (5.72), then supp $T_0 h \subset \overline{(\text{supp }h)^{3\delta k_0}}$ and (5.105) follows if, say, $4\delta_{k_0} \leq d$.

Let $g_j = (g\psi_j)(\Lambda_j^{(-1)})$. Since by Lemma (5.105) $\text{supp}(T_0 g_j)(\Lambda_j) \subset V_j \cap \Omega$ and supp $T_0 g_j \subset \Lambda_j(V_j \cap \Omega)$, by Lemma 16 of Chapter 4 we have

$$\|(T_0 g_j)(\Lambda_j)\|_{W_p^l(\mathbb{R}^n)} = \|(T_0 g_j)(\Lambda_j)\|_{W_p^l(V_j\cap\Omega)}$$

$$\leq M_3\|T_0 g_j\|_{W_p^l(\Lambda_j(V_j\cap\Omega))} = M_3\|T_0 g_j\|_{W_p^l(\mathbb{R}^n)},$$

where M_3 is independent of g and j.

By the proofs of the Theorems 3 and 5 g_j is a trace of $T_0 g_j$ on $\mathbb{R}^{n-1}$, hence $(g\psi_j)(\Lambda_j^{(-1)})$ is a trace of $T_0((g\psi_j)(\Lambda_j^{(-1)}))$ on $W_j^* = \Lambda_j(V_j\cap\partial\Omega)$. Consequently, $g\psi_j$ is a trace of $T_0((g\psi_j)(\Lambda_j^{(-1)}))(\Lambda_j)$ on $V_j \cap \partial\Omega$ and, by Definition 3, $g = \sum_{j=1}^{s} g\psi_j$ is a trace of Tg on $\partial\Omega$.

Suppose that $l > \frac{1}{p}$, the case $l = p = 1$ being similar. By Theorem 3 and Lemma 14 we get

$$\|T_0 g_j\|_{W_p^l(\mathbb{R}^n)} \leq M_4\|g_j\|_{B_p^{l-\frac{1}{p}}(\mathbb{R}^{n-1})}$$

$$\leq M_5\|(g\psi_j)(\Lambda_j^{(-1)})\|_{B_p^{l-\frac{1}{p}}(W_j^*)} = M_5\|g\psi_j\|_{B_p^{l-\frac{1}{p}}(V_j\cap\partial\Omega)},$$

where M_4 and M_5 are independent of g and j. So

$$\|Tg\|_{W_p^l(\mathbb{R}^n)} = \|\sum_{j=1}^{s}(T_0 g_j)(\Lambda_j)\|_{W_p^l(\mathbb{R}^n)} \leq M_3\sum_{j=1}^{s}\|T_0 g_j\|_{W_p^l(\mathbb{R}^n)}$$

$$\leq M_3 M_5\sum_{j=1}^{s}\|g\psi_j\|_{B_p^{l-\frac{1}{p}}(V_j\cap\partial\Omega)} = M_3 M_5\|g\|_{B_p^{l-\frac{1}{p}}(\partial\Omega)}. \quad \square$$

Remark 19 We note that in Theorem 8 the coverings $\{V_j\}$ and the partitions of unity $\{\psi_j\}$ could be different in the first and the second parts of the proof.

From this fact it follows that Definition 4 of the spaces $B_p^{l-\frac{1}{p}}(\partial\Omega)$, $l \in \mathbb{N}, l > \frac{1}{p}$, does not depend on $\{V_j\}$ and $\{\psi_j\}$ for the open sets with a C^l-boundary. Consider two coverings $\{V_{j,k}\}$ and partitions of unity $\{\psi_{j,k}\}$, $k = 1, 2$, and let $\|\cdot\|_{B_p^{l-\frac{1}{p}}(\partial\Omega)}^{(k)}$ be the norms defined with the help of $\{V_{j,k}\}, \{\psi_{j,k}\}$. Then by
(5.102) $\forall f \in W_p^l(\Omega)$

$$\|f\|_{B_p^{l-\frac{1}{p}}(\partial\Omega)}^{(1)} \leq M_1 \|T_2 f\|_{W_p^l(\Omega)} \leq M_2 \|f\|_{B_p^{l-\frac{1}{p}}(\partial\Omega)}^{(2)}, \qquad (5.106)$$

where T_2 is defined by (5.104) for $\{V_{j,2}\}$ and $\{\psi_{j,2}\}$ and M_1, M_2 are independent of f. Similarly we estimate $\|\cdot\|_{B_p^{l-\frac{1}{p}}(\partial\Omega)}^{(2)}$ via $\|\cdot\|_{B_p^{l-\frac{1}{p}}(\partial\Omega)}^{(1)}$. Hence the norms $\|\cdot\|_{B_p^{l-\frac{1}{p}}(\partial\Omega)}^{(1)}$ and $\|\cdot\|_{B_p^{l-\frac{1}{p}}(\partial\Omega)}^{(2)}$ are equivalent.

By this scheme it is also possible to prove, applying the trace theorem (5.68), the independence of Definition 4 of $\{V_j\}$ and $\{\psi_j\}$ for the spaces $B_p^l(\partial\Omega)$ with an arbitrary $l > 0$. In this case one should verify that an analogue of (5.68) and Theorem 8 holds for the spaces $W_{p,\varrho^s}^l(\Omega)$, where $\varrho(x) = \mathrm{dist}\,(x, \partial\Omega)$, and replace (5.106) by

$$\|f\|_{B_p^l(\partial\Omega)}^{(1)} \leq M_1 \|T_2 f\|_{W_{p,\varrho^s}^r(\Omega)} \leq M_2 \|f\|_{B_p^l(\partial\Omega)}^{(2)},$$

where $r \in \mathbb{N}, r \geq l + \frac{1}{p}, s = r - l - \frac{1}{p}$ and $\partial\Omega \in C^r$.

For an open set $\Omega \subset \mathbb{R}^n$ with a C^l-boundary let $\nu(x)$ be the unit vector of the outer normal at the point $x \in \partial\Omega$. Hence $\nu(x) = (\cos\gamma_1, ..., \cos\gamma_n)$, where γ_j are the angles between $\nu(x)$ and the unit coordinate vectors e_j. For $f \in W_p^l(\Omega)$ the traces of the weak derivatives $D_w^\alpha f$ exist on $\partial\Omega$ if $|\alpha| \leq l - 1$. We define the weak normal derivatives by

$$\left(\frac{\partial^s f}{\partial\nu^s}\right)_w = \sum_{j_1,...,j_s=1}^{n} \cos\gamma_{j_1} \cdots \cos\gamma_{j_s} \left(\frac{\partial^s f}{\partial x_{j_1} \cdots \partial x_{j_s}}\right)_w, \quad s = 1, ..., l - 1.$$

The total trace and the total trace space are defined by

$$\mathrm{Tr}_{\partial\Omega}\, f = \left(\left(\frac{\partial^s f}{\partial\nu^s}\right)_w\right)_{s=0,...,l-1},$$

respectively

$$\mathrm{Tr}_{\partial\Omega}\, W_p^l(\Omega) = \{\mathrm{Tr}_{\partial\Omega}\, f,\ f \in W_p^l(\Omega)\}.$$

Theorem 9 *Let $l \in \mathbb{N}, 1 \leq p \leq \infty$ and let $\Omega \subset \mathbb{R}^n$ be an open set with a C^l-boundary. Then*

$$\mathrm{Tr}_{\partial\Omega}\, W_p^l(\Omega) = \prod_{s=0}^{l-1} B_p^{l-s-\frac{1}{p}}(\partial\Omega), \quad 1 < p \leq \infty, \tag{5.107}$$

and

$$\mathrm{Tr}_{\partial\Omega}\, W_1^l(\Omega) = \prod_{s=0}^{l-2} B_p^{l-s-1}(\partial\Omega) \times L_1(\partial\Omega). \tag{5.108}$$

Idea of the proof. Combine the proofs of Theorems 7 and 8. $\square$

Remark 20 If $p > 1$, then as in Remark 15 one may state that there exists a bounded linear extension operator

$$T : \prod_{s=0}^{l-1} B_p^{l-s-\frac{1}{p}}(\partial\Omega) \to W_p^l(\Omega) \cap C^\infty(\Omega),$$

satisfying the inequalities

$$\left\| \varrho^{k-l} \frac{\partial^k (T\{g_s\})}{\partial \nu^k} \right\|_{L_p(\Omega)} \leq c_{24} \sum_{s=0}^{l-1} \|g_s\|_{B_p^{l-s-\frac{1}{p}}(\partial\Omega)}, \quad k \geq l, \tag{5.109}$$

and

$$\left\| \varrho^{k-l} \left(\frac{\partial^k (T\{g_s\})}{\partial \nu^k} - g_k \right) \right\|_{L_p(\Omega)} \leq c_{25} \sum_{s=0}^{l-1} \|g_s\|_{B_p^{l-s-\frac{1}{p}}(\partial\Omega)}, \quad 0 \leq k < l, \tag{5.110}$$

where $\varrho(x) = \mathrm{dist}\,(x, \partial\Omega)$ and $c_{24}, c_{25} > 0$ are independent of g_s.

In (5.109) the exponent $k - l$ cannot be replaced by $k - l - \varepsilon$ for any $\varepsilon > 0$.

If $p = 1$, then a similar statement holds. (We recall that in this case the extension operator T is nonlinear.)

Remark 21 The problem of the traces on smooth m-dimensional manifolds where $m < n - 1$ may be treated similarly, though technically this is more complicated. Suppose that $\Omega \subset \mathbb{R}^n$ is an open set such that $\Omega = \bigcup_{m=0}^{n-1} \Gamma_m$, where Γ_m are m-dimensional manifolds in the class C^l and $\Gamma_m \cap \Gamma_\mu = \varnothing$ if $m \neq \mu$. (Some of Γ_m may be absent.) Let, for example, $1 < p \leq \infty$. If $m \leq n - pl$, then, by Theorem 2, the traces on Γ_m of functions $f \in W_p^l(\Omega)$ may not exist. If $m > n - pl$, then for each $f \in W_p^l(\Omega)$ the trace of f on Γ_m exists.

Moreover, the traces of the weak derivatives $D_w^\alpha f$ also exist if $|\alpha| < l - \frac{n-m}{p}$. For this reason the total trace and the total trace space are defined by

$$\mathrm{Tr}_{\partial\Omega}\, f = \left(\left(D_{\nu,w}^\alpha \right)_{|\alpha| \leq l - \frac{n-m}{p}} \right)_{n-pl < m \leq n-1},$$

$$\mathrm{Tr}_{\partial\Omega}\, W_p^l(\Omega) = \{ \mathrm{Tr}_{\partial\Omega}, \ f \in W_p^l(\Omega) \}$$

respectively. Here $D_{\nu,w}^\alpha f = \left(\frac{\partial^{\alpha_1 + \ldots + \alpha_{n-m}} f}{\partial \nu_1^{\alpha_1} \ldots \partial \nu_{n-m}^{\alpha_{n-m}}} \right)_w$ are weak derivatives with respect to an orthonormal set of the normals $\nu_1, \ldots, \nu_{n-m}$ to Γ_m. The appropriate generalization of (5.107) has the form

$$\mathrm{Tr}_{\partial\Omega}\, W_p^l(\Omega) = \prod_{m=0}^{n-1} \ \prod_{|\alpha| < l - \frac{n-m}{p}} B_p^{l - |\alpha| - \frac{n-m}{p}} (\partial\Omega).$$

Similarly one may generalize (5.108).

This statement plays an important role in the theory of boundary-value problems for elliptic partial differential equations, because it explains what boundary values must be given and to which spaces they can belong.

Chapter 6

Extension theorems

The main aim of this chapter is to prove that under sertain assumptions on an open set $\Omega \subset \mathbb{R}^n$ there exists an extension [1] operator

$$T : W_p^l(\Omega) \longrightarrow W_p^l(\mathbb{R}^n),$$

which is linear and bounded. The existence of such an operator ensures that a number of properties of the space $W_p^l(\mathbb{R}^n)$ are inherited by the space $W_p^l(\Omega)$. Examples have been given in Section 4.2 (Remark 11 and the proof of Theorem 3) and Section 4.7 (Corollaries 20, 24 and the second proof of Theorem 13).

6.1 The one-dimensional case

We start with the simplest case of Sobolev spaces $W_p^l(a, b)$, in which it is possible to give sharp two-sided estimates of the minimal norm of an extension operator $T : W_p^l(a, b) \longrightarrow W_p^l(-\infty, \infty)$.

Lemma 1 *Let* $-\infty < a < b < \infty$. *If f is defined on $[a, c]$ and is absolutely continuous on $[a, b]$ and $[b, c]$, then f is absolutely continuous on $[a, c]$.* $\square$

Idea of the proof. Derive the statement directly from the definition of absolute continuity on $[a, b]$ and $[c, b]$.

Proof. Given $\varepsilon > 0$, there exists $\delta > 0$ such that for any finite system of disjoint intervals $(\alpha_i^{(1)}, \beta_i^{(1)}) \subset [a, b]$ and $(\alpha_i^{(2)}, \beta_i^{(2)}) \subset [b, c]$ satisfying the inequalities $\sum_i (\beta_i^{(j)} - \alpha_i^{(j)}) < \delta$, $j = 1, 2$, the inequalities $\sum_i |f(\alpha_i^{(j)}) - f(\beta_i^{(j)})| < \frac{\varepsilon}{2}$, $j = 1, 2$,

---

[1] This means that $(Tf)(x) = f(x)$, if $x \in \Omega$.

hold. Now let $(\alpha_i, \beta_i) \subset [a, b]$ be a finite system of disjoint intervals satisfying $\sum_i (\beta_i - \alpha_i) < \delta$. If one of them contains b, denote it by (α^*, β^*) . Then

$$\sum_i |f(\alpha_i) - f(\beta_i)| \leq \sum_{i:(\alpha_i,\beta_i) \subset [a,b]} |f(\alpha_i) - f(\beta_i)| + |f(\alpha^*) - f(b)|$$

$$+ |f(b) - f(\beta^*)| + \sum_{i:(\alpha_i,\beta_i) \subset [b,c]} |f(\alpha_i) - f(\beta_i)| < \varepsilon.$$

(If there is no such interval (α^*, β^*), then the summands $|f(\alpha^*) - f(\beta^*)|$ and $|f(b) - f(\beta^*)|$ must be omitted.) $\square$

Lemma 2 *Let $l \in \mathbb{N}, 1 \leq p \leq \infty, -\infty < a < b < \infty, f \in W_p^l(a, b)$ and $g \in W_p^l(b, c)$. Then the pasted function*

$$h = \begin{cases} f & \text{on } (a, b), \\ g & \text{on } (b, c). \end{cases} \tag{6.1}$$

belongs to $W_p^l(a, c)$ if, and only if,

$$f_w^{(s)}(b-) = g_w^{(s)}(b+), \;\; s = 0, 1, ..., l - 1, \tag{6.2}$$

where $f_w^{(s)}(b-)$ and $g_w^{(s)}(b+)$ are boundary values of $f_w^{(s)}$ and $g_w^{(s)}$ (see Remark 6 of Chapter 1).

If (6.2) is satisfied, then

$$\|h\|_{W_p^l(a,c)} \leq \|f\|_{W_p^l(a,b)} + \|g\|_{W_p^l(b,c)}. \tag{6.3}$$

Idea of the proof. Starting from Definition 4 and Remark 6 of Chapter 1, apply Lemma 1. $\square$

Proof. Let f_1 and g_1 be the functions, equivalent to f and g, whose derivatives $f_1^{(l-1)}$, $g_1^{(l-1)}$ exist and are absolutely continuous on $[a, b]$, $[b, c]$ respectively. Then $f_1^{(s)}(b) = f_w^{(s)}(b-)$ and $g_1^{(s)}(b) = g_w^{(s)}(b+), s = 0, 1, ..., l - 1$. If (6.2) is satisfied, then the function

$$h_1 = \begin{cases} f_1 & \text{on } [a, b], \\ g_1 & \text{on } [b, c] \end{cases}$$

is such that $h_1^{(l-1)}$ exists and is absolutely continuous on $[a, b]$. Consequently, the weak derivative $h_w^{(l)}$ exists on (a, b) and

$$h_w^{(l)} = \begin{cases} f_w^{(l)} & \text{on } (a, b), \\ g_w^{(l)} & \text{on } (b, c). \end{cases}$$

Hence, inequality (6.3) follows.

If (6.2) is not satisfied, then for any function h_2 defined on $[a, b]$, coinciding with f_1 on $[a, b)$ and with g_1 on $(b, c]$, the ordinary derivative $h_2^{(l-1)}(b)$ does not exist. Hence, the weak derivative $h_w^{(l-1)}$ does not exist on (a, c) and h is not in $W_p^{(l)}(a, c)$. $\square$

Lemma 3 *Let $l \in \mathbb{N}, 1 \le p \le \infty$. Then there exists a linear extension operator $T : W_p^l(\infty, 0) \longrightarrow W_p^l(-\infty, \infty)$, such that*

$$\|T\|_{W_p^l(-\infty,0) \to W_p^l(-\infty,\infty)} \le 8^l. \tag{6.4}$$

Idea of the proof. If $l = 1$, it is enough to consider the reflection operator, i.e., to set

$$(T_1 f)(x) = f(-x), \quad x > 0. \tag{6.5}$$

If $l \ge 2$, define $(T_2)(x)$ for $x > 0$ as a linear combination of reflection and dilations:

$$(T_2 f)(x) = \sum_{k=1}^{l} \alpha_k (T_1 f)(\beta_k x) = \sum_{k=1}^{l} \alpha_k f(-\beta_k x), \tag{6.6}$$

where $\beta_k > 0$ and α_k are chosen in such a way that

$$(T_2 f)_w^{(s)}(0+) = f_w^{(s)}(0-), \quad s = 0, 1, ..., l - 1. \tag{6.7}$$

Verify that $\|T_2\|_{W_p^l(\infty,0) \to W_p^l(-\infty,\infty)} < \infty$ and choose $\beta_k = \frac{k}{l}$, $k = 1, ..., l$, in order to prove (6.4). $\square$

Proof. Equalities (6.7) are equivalent to

$$\sum_{k=1}^{l} \alpha_k (-\beta_k)^s = 1, \quad s = 0, 1, ..., l - 1. \tag{6.8}$$

Consequently, by Cramer's rule and the formula for Van-der-Monde's determinant,

$$\alpha_k = \frac{\prod_{1 \le i < j \le l} (\beta_i - \beta_j) \big|_{\beta_k = -1}}{\prod_{1 \le i < j \le l} (\beta_i - \beta_j)}$$

$$= \frac{\prod_{1 \le i < k} (\beta_i + 1) \prod_{k < j \le l} (-1 - \beta_j)}{\prod_{1 \le i < k} (\beta_i - \beta_j) \prod_{k < j \le l} (\beta_k - \beta_j)} = \prod_{1 \le j \le l, j \ne k} \frac{1 + \beta_j}{\beta_j - \beta_k}, \quad k = 1, ..., l. \tag{6.9}$$

If $\beta_k = \frac{k}{l}, k = 1, ..., l$, then

$$\alpha_k = \frac{(-1)^{k-1}k}{l+k}\binom{2l}{l}\binom{l}{k}$$

and

$$|\alpha_k| \leq 4^l \frac{k}{l}\binom{l}{k}.$$

Therefore, setting $y = -\beta_k x$, we have

$$\|T_2 f\|_{W_p^l(0,\infty)} = \|T_2 f\|_{L_p(0,\infty)} + \|(T_2 f)_w^{(l)}\|_{L_p(0,\infty)}$$

$$\leq \Big(\sum_{k=1}^l |\alpha_k|\beta_k^{-\frac{1}{p}}\Big)\|f\|_{L_p(-\infty,0)} + \Big(\sum_{k=1}^l |\alpha_k|\beta_k^{l-\frac{1}{p}}\Big)\|f_w^{(l)}\|_{L_p(-\infty,0)}$$

$$\leq \Big(\sum_{k=1}^l |\alpha_k|\beta_k^{-\frac{1}{p}}\Big)\|f\|_{W_p^l(-\infty,0)} \leq 4^l \Big(\sum_{k=1}^l \Big(\frac{k}{l}\Big)^{1-\frac{1}{p}}\binom{l}{k}\Big)\|f\|_{W_p^l(-\infty,0)}$$

$$\leq (8^l - 1)\|f\|_{W_p^l(-\infty,0)}.$$

Hence, inequality (6.4) follows if we take into account Lemma 2 and, in particular, inequality (6.3). $\square$

Remark 1 It follows from the above proof that the inequalities

$$\|T_2\|_{w_p^m(-\infty,0)\to w_p^m(-\infty,\infty)} \leq 8^l, \quad m \in \mathbb{N}_0, \quad m \leq l,$$

also hold.

Corollary 1 *Let $l \in \mathbb{N}, 1 \leq p \leq \infty, -\infty < a < \infty$. Then there exists a linear extension operator $T : W_p^l(a, b) \longrightarrow W_p^l(2a - b, 2b - a)$, such that*

$$\|T\|_{W_p^l(a,b)\to W_p^l(2a-b,2b-a)} \leq 2 \cdot 8^l. \tag{6.10}$$

Idea of the proof. Define

$$(T_3 f)(x) = \begin{cases} \sum_{k=1}^l \alpha_k f(a + \beta_k(a - x)) & \text{for } x \in (2a - b, a), \\ f(x) & \text{for } x \in (a, b), \\ \sum_{k=1}^l \alpha_k f(b + \beta_k(b - x)) & \text{for } x \in (b, 2b - a), \end{cases} \tag{6.11}$$

where α_k and β_k are the same as in (6.6), observe that $T_3 f$ is defined on $(2a - b, 2b - a)$ since $0 < \beta_k \leq 1$, and apply the proof of Lemma 3. $\square$

Corollary 2 *Let $l \in \mathbb{N}, 1 \le p \le \infty, -\infty < a < b < \infty$. Then there exists a linear extension operator $T : W_p^l(a,b) \longrightarrow W_p^l(a-1,b+1)$ such that*

$$\|T\|_{W_p^l(a,b)\to W_p^l(a-1,b+1)} \le 2 \cdot 8^l(1 + (b-a)^{-l+\frac{1}{p'}}). \tag{6.12}$$

Idea of the proof. Let $\delta = \min\{1, b-a\}$ and define

$$(T_4 f)(x) = \begin{cases} \sum\limits_{k=1}^{l} \alpha_{k,\delta} f(a + \delta\beta_k(a-x)) & \text{for } x \in (a-1,a), \\ f(x) & \text{for } x \in (a,b), \\ \sum\limits_{k=1}^{l} \alpha_{k,\delta} f(b + \delta\beta_k(b-x)) & \text{for } x \in (b,b+1), \end{cases} \tag{6.13}$$

where β_k are the same as in (6.11) and $\alpha_{k,\delta}$ are such that $\sum\limits_{k=1}^{l} \alpha_{k,\delta}(-\delta\beta_k)^s = 1$, $s = 0, ..., l-1$. Observe that by (6.9) $|\alpha_{k,\delta}| \le (b-a)^{-l+1}|\alpha_k|$ and apply the proof of Lemma 3. $\square$

Proof. As in the proof of Lemma 3

$$\|T_4 f\|_{W_p^l(b,b+1)} \le \sum_{k=1}^{l} |\alpha_{k,\delta}| \cdot \|f(b + \delta\beta_k(b-x))\|_{L_p(b,b+1)}$$

$$+ \sum_{k=1}^{l} |\alpha_{k,\delta}|(\delta\beta_k)^l \|f_w^{(l)}(b + \delta\beta_k(b-x))\|_{L_p(b,b+1)}$$

$$\le \left(\sum_{k=1}^{l} |\alpha_{k,\delta}|(\delta\beta_k)^{-\frac{1}{p}} \right) \|f\|_{W_p^l(b-\delta\beta_k,b)}$$

$$\le \delta^{-l+\frac{1}{p'}}(8^l - 1)\|f\|_{W_p^l(a,b)} \le (8^l \delta^{-l+\frac{1}{p'}} - 1)\|f\|_{W_p^l(a,b)}$$

and

$$\|T_4 f\|_{W_p^l(a-1,b+1)} \le \|T_4 f\|_{W_p^l(a-1,a)} + \|T_4 f\|_{W_p^l(a,b)} + \|T_4 f\|_{W_p^l(b,b+1)}$$

$$\le 2 \cdot 8^l \delta^{-l+\frac{1}{p'}}\|f\|_{W_p^l(a,b)}. \quad \square$$

In order to estimate the norm of an extension operator $T : W_p^l(-\infty,0) \to W_p^l(-\infty,\infty)$ from below we prove the following statement, which reduces this problem to a certain type of extremal boundary-value problems.

For given $a_0, ..., a_{l-1} \in \mathbb{R}$ let

$$G_{p,l}^+(a_0, ..., a_{l-1}) = \inf_{\substack{f \in W_p^l(0,\infty): \\ f_w^{(k)}(0+)=a_k, k=0,...,l-1}} \|f\|_{W_p^l(0,\infty)}. \tag{6.14}$$

$G_{p,l}^-(a_0, ..., a_{l-1})$ is defined in a similar way with $(-\infty, 0)$ replacing $(0, \infty)$. Let

$$Q_{p,l} = \sup_{|a_0|+\cdots+|a_{l-1}|>0} \frac{G_{p,l}^+(a_0, a_1, ..., a_{l-1})}{G_{p,l}^-(a_0, a_1, ..., a_{l-1})}$$

$$= \sup_{|a_0|+\cdots+|a_{l-1}|>0} \frac{G_{p,l}^+(a_0, a_1, ..., a_{l-1})}{G_{p,l}^+(a_0, -a_1, ..., (-1)^{l-1}a_{l-1})}. \tag{6.15}$$

The latter equality follows if the argument x is replaced by $-x$ in the definition of $G_{p,l}^-$. Moreover, it follows from (6.15) that for $1 \le p \le \infty$

$$Q_{p,l} \ge 1, \quad l \in \mathbb{N}, \quad Q_{p,1} = 1. \tag{6.16}$$

Lemma 4 *Let $l \in \mathbb{N}, 1 \le p \le \infty$. Then*

$$\left(1 + Q_{p,l}^p\right)^{\frac{1}{p}} \le \inf_T \|T\|_{W_p^l(-\infty,0) \to W_p^l(-\infty,\infty)} \le 1 + Q_{p,l}. \tag{6.17}$$

(If $p = \infty$, then $(1 + Q_{p,l}^p)^{\frac{1}{p}}$ must be replaced by $Q_{\infty,l}$.)

Idea of the proof. Apply the inequality

$$\left(\|f\|_{W_p^l(-\infty,0)}^p + \|Tf\|_{W_p^l(0,\infty)}^p\right)^{\frac{1}{p}} \le \|Tf\|_{W_p^l(-\infty,\infty)} \le \|f\|_{W_p^l(-\infty,0)} + \|Tf\|_{W_p^l(0,\infty)}. \tag{6.18}$$

In order to prove the first inequality (6.17) apply also the inequality

$$\|Tf\|_{W_p^l(0,\infty)} \ge G_{p,l}^+(a_0, ..., a_{l-1}), \tag{6.19}$$

which, by the definition of $G_{p,l}^+$, holds for all $a_0, ..., a_{l-1}$ and for each extension operator T. In order to prove the second inequality (6.17) define, $\forall \varepsilon > 0$, the extension operator T_ε setting $T_\varepsilon f = g_\varepsilon$ for $x \in (0, \infty)$, where $g_\varepsilon \in W_p^l(0, \infty)$ is any function, which is such that $g_{\varepsilon,w}^{(k)}(0+) = f_w^{(k)}(0-), k = 0, ..., l-1$, and

$$\|g_\varepsilon\|_{W_p^l(0,\infty)} \le G_{p,l}^+(f(0-), ..., f_w^{(l-1)}(0-)) + \varepsilon \|f\|_{W_p^l(-\infty,0)}. \quad \Box \tag{6.20}$$

Proof. 1. The second inequality (6.18) is trivial since

$$\|h\|_{L_p(-\infty,\infty)} \le \|h\|_{L_p(-\infty,0)} + \|h\|_{L_p(0,\infty)}.$$

The first inequality (6.18) follows from Minkowski's inequality for finite sums, because

$$\|h\|_{W_p^l(-\infty,\infty)} = \left(\|h\|_{L_p(-\infty,0)}^p + \|h\|_{L_p(0,\infty)}^p \right)^{\frac{1}{p}}$$

$$+ \left(\|h_w^{(l)}\|_{L_p(-\infty,0)}^p + \|h_w^{(l)}\|_{L_p(0,\infty)}^p \right)^{\frac{1}{p}} \ge \left\{ \left(\|h\|_{L_p(-\infty,0)} + \|h_w^{(l)}\|_{L_p(-\infty,0)} \right)^p \right.$$

$$\left. + \left(\|h\|_{L_p(0,\infty)} + \|h_w^{(l)}\|_{L_p(0,\infty)} \right)^p \right\}^{\frac{1}{p}} = \left(\|h\|_{W_p^l(-\infty,0)}^p + \|h\|_{W_p^l(0,\infty)}^p \right)^{\frac{1}{p}}.$$

2. It follows from (6.18) and (6.19) that for each $a_0, ..., a_{l-1} \in \mathbb{R}$ such that $|a_0| + \cdots + |a_{l-1}| > 0$

$$\|T\|_{W_p^l(-\infty,0)\to W_p^l(-\infty,\infty)} = \sup_{f\in W_p^l(0,\infty), f\not\approx 0} \frac{\|Tf\|_{W_p^l(-\infty,\infty)}}{\|f\|_{W_p^l(-\infty,0)}}$$

$$\ge \left(1 + \sup_{\substack{f\in W_p^l(-\infty,0):\\ f_w^{(k)}(0-)=a_k, k=0,...,l-1}} \left(\frac{\|Tf\|_{W_p^l(0,\infty)}}{\|f\|_{W_p^l(-\infty,0)}} \right)^p \right)^{\frac{1}{p}}$$

$$\ge \left(1 + \left(G_{p,l}^+(a_0, \dots, a_{l-1}) \right)^p \sup_{\substack{f\in W_p^l(-\infty,0):\\ f_w^{(k)}(0-)=a_k, k=0,...,l-1}} \frac{1}{\|f\|_{W_p^l(-\infty,0)}^p} \right)^{\frac{1}{p}}$$

$$= \left(1 + \left(\frac{G_{p,l}^+(a_0, \dots, a_{l-1})}{G_{p,l}^-(a_0, \dots, a_{l-1})} \right)^p \right)^{\frac{1}{p}},$$

and we arrive at the first inequality (6.17).

3. Given $\varepsilon > 0$ by (6.18) and (6.20) we have

$$\|T_\varepsilon\| \le 1 + \sup_{f\in W_p^l(-\infty,0), f\not\approx 0} \frac{\|g_\varepsilon\|_{W_p^l(0,\infty)}}{\|f\|_{W_p^l(-\infty,0)}}$$

$$\le 1 + \varepsilon + \sup_{\substack{a_0,...,a_{l-1}\in\mathbb{R}:\\ |a_0|+...+|a_{l-1}|>0}} \sup_{\substack{f\in W_p^l(-\infty,0):\\ f_w^{(k)}(0-)=a_k, k=0,...,l-1}} \frac{G_{p,l}^+(a_0, \dots, a_{l-1})}{\|f\|_{W_p^l(-\infty,0)}} = 1 + Q_{p,l} + \varepsilon$$

and the second inequality of (6.17) follows. $\square$

Corollary 3 *Let $1 < p \leq \infty$. Then*

$$\inf_T \|T\|_{W_p^1(-\infty,0)\to W_p^1(-\infty,\infty)} = 2^{\frac{1}{p}}.$$

Idea of the proof. By (6.15) and (6.16) $\|T\|_{W_p^1(-\infty,0)\to W_p^1(-\infty,\infty)} \geq 2^{\frac{1}{p}}$ for each extension operator T. On the other hand it is clear that for the extension operator T_1 defined by (6.5) $\|T_1\|_{W_p^1(-\infty,0)\to W_p^1(-\infty,\infty)} = 2^{\frac{1}{p}}$. $\square$

Remark 2 Note also that if the norm in the space $W_p^l(a,b)$ is defined by

$$\|f\|_{W_p^l(a,b)}^{(1)} = \left(\int_a^b (|f(x)|^p + |f_w^{(l)}(x)|^p)\,dx \right)^{\frac{1}{p}}$$

(see Remark 8 of Section 1.3), then

$$\inf_T \|T\|_{W_p^l(-\infty,0)\to W_p^l(-\infty,\infty)} = (1 + (Q_{p,l}^{(1)})^p)^{\frac{1}{p}},$$

where $Q_{p,l}^{(1)}$ is defined by (6.14) – (6.15) with $\|\cdot\|^{(1)}$ replacing $\|\cdot\|$. This follows from the proof of Lemma 4 and the equality

$$\|Tf\|_{W_p^l(-\infty,\infty)}^{(1)} = \left((\|Tf\|_{W_p^l(\infty,0)}^{(1)})^p + (\|Tf\|_{W_p^l(0,\infty)}^{(1)})^p \right)^{\frac{1}{p}}.$$

Lemma 5 *Let $l \in \mathbb{N}, 1 \leq p \leq \infty$ and $f \in W_p^l(0,\infty)$. Then*

$$\|f\|_{W_p^l(0,\infty)} \geq \left\| \sum_{k=0}^{l-1} \frac{f^{(k)}(0+)}{k!} x^k \right\|_{L_p(0,\sqrt[l]{l!})}. \tag{6.21}$$

Idea of the proof. Apply Taylor's formula and Hölder's inequality. $\square$
Proof. Let $f \in W_p^l(0,\infty)$. Then for almost every $x \in (0,\infty)$

$$f(x) = \sum_{k=0}^{l-1} \frac{f_w^{(k)}(0+)\, x^k}{k!} + \frac{1}{(l-1)!} \int_0^x (x-u)^{l-1} f_w^{(l)}(u)\,du,$$

where the $f_w^{(k)}(0+), k = 0, 1, \ldots, l-1$, are the boundary values of the weak derivatives $f_w^{(k)}$. (See formula (3.10) and comments on it in Section 3.1). Hence, by the triangle inequality for each $a > 0$

$$\left\| \sum_{k=0}^{l-1} \frac{f_w^{(k)}(0+)\, x^k}{k!} \right\|_{L_p(0,a)} \leq \|f\|_{L_p(0,a)} + \frac{1}{(l-1)!} \left\| \int_0^x (x-u)^{l-1} f_w^{(l)}(u)\,du \right\|_{L_p(0,a)}.$$

By Hölder's inequality

$$\left\| \int_0^x (x-u)^{l-1} f_w^{(l)}(u)\, du \right\|_{L_p(0,a)} \leq \left\| \left(\frac{x^{(l-1)p'+1}}{(l-1)p'+1} \right)^{\frac{1}{p'}} \left\| f_w^{(l)} \right\|_{L_p(0,x)} \right\|_{L_p(0,a)}$$

$$\leq ((l-1)p'+1)^{-\frac{1}{p'}} \| x^{l-\frac{1}{p}} \|_{L_p(0,a)} \| f_w^{(l)} \|_{L_p(0,a)}$$

$$= a^l (lp)^{-\frac{1}{p}} ((l-1)p'+1)^{-\frac{1}{p'}} \| f_w^{(l)} \|_{L_p(0,a)} \leq \frac{a^l}{l} \| f_w^{(l)} \|_{L_p(0,a)}.$$

Consequently,

$$\left\| \sum_{k=0}^{l-1} \frac{f_w^{(k)}(0+) x^k}{k!} \right\|_{L_p(0,a)} \leq \| f \|_{L_p(0,a)} + \frac{a^l}{l!} \| f_w^{(l)} \|_{L_p(0,a)}.$$

Setting $a = \sqrt[l]{l!}$, we get (6.21). $\square$

Corollary 4 *For all* $l \in \mathbb{N}, 1 \leq p \leq \infty, a_0, \ldots, a_{l-1} \in \mathbb{R}$

$$G_{p,l}^+(a_0, \ldots, a_{l-1}) \geq \left\| \sum_{k=0}^{l-1} \frac{a_k}{k!} x^k \right\|_{L_p(0,\sqrt[l]{l!})}. \tag{6.22}$$

Idea of the proof. Apply (6.14) and (6.21). $\square$

Lemma 6 *Let* $l \in \mathbb{N}, 1 \leq p \leq \infty$. *Then for every extension operator* $T : W_p^l(-\infty, 0) \to W_p^l(-\infty, \infty)$

$$\| T \|_{W_p^l(-\infty,0) \to W_p^l(-\infty,\infty)} \geq 2^{l-2}\, l^{-\frac{1}{p}}. \tag{6.23}$$

Idea of the proof. For $l = 1, 2$ inequality (6.23) is trivial since $\| T \| \geq 1$ for each extension operator T. Assume that $l \geq 3$ and set

$$f_l(x) = \begin{cases} 0 & \text{for } -\infty < x \leq -a, \\ (x+a)^l & \text{for } -a \leq x \leq 0, \end{cases}$$

where $a = \sqrt[l]{l!}$. $\square$

Proof. By (6.14), (6.15), (6.22) and the triangle inequality we have

$$Q_{p,l} \geq \frac{G_{p,l}^+(f_l(0), \ldots, f_l^{(l-1)}(0))}{\| f_l \|_{W_p^l(-\infty,0)}} \geq \frac{\left\| \sum_{k=0}^{l-1} \frac{f_l^{(k)}(0) x^k}{k!} \right\|_{L_p(0,a)}}{\| f_l \|_{W_p^l(-\infty,0)}}$$

$$= \frac{\|(x+a)^l - x^l\|_{L_p(0,a)}}{\|(x+a)^l\|_{W_p^l(-a,0)}} \geq \frac{\|(x+a)^l\|_{L_p(0,a)} - \|x^l\|_{L_p(0,a)}}{\|x^l\|_{W_p^l(0,a)}}$$

$$= \frac{(2^{lp+1}-1)^{\frac{1}{p}} - 1}{1 + (lp+1)^{\frac{1}{p}}} \geq \frac{2^l - 1}{l^{\frac{1}{p}}(l^{-\frac{1}{p}} + (p+\frac{1}{l})^{\frac{1}{p}})} \geq \frac{2^l - 1}{3} l^{-\frac{1}{p}} \geq 2^{l-2} l^{-\frac{1}{p}}.$$

Hence by (6.17) inequality (6.23) follows. $\square$

Remark 3 Note also that there exists a constant $c_1 > 1$ such that [2]

$$\|T\|_{W_p^l(-\infty,0) \to W_p^l(-\infty,\infty)} \geq c_1^l, \quad l \geq 2, \quad 1 \leq p \leq \infty, \tag{6.24}$$

for every extension operator T. For $c_1 = \frac{2}{\sqrt{3}}$ this follows from the inequality $(2^{lp+1} - 1)^{\frac{1}{p}} - 1 - (\frac{2}{\sqrt{3}})^l(1 + (lp+1)^{\frac{1}{p}}) \geq 0$ for $l \geq 2, \ 1 \leq p \leq \infty$.

Lemma 7 *Let $l \in \mathbb{N}, -\infty < a < b < \infty, \varepsilon > 0$ Then there exists a "cap-shaped" function $\eta \in C_0^\infty(\mathbb{R})$ such that $0 \leq \eta \leq 1, \eta = 1$ on $(a,b), \mathrm{supp}\, \eta \subset (a - \varepsilon, b + \varepsilon)$ and*

$$|\eta^{(k)}(x)| \leq (4l)^k \varepsilon^{-k}, \quad x \in \mathbb{R}, \quad k = 0,\ldots,l. \tag{6.25}$$

Idea of the proof. Set

$$\eta = \widetilde{\omega}_{\frac{\gamma\varepsilon}{4(l+\gamma)}} * \underbrace{\omega_{\frac{\varepsilon}{2(l+\gamma)}} * \ldots * \omega_{\frac{\varepsilon}{2(l+\gamma)}}}_{l \text{ times}} * \chi_{(a-\frac{\varepsilon}{2}, b+\frac{\varepsilon}{2})}, \tag{6.26}$$

where $\chi_{(a-\frac{\varepsilon}{2}, b+\frac{\varepsilon}{2})}$ is the characteristic function of the interval $(a - \frac{\varepsilon}{2}, b + \frac{\varepsilon}{2})$, $\omega(x) = 1 - |x|$ if $|x| \leq 1$, $\omega(x) = 0$ if $|x| > 1$, $\widetilde{\omega}$ is any nonnegative infinitely differentiable kernel of mollification (see Section 1.1) and γ is a sufficiently small positive number. Apply Young's inequality (4.138) and the equality

$$\left\| \left(\omega_{\frac{\varepsilon}{2(l+\gamma)}} * \chi_{(a-\frac{\varepsilon}{2}, b+\frac{\varepsilon}{2})} \right)' \right\|_{L_\infty(\mathbb{R})} = \left\| \omega_{\frac{\varepsilon}{2(l+\gamma)}} \right\|_{L_\infty(\mathbb{R})}. \quad \square \tag{6.27}$$

Proof. Let $\sigma = \frac{\varepsilon}{2}(l + \frac{\gamma}{4})(l + \gamma)^{-1}$. By Section 1.1 $\eta \in C_0^\infty(\mathbb{R}), 0 \leq \eta \leq 1, \eta = 1$ on $(a - \frac{\varepsilon}{2} + \sigma, b + \frac{\varepsilon}{2} - \sigma) \supset (a,b)$ and $\mathrm{supp}\, \eta \subset [a - \frac{\varepsilon}{2} - \sigma, b + \frac{\varepsilon}{2} + \sigma] \subset (a - \frac{\varepsilon}{2}, b + \frac{\varepsilon}{2})$. Moreover,

$$\|\eta^{(k)}\|_{L_\infty(\mathbb{R})} \leq \left\| (\omega_{\frac{\varepsilon}{2(l+\gamma)}} * \ldots * \omega_{\frac{\varepsilon}{2(l+\gamma)}} * \chi_{(a-\frac{\varepsilon}{2}, b+\frac{\varepsilon}{2})})_w^{(k)} \right\|_{L_\infty(\mathbb{R})}$$

---

[2] Inequailty (6.24) does not hold $\forall l \in \mathbb{N}$ because of Corollary 4 for $p = \infty$.

$$= \Big\| \underbrace{(\omega_{\frac{\varepsilon}{2(l+\gamma)}})' * \ldots * (\omega_{\frac{\varepsilon}{2(l+\gamma)}})'}_{k-1 \text{ times}} * \underbrace{\omega_{\frac{\varepsilon}{2(l+\gamma)}} * \ldots * \omega_{\frac{\varepsilon}{2(l+\gamma)}}}_{l-k \text{ times}} * (\omega_{\frac{\varepsilon}{2(l+\gamma)}} * \chi_{(a-\frac{\varepsilon}{2}, b+\frac{\varepsilon}{2})})' \Big\|_{L_\infty(\mathbb{R})}$$

$$\leq \Big(\frac{2(l+\gamma)}{\varepsilon} \Big)^{k-1} \|\omega'\|_{L_1(\mathbb{R})}^{k-1} \|\omega\|_{L_1(\mathbb{R})}^{l-k} \|\omega_{\frac{\varepsilon}{2(l+\gamma)}}\|_{L_\infty(\mathbb{R})} \leq 2 \cdot 4^{k-1} (l+\gamma)^k \varepsilon^{-k}.$$

Choose $\gamma > 0$ satisfying $e^\gamma \leq 2$, then $(l + \gamma)^k \leq e^k (1 + \frac{\gamma}{e})^l \leq 2 \cdot l^k$ and so obtain (6.25).

Finally we note that (6.27) follows from

$$(\omega_{\frac{\varepsilon}{2(l+\gamma)}} * \chi_{(a-\frac{\varepsilon}{2}, b+\frac{\varepsilon}{2})})'(x) = \Big(\int_{a-\frac{\varepsilon}{2}}^{b+\frac{\varepsilon}{2}} \omega_{\frac{\varepsilon}{2(l+\gamma)}}(x - y)\, dy \Big)'$$

$$= \Big(\int_{x-a-\frac{\varepsilon}{2}}^{x-b-\frac{\varepsilon}{2}} \omega_{\frac{\varepsilon}{2(l+\gamma)}}(z)\, dz \Big)' = \omega_{\frac{\varepsilon}{2(l+\gamma)}}\Big(x - b - \frac{\varepsilon}{2}\Big) - \omega_{\frac{\varepsilon}{2(l+\gamma)}}\Big(x - a + \frac{\varepsilon}{2}\Big)$$

since the terms of the right-hand side have disjoint supports. $\square$

Corollary 5 *In the one-dimensional case* $\forall l \in \mathbb{N}$ *there exists a nonnegative infinitely differentiable kernel of mollification* μ *satisfying* (1.1) *such that*

$$|\mu^{(k)}(x)| \leq (4l)^k, \quad x \in \mathbb{R}, \quad k = 0, \ldots, l. \tag{6.28}$$

Idea of the proof. Define η by (6.26), where $a = b = 0$ and $\varepsilon = 1$, and apply the equality $\|f * g\|_{L_1(\mathbb{R})} = \|f\|_{L_1(\mathbb{R})} \cdot \|g\|_{L_1(\mathbb{R})}$ for non-negative $f, g \in L_1(\mathbb{R})$. $\square$

Lemma 8 *There exists* $c_2 > 0$ *such that for all* $l, m \in \mathbb{N}, m < l, 1 \leq p, q \leq \infty, -\infty < a < b < \infty$ *and* $\forall f \in W_p^l(a, b)$

$$\|f_w^{(m)}\|_{L_q(a,b)} \leq c_2^l (b-a)^{\frac{1}{q}-\frac{1}{p}} \Big(\Big(\frac{l}{b-a} \Big)^m \|f\|_{L_p(a,b)} + \Big(\frac{b-a}{l} \Big)^{l-m} \|f_w^{(l)}\|_{L_p(a,b)} \Big). \tag{6.29}$$

Idea of the proof. Apply the integral representation (3.17) with $(\alpha, \beta) = (a, b)$ and $\omega(x) = \frac{2}{b-a}\, \mu\Big(\frac{2(x - \frac{a+b}{2})}{b-a} \Big)$, where the function μ is a function constructed in Corollary 5. $\square$

Proof. The numbers $\sigma_{s,m}$ defined by (3.20) satisfy the following inequality

$$|\sigma_{s,m}| \leq \frac{1}{(s-m)!} \sum_{k=0}^{l-1} \binom{l-1}{k} = \frac{2^{l-1}}{(s-m)!}.$$

Consequently, taking into account (3.5) and Remark 5 in Chapter 3, we have that for almost all $x \in [a, b]$

$$|f_w^{(m)}(x)| \leq \int_a^b \left(\sum_{s=m}^{l-1} \frac{2^{l-1}}{(s-m)!} (b-a)^{s-m} (4l)^s \left(\frac{2}{b-a}\right)^{s+1} (4l)^s \right) |f| \, dy$$

$$+ \frac{(b-a)^{l-m-1}}{(l-m-1)!} \int_a^b |f_w^{(l)}| \, dy \leq (b-a)^{-m-1} 16^l l^m \sum_{s=m}^{l-1} \frac{l^{s-m}}{(s-m)!} \int_a^b |f| \, dy$$

$$+ (b-a)^{l-m-1} l^{m-l-1} \left(\frac{l^{l-m-1}}{(l-m-1)!} \right) \int_a^b |f_w^{(l)}| \, dy$$

$$\leq (16e)^l (b-a)^{-\frac{1}{p}} \left(\left(\frac{l}{b-a}\right)^m \|f\|_{L_p(a,b)} + \left(\frac{b-a}{l}\right)^{l-m} \|f_w^{(l)}\|_{L_p(a,b)} \right) \qquad (6.30)$$

and inequality (6.29) follows with $c_2 = 16\,e$. $\square$

Remark 4 Inequality (6.29) is an improved version of inequality (4.55). $\square$

Corollary 6 *If, in addition to the assumptions of Lemma 8, $b - a \leq l$, then*

$$\|f_w^{(m)}\|_{L_q(a,b)} \leq c_2{}^l \, l^m \, (b-a)^{-m+\frac{1}{q}-\frac{1}{p}} \, \|f\|_{W_p^l(a,b)}. \qquad (6.31)$$

If, in addition to the assuptions of Lemma 8, $b - a \geq 1$ and $q \geq p$, then

$$\|f_w^{(m)}\|_{L_q(a,b)} \leq 2^{\frac{1}{q}} c_2{}^l \, l^m \, \|f\|_{W_p^l(a,b)}. \qquad (6.32)$$

Idea of the proof. Inequality (6.31) is a direct corollary of (6.29). In order to prove (6.32) apply (6.29) and Lemma 7 of Chapter 4. $\square$

Proof. Let $b - a \geq 1$ and $q \geq p$. Choose intervals (a_k, b_k), $k = 1, \ldots, s$, in such a way that $b_k - a_k = 1, (a, b) = \bigcup_{k=1}^s (a_k, b_k)$ and the multiplicity of the covering $\{(a_k, b_k)\}_{k=1}^s$ is equal to 2. By (6.29)

$$\|f_w^{(m)}\|_{L_q(a_k,b_k)} \leq c_2{}^l \, (l^m \, \|f\|_{L_p(a_k,b_k)} + l^{m-l} \, \|f_w^{(l)}\|_{L_p(a_k,b_k)}).$$

Hence, by Lemma 7 of Chapter 4

$$\|f_w^{(m)}\|_{L_q(a,b)} \leq 2^{\frac{1}{q}} c_2{}^l \, (l^m \, \|f\|_{L_p(a,b)} + l^{m-l} \, \|f_w^{(l)}\|_{L_p(a,b)})$$

and (6.32) follows. $\square$

Lemma 9 *Let $l \in \mathbb{N}$, $1 \le p \le \infty$, $-\infty < a < b < \infty$, $b - a \le 1$. There exists a linear operator $T : W_p^l(a,b) \to W_p^l(-\infty,\infty)$, such that*

$$\|T\|_{W_p^l(a,b) \to W_p^l(-\infty,\infty)} \le \frac{c_3{}^l l^l}{(b-a)^{l-\frac{1}{p'}}}, \tag{6.33}$$

where c_3 is a constant greater than 1.

Idea of the proof. Consider the operator

$$(T_5 f)(x) = (T_4 f)(x)\eta(x), \quad x \in \mathbb{R}, \tag{6.34}$$

where η is the function constructed in Lemma 7 for $\varepsilon = 1$ and T_4 is defined by (6.13), assuming that $(T_5 f)(x) = 0$ for $x \notin (a-1, b+1)$ and apply Corollary 6. $\square$

Proof. It follows from the Leibnitz formula, (6.25), (6.32) and (6.12) that

$$\|T_5 f\|_{W_p^l(-\infty,\infty)} = \|\eta T_4 f\|_{L_p(a-1,b+1)} + \|(\eta T_4 f)_w^{(l)}\|_{L_p(a-1,b+1)}$$

$$\le \|T_4 f\|_{L_p(a-1,b+1)} + \sum_{m=0}^{l} \binom{l}{m} \|\eta^{(l-m)}\|_{L_\infty(-\infty,\infty)} \|(T_4 f)_w^{(l)}\|_{L_p(a-1,b+1)}$$

$$\le \|T_4 f\|_{L_p(a-1,b+1)} + \left(\sum_{m=0}^{l} \binom{l}{m} (4l)^{l-m} (2c_2)^l l^m \right) \|T_4 f\|_{W_p^l(a-1,b+1)}$$

$$\le (1 + (16\,c_2\,l)^l) \|T_4 f\|_{W_p^l(a-1,b+1)} \le 4\,(1 + (16\,c_2\,l)^l)\,8^l\,(b-a)^{-l+\frac{1}{p'}} \|f\|_{W_p^l(a,b)}$$

$$\le c_3^l\,l^l\,(b-a)^{-l+\frac{1}{p'}} \|f\|_{W_p^l(a,b)} ,$$

where $c_3 = 32\,(1 + 16\,c_2)$. Hence we obtain (6.33). $\square$

Lemma 10 *Let $l \in \mathbb{N}$, $1 \le p \le \infty$, $-\infty < a < b < \infty$, $b - a \ge 1$. There exists a linear extension operator $T : W_p^l(a,b) \to W_p^l(-\infty,\infty)$ such that*

$$\|T\|_{W_p^l(a,b) \to \dot{W}_p^l(-\infty,\infty)} \le c_4^l \left(1 + \frac{l^l}{(b-a)^{l-\frac{1}{p'}}} \right), \tag{6.35}$$

where c_4 is a constant greater than 1.

Idea of the proof. Consider the operator

$$(T_6 f)(x) = (T_3 f)(x)\,\eta(x), \tag{6.36}$$

where η is the function constructed in Lemma 7 for $\varepsilon = b - a$ and T_3 is defined by (6.11), and apply Lemma 8. $\square$

Proof. It follows from the Leibnitz formula, (6.25), (6.29) and (6.10) that

$$\|T_6 f\|_{W_p^l(-\infty,\infty)} = \|\eta T_3 f\|_{L_p(2a-b,2b-a)} + \|(\eta T_3 f)_w^{(l)}\|_{L_p(2a-b,2b-a)}$$

$$\leq \|T_3 f\|_{L_p(2a-b,2b-a)} + \sum_{m=0}^{l} \binom{l}{m} \|\eta^{(l-m)}\|_{L_\infty(-\infty,\infty)} \|(T_3 f)_w^{(m)}\|_{L_p(2a-b,2b-a)}$$

$$\leq \|T_3 f\|_{L_p(2a-b,2b-a)}$$

$$+ \sum_{m=0}^{l} \binom{l}{m} (4l)^{l-m} (b-a)^{m-l} c_2^{\,m} \left(\left(\frac{l}{b-a} \right)^m \|T_3 f\|_{L_p(2a-b,2b-a)} \right.$$

$$\left. + \left(\frac{b-a}{l} \right)^{l-m} \|(T_3 f)_w^{(l)}\|_{L_p(2a-b,2b-a)} \right)$$

$$\leq \|T_3 f\|_{L_p(2a-b,2b-a)} + (4l)^l \left(\sum_{m=0}^{l} \binom{l}{m} c_2^{\,m} \right) (b-a)^{-l} \|T_3 f\|_{L_p(2a-b,2b-a)}$$

$$+ 4^l \sum_{m=0}^{l} \binom{l}{m} c_2^{\,m} \|(T_3 f)_w^{(l)}\|_{L_p(2a-b,2b-a)}$$

$$\leq (1 + (4\,(1+c_2))^l\,(1 + l^l (b-a)^{-l})\|T_3 f\|_{W_p^l(2a-b,2b-a)}$$

$$\leq 2\,(1 + (4\,(1+c_2))^l\, 8^l\,(1 + l^l (b-a)^{-l})\|f\|_{W_p^l(a,b)}$$

$$\leq c_4^l\,(1 + l^l (b-a)^{-l})\|f\|_{W_p^l(a,b)} \leq c_4^l \left(1 + l^l (b-a)^{-l+\frac{1}{p'}} \right)\|f\|_{W_p^l(a,b)},$$

where $c_4 = 16\,(1 + 4\,(1 + c_2))$. Hence we obtain (6.35). $\square$

Remark 5 It follows from the proofs of Lemmas 9 and 10 that for all $-\infty < a < b < \infty$ there exists an extension operator T such that

$$\|T\|_{W_p^m(a,b)\to W_p^m(-\infty,\infty)} \leq c_5^l \left(1 + \frac{m^m}{(b-a)^{m-\frac{1}{p'}}} \right), \quad m \in N_0, \ \ m \leq l, \tag{6.37}$$

where c_5 is a constant greater than 1.

Now we consider estimates from below for the minimal norm of an extension operator.

Lemma 11 *Let $l \in \mathbb{N}, 1 \leq p \leq \infty, \infty < a < b < \infty$. Then for every extension operator $T : W_p^l(a,b) \to W_p^l(-\infty,\infty)$*

$$\|T\|_{W_p^l(a,b) \to W_p^l(-\infty,\infty)} \geq \frac{1}{8\sqrt{l}}\left(\frac{4}{e}\right)^l l^l (b-a)^{-l+\frac{1}{p'}}. \tag{6.38}$$

Remark 6 We shall give two proofs of Lemma 11. The first of them is a direct one: as in the proof of Lemma 6 it is based on the choice of a function $f \in W_p^l(a,b)$, which is the "worst" for extension. The second one is based on Lemma 12 below, in which a lower bound for the norm of an arbitrary extension operator via the best constants in the inequalities for the norms of intermediate derivatives is given. In both proofs the polynomials $Q_{l-1;p}$ of degree $l-1$ closest to zero in $L_p(0,1)$ are involved, i.e., $Q_{l-1;p} = x^{l-1} + a_{l-2}x^{l-2} + \ldots + a_0$ and

$$\|Q_{l-1;p}\|_{L_p(0,1)} = \inf_{b_0,\ldots,b_{l-2} \in \mathbb{R}} \|x^{l-1} + b_{l-2}x^{l-2} + \ldots + b_0\|_{L_p(0,1)}.$$

We recall that $Q_{l-1;\infty}(x) = 2^{-l+1}R_{l-1}(2x - 1)$, where R_m is the Chebyshev polynomial of the 1-st type: $R_m(x) = 2^{-m+1}\cos(m \arccos x)$. Moreover,

$$\|Q_{l-1;p}\|_{L_p(0,1)} \leq \|Q_{l-1;\infty}\|_{L_p(0,1)} \leq \|Q_{l-1;\infty}\|_{L_\infty(0,1)} = 8 \cdot 4^{-l}. \tag{6.39}$$

Idea of the first proof of Lemma 11. In the inequality

$$\|T\| = \|T\|_{W_p^l(a,b) \to W_p^l(-\infty,\infty)} \geq \frac{\|Tf\|_{W_p(-\infty,\infty)}}{\|f\|_{W_p^l(a,b)}} \tag{6.40}$$

set

$$f(x) = \frac{(b-a)^{l-1}}{(l-1)!}Q_{l-1;p}\left(\frac{x-a}{b-a}\right), \tag{6.41}$$

apply inequality (4.50) and the relation

$$\inf_{\substack{h \in W_p^1(-\infty,a): \\ h(a-)=1}} \|h\|_{W_p^1(-\infty,a)} \geq 1. \; \square \tag{6.42}$$

First proof. It follows from (6.40), (6.41) and (6.39) that

$$\|T\| \geq \frac{(l-1)!(b-a)^{-l+\frac{1}{p'}}}{\|Q_{l-1;p}\|_{L_p(0,1)}} \|g\|_{W_p^l(-\infty,\infty)}$$

$$\geq \frac{1}{2} 4^{l-1}(l-1)!\,(b-a)^{-l+\frac{1}{p'}}\, \|g\|_{W_p^l(-\infty,\infty)}$$

where $g = Tf$. By inequality (4.50)

$$\|g_w^{(l-1)}\|_{L_p(-\infty,\infty)} \leq \frac{\pi}{2}\,\|g\|_{L_p(-\infty,\infty)}^{\frac{1}{l}}\,\|g_w^{(l)}\|_{L_p(-\infty,\infty)}^{1-\frac{1}{l}} \leq \frac{\pi}{2}\,\|g\|_{W_p^l(-\infty,\infty)}.$$

Consequently

$$\|g_w^{(l-1)}\|_{W_p^1(-\infty,\infty)} = \|g_w^{(l-1)}\|_{L_p(-\infty,\infty)} + \|g_w^{(l)}\|_{L_p(-\infty,\infty)} \leq \left(\frac{\pi}{2}+1\right)\|g\|_{W_p^l(-\infty,\infty)}$$

and

$$\|T\| \geq \frac{4^{l-1}(l-1)!}{\pi+2}\,\|g_w^{(l-1)}\|_{W_p^1(-\infty,\infty)}$$

Since $f_w^{(l-1)} \equiv 1$ and $g \in W_p^l(-\infty,\infty)$, by Lemma 2, $g_w^{(l-1)}(a-) = 1$. Hence by (6.42)

$$\|g_w^{(l-1)}\|_{W_p^1(-\infty,\infty)} \geq \inf_{\substack{h\in W_p^1(-\infty,a):\\ h(a-)=1}} \|h\|_{W_p^1(-\infty,a)} \geq 1.$$

Thus by Stirling's formula

$$\|T\| \geq \frac{4^{l-1}(l-1)!\,(b-a)^{-l+\frac{1}{p'}}}{\pi+2} \geq \frac{\sqrt{2\pi(l-1)}}{\pi+2}\left(\frac{4}{e}\right)^{l-1}(l-1)^{l-1}(b-a)^{-l+\frac{1}{p'}}$$

$$= \frac{e\sqrt{2\pi}}{4\,(\pi+2)\sqrt{l-1}}\left(1-\frac{1}{l}\right)^l\left(\frac{4}{e}\right)^l l^l(b-a)^{-l+\frac{1}{p'}}$$

$$\geq \frac{\sqrt{2\pi}}{4(\pi+2)\sqrt{l}}\left(\frac{4}{e}\right)^l l^l(b-a)^{-l+\frac{1}{p'}} \geq \frac{0.12}{\sqrt{l}}\left(\frac{4}{e}\right)^l l^l(b-a)^{-l+\frac{1}{p'}}$$

and we obtain (6.38) with 0.12 replacing $\frac{1}{8}$.

Finally we note that (6.42), by Hölder's inequality, follows from (3.8):

$$1 = |h(a-)| \leq \int_a^{a+1} |h|\,dy + \int_a^{a+1} |h'_w|\,dy \leq \|h\|_{W_p^1(-\infty,\infty)}. \quad \square$$

Now for $l,n \in \mathbb{N}$ and $1 \leq p \leq \infty$ we shall denote by $M_{l,n,p}$ the set of q,β satisfying $1 \leq q \leq \infty, \beta \in \mathbb{N}_0^n$, which are such that for some $A > 0$ and $\forall f \in W_p^l(\mathbb{R}^n)$

$$\|D_w^\beta f\|_{L_q(\mathbb{R}^n)} \leq A\,\|f\|_{W_p^l(\mathbb{R}^n)}. \tag{6.43}$$

It follows from Chapter 4 that $p \le q < \infty$ and $|\beta| \le l - n(\frac{1}{p} - \frac{1}{q})$ or $q = \infty$ and $|\beta| \le l$ for $p = \infty$, $|\beta| < l - \frac{n}{p}$ for $1 < p < \infty$, $|\beta| \le l - n$ for $p = 1$. Furthermore, for an open set $\Omega \subset \mathbb{R}^n$ and $(q, \beta) \in M_{l,n,p}$ we denote by $C^*(\Omega, p, q, l, \beta)$ the best (minimal possible) value of C, for which $\forall f \in W_p^l(\Omega)$

$$\|D_w^\beta f\|_{L_q(\Omega)} \le C \|f\|_{W_p^l(\Omega)}. \tag{6.44}$$

Lemma 12 *Let $l, n \in \mathbb{N}$, $1 \le p \le \infty$, $(q, \beta) \in M_{l,n,p}$ and let $\Omega \subset \mathbb{R}^n$ be an open set. Then for every extension operator $T : W_p^l(\Omega) \to W_p^l(\mathbb{R}^n)$*

$$\|T\|_{W_p^l(\Omega) \to W_p^l(\mathbb{R}^n)} \ge \sup_{(q,\beta) \in M_{l,n,p}} \frac{C^*(\Omega, p, q, l, \beta)}{C^*(\mathbb{R}^n, p, q, l, \beta)}. \tag{6.45}$$

Idea of the proof. Prove (6.44) by applying an arbitrary extension operator T and inequality (6.43) where $A = C^*(\mathbb{R}^n, p, q, l, \beta)$. $\square$

Proof. For all $(q, \beta) \in M_{l,n,p}$

$$\|D_w^\beta f\|_{L_q(\Omega)} \le \|D^\beta(Tf)\|_{L_q(\mathbb{R}^n)} \le C^*(\mathbb{R}^n, p, q, l, \beta) \|Tf\|_{W_p^l(\mathbb{R}^n)}$$

$$\le C^*(\mathbb{R}^n, p, q, l, \beta) \|T\|_{W_p^l(\Omega) \to W_p^l(\mathbb{R}^n)} \|f\|_{W_p^l(\Omega)}.$$

Hence,

$$C^*(\Omega, p, q, l, \beta) \le C^*(\mathbb{R}^n, p, q, l, \beta) \|T\|_{W_p^l(\Omega) \to W_p^l(\mathbb{R}^n)}$$

and (6.45) follows. $\square$

Idea of the second proof of Lemma 11. Apply Lemma 12 with $\beta = l - 1, q = \infty$ and inequality (4.53). Use the function f, defined by (6.41) to obtain a lower bound for $C^*((a, b), p, \infty, l - 1, l)$. $\square$

Second proof. By (6.45) for every extension operator $T : W_p^l(a, b) \to W_p^l(-\infty, \infty)$

$$\|T\| \equiv \|T\|_{W_p^l(a,b) \to W_p^l(-\infty,\infty)} \ge \frac{C^*((a, b), p, \infty, l, l - 1)}{C^*((-\infty, \infty), p, \infty, l, l - 1)}.$$

It follows from (6.44), with f defined by (6.41), and (6.39) that

$$C^*((a, b), p, \infty, l, l - 1) \ge \frac{\|1\|_{L_\infty(a,b)}}{\frac{(b-a)^{l-1}}{(l-1)!} \|Q_{l-1;p}(\frac{x-a}{b-a})\|_{L_p(a,b)}}$$

$$= \frac{(l-1)! (b-a)^{-l+\frac{1}{p'}}}{\|Q_{l-1;p}\|_{L_p(0,1)}} \ge \frac{1}{8} 4^l (l-1)! (b-a)^{-l+\frac{1}{p'}}.$$

From (4.53) $C^*((-\infty,\infty), p, \infty, l, l-1) \leq \sqrt{2\pi}$. Hence, applying Stirling's formula as in the first proof of Lemma 12 , we get

$$\|T\| \geq \frac{4^{l-1}(l-1)!}{2\sqrt{2\pi}}(b-a)^{-l+\frac{1}{p'}} \geq \frac{1}{8\sqrt{l}}\left(\frac{4}{e}\right)^l l^l (b-a)^{-l+\frac{1}{p'}}. \quad \Box$$

Finally, we give a formulation of the main result of Section 6.1.

Theorem 1 *There exist constants $c_6, c_7 > 0$ such that for all $l \in \mathbb{N}, 1 \leq p \leq \infty$ and $-\infty \leq a < b \leq \infty$*

$$c_6^l\left(1 + \frac{l^l}{(b-a)^{l-\frac{1}{p'}}}\right) \leq \inf_T \|T\|_{W_p^l(a,b) \to W_p^l(-\infty,\infty)} \leq c_7^l\left(1 + \frac{l^l}{(b-a)^{l-\frac{1}{p'}}}\right). \quad (6.46)$$

Idea of the proof. Apply Lemmas 3, 6, 9, 10 and 11. $\Box$
Proof. If $b - a = \infty$, then (6.47) follows from (6.4) and (6.24). If $b - a < \infty$, then (6.47) follows from (6.33), (6.35) and (6.38). $\Box$

Remark 7 If $p = \infty$, then the statement of the Theorem is also valid for the spaces $\overline{C}^l(a,b)$, i.e., there exist $c_8, c_9 > 0$ such that

$$c_8^l\left(1 + \frac{l^l}{(b-a)^{l-1}}\right) \leq \inf_T \|T\|_{\overline{C}^l(a,b) \to \overline{C}^l(-\infty,\infty)} \leq c_9^l\left(1 + \frac{l^l}{(b-a)^{l-1}}\right). \quad (6.47)$$

The estimate from below is proved in the same manner as for the space $W_\infty^l(a,b)$. When proving estimates from above, the operator T_2 defined by (6.6) must be replaced by $\widetilde{T}_2$ defined by $(\widetilde{T}_2 f)(0) = f(0-)$ and $(\widetilde{T}_2 f)(x) = \sum_{k=1}^{l+1} \alpha_k f(-\beta_k x), \ x > 0$, where $\beta_k > 0$ and $\sum_{k=1}^{l+1} \alpha_k(-\beta_k)^s = 1, \ s = 0, 1, \ldots, l$. In that case $(\widetilde{T}_2 f)^{(s)}(0+) = f^{(s)}(0-), \ s = 0, 1, \ldots, l$, which ensures that $\widetilde{T}_2 f \in \overline{C}^l(-\infty,\infty)$ for each $f \in \overline{C}^l(-\infty,0)$. Moreover, $\|\widetilde{T}_2\|_{\overline{C}^l(-\infty,0) \to \overline{C}^l(-\infty,\infty)} \leq 16^l$. The rest of the proof is the same as for the space $W_\infty^l(a,b)$.

6.2   Pasting local extensions

We pass to the multidimensional case and start by reducing the problem of extensions to the problem of local extensions.

Lemma 13 *Let $l \in N$, $1 \le p \le \infty$ and let $\Omega \subset \mathbb{R}^n$ be an open set with a quasi-resolved boundary. Moreover, let $U_j \subset \mathbb{R}^n$, $j = \overline{1,s}$, where $s \in \mathbb{N}$ or $s = \infty$, be open sets such that*

$$\Omega \subset \bigcup_{j=1}^{s} (U_j)_\delta$$

for some $\delta > 0$. If $s = \infty$, suppose, in addition, that the multiplicity of the covering $\varkappa \equiv \varkappa(\{U_j\}_{j=1}^{s})$ is finite.

Suppose that for all $j = \overline{1,s}$ there exist bounded extension operators

$$T_j : \widehat{W}_p^l(\Omega \cap U_j) \to W_p^l(U_j), \tag{6.48}$$

where $\widehat{W}_p^l(\Omega \cap U_j) = \{f \in W_p^l(\Omega \cap U_j) : \operatorname{supp} f \subset \overline{\Omega} \cap U_j\}$. If $s = \infty$, suppose also that $\sup\limits_{j \in \mathbb{N}} \|T_j\| < \infty$. Then there exists a bounded extension operator

$$T : W_p^l(\Omega) \to W_p^l(\mathbb{R}^n). \tag{6.49}$$

Moreover,

$$\|T\| \le c_{10} \sup_{j=\overline{1,s}} \|T_j\|, \tag{6.50}$$

where $c_{10} > 0$ depends only on n, l, δ and $\varkappa$.

If all the T_j are linear, then T is also linear.

Idea of the proof. Assuming, without loss of generality, that $(U_j)_\delta \cap \Omega \ne \varnothing$ construct functions $\psi_j \in C^\infty(\mathbb{R}^n)$, $j = \overline{1,s}$ such that the collection $\{\psi_j^2\}_{j=1}^{s}$ is a partition of unity corresponding to the covering $\{U_j\}_{j=1}^{s}$, i.e., the following properties hold: $0 \le \psi_j \le 1$, $\operatorname{supp} \psi_j \subset U_j$, $\sum\limits_{j=1}^{s} \psi_j^2 = 1$ on Ω and $\forall \alpha \in \mathbb{N}_0^n$ satisfying $|\alpha| \le l$, $\|D^\alpha \psi_j\|_{L_\infty(\mathbb{R}^n)} \le M_1$, where M_1 depends only on n, l and δ. For $f \in W_p^l(\Omega)$ set

$$Tf = \sum_{j=1}^{s} \psi_j \, T_j(f\psi_j) \quad \text{on} \quad \mathbb{R}^n. \tag{6.51}$$

(Assume that $\psi_j T_j(f\psi_j) = 0$ on ${}^c(U_j)$.) $\square$

Proof. 1. Let $\eta_j \in C^\infty(\mathbb{R}^n)$ be "cap-shaped" functions satisfying $0 \le \eta_j \le 1$, $\eta_j = 1$ on $(U_j)_{\frac{\delta}{2}}$, $\eta_j = 0$ on ${}^c((U_j)_{\frac{\delta}{4}})$ and $|D^\alpha \eta_j(x)| \le M_2 \delta^{-|\alpha|}$, $\alpha \in \mathbb{N}_0^n$, where M_2 depends only on n and α. (See Section 1.1.) Then $1 \le \sum\limits_{j=1}^{s} \eta_j^2 \le \varkappa$

on $\bigcup_{j=1}^{s}(U_j)_{\frac{\delta}{2}}$. Further, let $\eta \in C_b^{\infty}(\mathbb{R}^n)$, $\eta = 1$ on Ω, $\eta = 0$ on $^c(\bigcup_{j=1}^{s}(U_j)_{\frac{\delta}{2}})$. One

can construct functions ψ_j by setting $\psi_j = \eta_j\,\eta\,(\sum_{i=1}^{s}\eta_i^2)^{-\frac{1}{2}}$ on $\bigcup_{i=1}^{s}(U_i)_{\frac{\delta}{2}}$ assuming

that $\psi_j = 0$ on $^c(\bigcup_{i=1}^{s}(U_i)_{\frac{\delta}{2}})$.

2. The operator T defined by (6.51) is an extension operator. For, let $x \in \Omega$. If $x \in \operatorname{supp}\psi_j$ for some j, then $\psi_j(x)(T_j(f\psi_j))(x) = \psi_j^2(x)\,f(x)$. If $x \notin \operatorname{supp}\psi_j$, then $\psi_j(x)(T_j(f\,\psi_j))(x) = 0 = \psi_j^2(x)\,f(x)$. So $(Tf)(x) = \sum_{j=1}^{s}\psi_j^2(x)\,f(x) = f(x)$.

3. Let $\alpha \in \mathbb{N}_0^n$ and $|\alpha| = l$. If $s \in \mathbb{N}$, then

$$D_w^{\alpha}(Tf) = \sum_{j=1}^{s} D_w^{\alpha}(\psi_j\,T_j(f\,\psi_j)) \quad \text{on } \mathbb{R}^n. \tag{6.52}$$

If $s = \infty$, then (6.52) still holds, because on $^c(\bigcup_{j=1}^{s}(U_j)_{\frac{\delta}{2}})$ both sides of (6.52)

are equal to 0 and $\forall x \in \bigcup_{j=1}^{s}(U_j)_{\frac{\delta}{2}}$ the number of sets $(U_j)_{\frac{\delta}{2}}$ intersecting the

ball $B(x,\frac{\delta}{2})$ is finite. Otherwise there exists a countable set of $U_{j_s}, s \in \mathbb{N}$, satisfying $(U_{j_s})_{\frac{\delta}{2}} \cap B(x,\frac{\delta}{2}) \neq \varnothing$. Hence $x \in U_{j_s}$, and we arrive to a contradiction since $\varkappa(\{U_j\}_{j=1}^{\infty}) < \infty$. Consequently, there exists $s_x \in \mathbb{N}$ such that $\operatorname{supp}(\psi_j T_j(f\psi_j)) \cap \overline{B(x,\frac{\delta}{2})} \neq \varnothing$ for $j > s_x$. So

$$Tf = \sum_{j=1}^{s_x} \psi_j\,T_j(f\,\psi_j) \quad \text{on } B(x,\tfrac{\delta}{2}).$$

Hence,

$$D_w^{\alpha}(Tf) = \sum_{j=1}^{s_x} D_w^{\alpha}(\psi_j T_j(f\,\psi_j)) = \sum_{j=1}^{\infty} D_w^{\alpha}(\psi_j T_j(f\,\psi_j)) \quad \text{on } B(x,\tfrac{\delta}{2}).$$

Therefore by the appropriate properties of weak derivatives (see Section 1.2) (6.52) with $s = \infty$ follows.

4. Let $\alpha \in \mathbb{N}_0^n$ and $\alpha = 0$ or $|\alpha| = l$. In (6.51), for all $x \in \mathbb{R}^n$, and in (6.52), for almost all $x \in \mathbb{R}^n$, the number of nonzero summands does not exceed $\varkappa$. Hence, by Hölder's inequality for finite sums,

$$|D_w^{\alpha}(Tf)|^p \leq \varkappa^{p-1} \sum_{j=1}^{s} |D_w^{\alpha}(\psi_j\,T_j\,(f\psi_j))|^p$$

almost everywhere on $\mathbb{R}^n$ and

$$\int\limits_{\mathbb{R}^n} |D_w^\alpha(Tf)|^p\, dx \le \varkappa^{p-1} \sum_{j=1}^{s} \int\limits_{\mathbb{R}^n} |D_w^\alpha(\psi_j\, T_j(f\psi_j))|^p\, dx.$$

Therefore, taking into account Remark 8 of Chapter 1, we have

$$\|Tf\|_{W_p^l(\mathbb{R}^n)} \le M_3 \left(\sum_{j=1}^{s} \|\psi_j\, T_j(f\psi_j)\|_{W_p^l(\mathbb{R}^n)}^p \right)^{\frac{1}{p}},$$

where M_3 depends only on n, l and $\varkappa$. Since $\operatorname{supp}\psi_j \subset U_j$, applying Corollary 18 of Chapter 4, we have

$$\|\psi_j\, T_j(f\psi_j)\|_{W_p^l(\Omega)} \le M_4\, \|T_j(f\psi_j)\|_{W_p^l(U_j)} \le M_4\, \|T_j\|\, \|f\psi_j\|_{W_p^l(\Omega\cap U_j)}$$

$$\le M_5\, \|T_j\|\, \|f\|_{W_p^l(\Omega\cap U_j)},$$

where M_4 and M_5 depend only on n, l and δ. Now it follows, by (2.59), that

$$\|Tf\|_{W_p^l(\mathbb{R}^n)} \le M_6 \sup_{j} \|T_j\| \left(\sum_{j=1}^{s} \|f\|_{W_p^l(\Omega\cap U_j)}^p \right)^{\frac{1}{p}}$$

$$\le M_7 \sup_{j} \|T_j\| \left(\sum_{j=1}^{s} \int\limits_{\Omega\cap U_j} \Big(|f|^p + \sum_{|\alpha|=l} |D_w^\alpha f|^p \Big)\, dx \right)^{\frac{1}{p}}$$

$$\le M_8 \sup_{j} \|T_j\|\, \|f\|_{W_p^l(\Omega)},$$

where M_6, M_7 and M_8 depend only on n, l, δ and $\varkappa$.

Remark 8 Suppose that in Lemma 13 the operators T_j satisfy the additional condition

$$f \in \widehat{W}_p^l(\Omega\cap U_j) \implies \operatorname{supp} T_j f \subset U_j. \tag{6.53}$$

In this case the operator T may be constructed in a simpler way with the help of a standard partition of unity $\{\psi_j\}_{j=1}^{s}$, i.e., $\sum_{j=1}^{s}\psi_j = 1$ on Ω. We assume that $T_j(f\psi_j)(x) = 0$ if $x \in U_j$ and set

$$Tf = \sum_{j=1}^{s} T_j(f\psi_j) \quad \text{on } \mathbb{R}^n. \tag{6.54}$$

The operator T is an extension operator. For, let $x \in \Omega$. If $x \in U_j$, then $(T_j(f\psi_j))(x) = \psi_j(x)\, f(x)$, and if $x \notin U_j$, then $(T_j(f\psi j))(x) = 0 = \psi_j(x)\, f(x)$. Thus $(Tf)(x) = \sum_{j=1}^{s} \psi_j(x)\, f(x) = f(x)$. Note also that for $f \in W_p^l(\Omega)$, because of (6.53), we have $T_j(f\psi_j) \in W_p^l(\mathbb{R}^n)$ and $\| T_j(f\psi_j) \|_{W_p^l(\mathbb{R}^n)} = \| T_j(f\psi_j) \|_{W_p^l(U_j)}$.

Further we consider a bounded elementary domain $H \subset \mathbb{R}^n$ with a C^l- or Lipschitz boundary with the parameters $0 < d \le D < \infty$, $0 \le M < \infty$, which by Section 4.3 means that

$$H = \{x \in \mathbb{R}^n : a_n < x_n < \varphi(\bar{x}),\ \bar{x} \in W\}, \tag{6.55}$$

where $\bar{x} = (x_1, \ldots, x_{n-1})$, $W = \{\bar{x} \in \mathbb{R}^{n-1},\ a_i < x_i < b_i,\ i = 1, \ldots, n-1\}$, $-\infty < a_i < b_i < \infty$, $\operatorname{diam} H \le D$,

$$a_n + d \le \varphi(\bar{x}),\quad \bar{x} \in W, \tag{6.56}$$

and

$$\max_{1 \le |\alpha| \le l} \| D^\alpha \varphi \|_{C(\bar{W})} \le M \tag{6.57}$$

or

$$|\varphi(\bar{x}) - \varphi(\bar{y})| \le M\, |\bar{x} - \bar{y}|,\quad \bar{x}, \bar{y} \in \overline{W}, \tag{6.58}$$

respectively. Moreover, let $V = \{x \in \mathbb{R}^n : a_i < x_i < b_i,\ i = 1, \ldots, n-1,\ a_n < x_n < \infty\}$.

Lemma 14 *Let $l \in \mathbb{N}$, $1 \le p \le \infty$. Suppose that for each bounded elementary domain $H \subset \mathbb{R}^n$ with a C^l- or Lipschitz boundary with the parameters d, D and M there exists a bounded linear extension operator*

$$T: \widehat{W}_p^l(H) \to W_p^l(V), \tag{6.59}$$

where $\widehat{W}_p^l(H) = \{f \in W_p^l(\Omega) : \operatorname{supp} f \subset \overline{H} \cap V\}$ and $\|T\| \le c_{11}$, where $c_{11} > 0$ depends only on n, l, p, d, D and M.

Then for each open set $\Omega \subset \mathbb{R}^n$ with a C^l-, Lipschitz respectively, boundary there exists a bounded linear extension operator (6.49).

Idea of the proof. Apply Lemma 13 with $U_j = V_j$, where V_j, $j = \overline{1, s}$ are open parallelepipeds as in the definition of an open set with a C^l- or a Lipschitz boundary.

Proof. By the assumptions of the lemma for all $j = \overline{1,s}$ there exist bounded extension operators

$$T_j : \widehat{W}_p^l(\lambda_j(\Omega \cap V_j)) \to W_p^l(\lambda_j(V_j)).$$

Let $(\Lambda_j f)(x) = f(\lambda_j(x))$ and define

$$T_j^{(1)} = \Lambda_j^{-1} T_j \Lambda_j.$$

It follows from the proof of Lemma 16 of Chapter 4 that $\Lambda_j : \widehat{W}_p^l(\Omega \cap V_j) \to \widehat{W}_p^l(\lambda(\Omega \cap V_j))$, $\Lambda_j^{-1} : \widehat{W}_p^l(\lambda_j(V_j)) \to \widehat{W}_p^l(V_j)$ and $\| \Lambda_j \|$, $\| \Lambda_j \|^{-1}$ do not exceed some quantity depending only on n and l. Hence,

$$T_j^{(1)} : \widehat{W}_p^l(\Omega \cap V_j) \to W_p^l(V_j)$$

and

$$\| T_j^{(1)} \| \leq \| \Lambda^{-1} \| \cdot \| T_j \| \cdot \| \Lambda_j \| \leq M_1 \| T_j \|,$$

where M_1 depends only on n and l.

If Ω is bounded, then $s \in \mathbb{N}$ and by Lemma 13 there exists a bounded extension operator (6.49). If Ω is unbounded, then $s = \infty$ and by the definition of an open set with a C^l- or Lipschitz boundary each bounded elementary domain $\lambda_j(\Omega \cap V_j)$ has the same parameters d, D, M. Hence, by the assumptions of the lemma $\|T_j\| \leq c_{11}$. Moreover, in this case the multiplicity of the covering $\{V_j\}_{j=1}^\infty$ is finite. Thus Lemma 13 is applicable, which ensures the existence of a bounded linear operator (6.49). $\square$

6.3 Extensions for sufficiently smooth boundaries

Lemma 15 *Let $l \in \mathbb{N}$, $1 \leq p \leq \infty$ and $\Omega = \{x \in \mathbb{R}^n : a_i < x_i < b_i, i = 1,\ldots,n\}$, where $-\infty \leq a_i < b_i \leq \infty$. Then there exists a bounded linear extension operator* (6.49).

Idea of the proof. Apply Lemmas $9-10$ n times. $\square$

Lemma 16 *Let $l \in \mathbb{N}, 1 \leq p \leq \infty$. Then for each bounded elementary domain $H \subset \mathbb{R}^n$ with a C^l-boundary with the parameters d, D and M there exists a bounded linear extension operator* (6.59), *which is such that $\|T\| \leq c_{12}$, where $c_{12} > 0$ depends only on n, l and M.*

Idea of the proof. Let $H^- = \{x \in R^n; -\infty < x_n < \varphi(\bar{x}), \bar{x} \in W\}$, $(T_0 f)(x) = f(x)$ for $x \in H$, $(T_0 f)(x) = 0$ for $x \in H^- \setminus H$. Moreover, let $(Af)(x) = f(a(x))$, where $(a(x))_k = x_k, k = 1, \ldots, n-1, (a(x))_n = x_n + \varphi(\bar{x})$ and $(T_2 f)(x) = \sum_{k=1}^{l} \alpha_k f(\bar{x}, -\beta_k x_n)$ for $\bar{x} \in W$, $x_n > 0$, where $\beta_k > 0$ and α_k are defined by (6.8). Set

$$T = A^{-1} T_2 A T_0 \tag{6.60}$$

and apply Lemma 16 and Remark 25 of Chapter 4. $\square$

Proof. If $f \in \widehat{W}_p^l(H)$, then $T_0 f \in W_p^l(H^-)$ and $\|T_0 f\|_{W_p(H^-)} = \|f\|_{W_p^l(H)}$. Hence, $\|T_0\|_{\widehat{W}_p^l(H) \to W_p^l(H^-)} = 1$. Since $A(H^-) = Q^- = \{x \in \mathbb{R}^n : \bar{x} \in W, x_n < 0\}$ and $\frac{Da}{Dx}(x) \equiv 1$, by (4.126) and (4.148) we have

$$\|A\|_{W_p^l(H^-) \to W_p^l(Q^-)} \leq M_1 \max_{1 \leq |\alpha| \leq l} \|D^\alpha \varphi\|_{C(\overline{W})} \leq M_1 M,$$

where M_1 depends only on n and l.

Since $(a^{(-1)}(x))_k = x_k, k = 1, \ldots, n-1, (a^{(-1)}(x))_n = x_n - \varphi(\bar{x})$, the same estimate holds for $\|A^{-1}\|_{W_p^l(\widehat{Q}) \to W_p^l(\widehat{Q})}$ where $\widehat{Q} = W \times \mathbb{R}$. Finally by Lemma 3, $\|T_2\|_{W_p^l(Q^-) \to W_p^l(\widehat{Q})} \leq 8^l$. Thus,

$$\|T\|_{\widehat{W}_p^l(H) \to W_p^l(V)} \leq \|A^{-1}\| \cdot \|T_2\| \cdot \|A\| \cdot \|T_0\| \leq c_{12},$$

where c_{12} depends only on n, l and M. $\square$

Remark 9 Note that

$$(T_2 A f)(x) = \sum_{k=1}^{l} \alpha_k f(\bar{x}, x_n - (1 + \beta_k)(x_n - \varphi(\bar{x}))) \tag{6.61}$$

on $H^+ = \{x \in \mathbb{R}^n : \bar{x} \in W, x_n > \varphi(\bar{x})\}$, where $\beta_k > 0$ and α_k satisfy (6.8).

Theorem 2 *Let $l \in \mathbb{N}, 1 \leq p \leq \infty$ and let $\Omega \subset \mathbb{R}^n$ be an open set with a C^l-boundary. Then there exists a bounded linear extension operator* (6.49).

Idea of the proof. Apply Lemmas 14 and 17. $\square$

Remark 10 If $p = \infty$ then Lemmas 13–16 and Theorem 2 are also valid for the space $\overline{C}^l(\Omega)$. Thus, for each open set with a C^l-boundary there exists a bounded linear extension operator $T : \overline{C}^l(\Omega) \to \overline{C}^l(\mathbb{R}^n)$. (See also Remark 7.)

6.4 Extensions for Lipschitz boundaries

Let

$$\Omega = \{x \in \mathbb{R}^n : x_n < \varphi(\bar{x}), \bar{x} \in \mathbb{R}^{n-1}\}, \tag{6.62}$$

where φ satisfies a Lipschitz condition on $\mathbb{R}^{n-1}$:

$$|\varphi(\bar{x}) - \varphi(\bar{y})| \leq M|\bar{x} - \bar{y}|, \quad \bar{x}, \bar{y} \in \mathbb{R}^{n-1}. \tag{6.63}$$

Lemma 17 *Let $l \in \mathbb{N}, 1 \leq p \leq \infty$. Suppose that for each domain Ω defined by* (6.62) – (6.63) *there exists a bounded linear extension operator*

$$T : \widehat{W}_p^l(\Omega) \longrightarrow W_p^l(\mathbb{R}^n), \tag{6.64}$$

where $\widehat{W}_p^l(\Omega) = \{f \in W_p^l(\Omega) : \operatorname{supp} f$ is compact in $\mathbb{R}^n\}$ and $\|T\| \leq c_{13}$, where $c_{13} > 0$ depends only on n, p, l and M.

Then for each open set Ω with a Lipshitz boundary there exists a bounded linear extension operator (6.49).

Idea of the proof. Prove that for each bounded elementary domain $H \subset \mathbb{R}^n$ with a Lipschitz boundary there exists a bounded linear extension operator (6.59) such that $\|T\| \leq c_{13}$ and apply Lemma 14. $\square$

Proof. 1. Let H be defined by (6.55), (6.56) and (6.58). Denote by ψ the following extension of the function φ in (6.55):

$$\psi(x_1, x_2, ..., x_{n-1}) = \begin{cases} \varphi(a_1, x_2, ..., x_{n-1}) & \text{for } x_1 < a_1, \\ \varphi(x_1, x_2, ..., x_{n-1}) & \text{for } a_1 \leq x_1 \leq b_1, \\ \varphi(b_1, x_2, ..., x_{n-1}) & \text{for } b_1 < x_1. \end{cases} \tag{6.65}$$

Then ψ satisfies a Lipschitz condition on $W_1 = \{x \in \mathbb{R}^{n-1} : \infty < x_1 < \infty,$ $a_i < x_i < b_i, i = 2, ..., n-1\}$ with the same constant M as the function φ. For, if, say $\bar{x} \in W, \bar{y} \in W_1$ and $y_1 > b_1$, we have

$$|\psi(x_1, x_2, ..., x_{n-1}) - \psi(y_1, y_2, ..., y_{n-1})| = |\varphi(x_1, x_2, ..., x_{n-1}) - \varphi(b_1, y_2, ..., y_{n-1})|$$

$$\leq |\varphi(x_1, x_2, ..., x_{n-1}) - \varphi(b_1, x_2, ..., x_{n-1})| + |\varphi(b_1, x_2, ..., x_{n-1}) - \varphi(b_1, y_2, ..., y_{n-1})|$$

$$\leq M(y_1 - x_1) + M\left((x_2 - y_2)^2 + \cdots + (x_{n-1} - y_{n-1})^2\right)^{\frac{1}{2}} \leq M|\bar{x} - \bar{y}|.$$

Repeating this procedure with respect to the variables $x_2, ..., x_n$ we obtain a function, which coincides with φ on $\overline{W}$ and satisfies a Lipschitz condition on

$\mathbb{R}_{n-1}$ with the same constant M as the function φ. We denote it also by φ and consider the domain Ω defined by (6.62) and the operator T satisfying (6.64).

2. For $f \in \widehat{W}_p^l(H)$ let $T_0 f$ be the extension of f by zero to Ω. Since $\operatorname{supp} f \setminus \overline{H} \cap V$, we have $T_0 f \in \widehat{W}_p^l(\Omega)$ and $\|T_0\|_{W_p^l(\Omega)} = \|f\|_{W_p^l(H)}$. Hence $\|T_0\|_{\widehat{W}_p^l(H) \to W_p^l(\Omega)} = 1$. Next we observe that $TT_0 : \widehat{W}_p^l(H) \to W_p^l(\mathbb{R}^n)$ and

$$\|TT_0\|_{\widehat{W}_p^l(H) \to W_p^l(V)} \leq \|T\|_{\widehat{W}_p^l(\Omega) \to W_p^l(\mathbb{R}^n)} \leq c_{13}.$$

Thus Lemma 14 is applicable and the statement of Lemma 17 follows. $\square$

Our next aim is to construct a bounded linear extension operator (6.64) for Ω defined by (6.62), (6.63).

Let $G = \mathbb{R}^n \setminus \overline{\Omega} = \{x \in \mathbb{R}_n : x_n > \varphi(x)\}$. We set

$$G_k = \{x \in G : 2^{-k-1} < \varrho_n(x) \leq 2^{-k}\}, \quad k \in \mathbb{Z},$$

where

$$\varrho_n(x) = x_n - \varphi(\bar{x})$$

is the distance from $x \in G$ to $\partial G = \partial \Omega$ in the direction of the axis Ox_n.

First we need an appropriate partition of unity. [3]

Lemma 18 *There exists a sequence of nonnegative functions ψ_k satisfying the following conditions:*

$$1) \quad \sum_{k=-\infty}^{\infty} \psi_k = \begin{cases} 1 & \text{for } x \in G, \\ 0 & \text{for } x \notin G, \end{cases} \tag{6.66}$$

$$2) \quad G = \bigcup_{k=-\infty}^{\infty} \operatorname{supp} \psi_k \tag{6.67}$$

and the multiplicity of the covering $\{\operatorname{supp} \psi_k\}_{k \in \mathbb{Z}}$ is equal to 2,

$$3) \quad G_k \subset \operatorname{supp} \psi_k \subset G_{k-1} \cup G_k \cup G_{k+1}, \quad k \in \mathbb{Z}, \tag{6.68}$$

$$4) \quad |D^\alpha \psi_k(x)| \leq c_{14}(\alpha) 2^{k|\alpha|}, \quad x \in \mathbb{R}^n, k \in \mathbb{Z}, \alpha \in \mathbb{N}_0^n, \tag{6.69}$$

where $c_{14}(\alpha) > 0$ depends only on α.

---

[3] In Lemmas 18–25 below Ω is always a domain defined by (6.62)–(6.63) and $G = \mathbb{R}^n \setminus \overline{\Omega}$.

Idea of the proof. Apply the proof of Lemma 5 of Chapter 2. $\square$

With the help of the partition of unity constructed in Lemma 18 we define an extension operator in the following way:

$$(Tf)(x) = \begin{cases} f(x) & \text{for } x \in \Omega, \\ \sum_{k=-\infty}^{\infty} \psi_k(x) f_k(x) & \text{for } x \in G, \end{cases} \tag{6.70}$$

where

$$f_k(x) = \int_{\mathbb{R}^n} f(\bar{x} - 2^{-k}\bar{z}, x_n - A2^{-k}z_n)\,\omega(z)\,dz$$

$$= A^{-1}2^{kn} \int_{\mathbb{R}^n} \omega(2^k(\bar{x} - \bar{y}), A^{-1}2^k(x_n - y_n)) f(y)\,dy. \tag{6.71}$$

Here [4]

$$A = 16\,(M+1) \tag{6.72}$$

and $\omega \in C_0^{\infty}(\mathbb{R}^n)$ is a kernel of mollification satisfying

$$\operatorname{supp}\omega \subset \{x \in \overline{B(0,1)} : x_n \geq \tfrac{1}{2}\} \tag{6.73}$$

and

$$\int_{B(0,1)} \omega(z)\,dz = 1; \quad \int_{B(0,1)} \omega(z)z^\alpha\,dz = 0, \ \alpha \in \mathbb{N}_0^n, \ 0 < |\alpha| \leq l. \tag{6.74}$$

Now let us show that the operator T is well defined. First, we assume that $\psi_k(x)f_k(x) = 0$ for $x \notin \operatorname{supp}\psi_k$ even if $f_k(x)$ is not defined. On the other hand, if $x \in \operatorname{supp}\psi_k$, $f_k(x)$ is defined. This is a consequence of the following inequality, which holds for $x \in \operatorname{supp}\psi_k$ and $z \in \operatorname{supp}\omega$ since by (6.68) $\varrho_n(x) \leq 2^{-k+1}$ and by (6.73) $|z| \leq 1, z_n \geq \tfrac{1}{2}$:

$$x_n - A2^{-k}z_n - \varphi(\bar{x} - 2^{-k}\bar{z}) = x_n - \varphi(\bar{x}) + \varphi(\bar{x} - 2^{-k}\bar{z}) - A2^{-k}z_n$$

$$\leq 2^{-k+1} + M2^{-k}|\bar{z}| - A2^{-k}z_n 2^{-k+1} + M2^{-k}|\bar{z}| - A2^{-k}z_n \leq 2^{-k}\left(2 + M - \tfrac{A}{2}\right) < 0.$$

(This means that the point $(\bar{x} - 2^{-k}\bar{z}, x_n - A2^{-k}z_n) \in \Omega$.)

---

[4] One can choose any larger fixed quantity depending only on M.

Furthermore, by Lemma 18, $\forall x \in G$ the sum in (6.70) is in fact finite: for each $x \in G$ it contains at most two nonzero terms. Moreover,

$$Tf = \sum_{k=m-1}^{m+1} \psi_k f_k \quad \text{on } G_m. \tag{6.75}$$

Thus T is a linear extension operator defined for functions $f \in L_1^{loc}(\Omega)$.

If $x \in \partial\Omega$ the values $(Tf)(x)$ are not defined by (6.70). When considering the spaces $W_p^l(\mathbb{R}^n)$ this is of no importance, because $\text{meas}_n\, \partial\Omega = 0$. In those cases, in which the functions f are defined and continuous in $\overline{\Omega}$, we shall naturally assume that $(Tf)(x) = f(x)$ for $x \in \overline{\Omega}$.

Remark 11 Because of the factor A in (6.71), f_k is an inhomogeneous mollification of f with the steps $2^{-k}, ..., 2^{-k}, A2^{-k}$ with respect to the variables $x_1, ..., x_{n-1}, x_n$. For $x \in \mathbb{R}^n, r > 0, h > 0$ consider an open cylinder centered at the point x of radius r and height h

$$C(x, r, h) = \{y \in \mathbb{R}^n : \bar{y} \in B(\bar{x}, r), |x_n - y_n| < \tfrac{h}{2}\}.$$

Because of (6.73) the value $f_k(x)$ is determined by the values $f(y)$ for y belonging to the cylinder

$$C_{x,k} \equiv C\left((\bar{x}, x_n - \tfrac{3}{4}A2^{-k}), 2^{-k}, \tfrac{1}{4}A2^{-k}\right),$$

which is centered at the point $(\bar{x}, x_n - \tfrac{3}{4}A2^{-k})$ translated with respect to x in the direction of the set Ω. This follows since in (6.71) $\omega(2^k(\bar{x} - \bar{y}), A^{-1}2^k(x_n - y_n))$ can be nonzero only if $2^k|\bar{x} - \bar{y}| < 1$ and $\tfrac{1}{2} \leq A^{-1}2^k(x_n - y_n) \leq 1$. For this reason $(Tf)(x), x \in G$, can be looked at as an inhomogeneous mollification of the function f, for which both the step and the translation are variable. Thus the extension operator T is closely related to the mollifiers with variable steps considered in Chapter 2.

Note also that on G the operator T is an integral operator:

$$(Tf)(x) = \int_\Omega K(x, y) f(y)\, dy, \quad x \in G,$$

with kernel

$$K(x, y) = A^{-1} \sum_{k=-\infty}^{\infty} \psi_k(x) 2^{kn} \omega(2^k(\bar{x} - \bar{y}), A^{-1}2^k(x_n - y_n)). \tag{6.76}$$

Lemma 19 *Let $f \in L_1^{loc}(\Omega), x \in G$ and $x^* = (\bar{x}, x_n - \frac{9}{4}A\varrho_n(x))$. Then the value $(Tf)(x)$ is determined by the values $f(y)$ for y belonging to the cylinder*

$$C_x \equiv C(x^*, 4\varrho_n(x), 4A\varrho_n(x))) \subset \overline{C_x} \subset \Omega. \tag{6.77}$$

Idea of the proof. Apply (6.76) and Remark 11. $\square$

Proof. Let $x \in G$. Choose the unique $m \in \mathbb{N}$ such that $x \in G_m$. Then $\psi_k(x) = 0$ if $k \notin \{m-1, m, m+1\}$ and the value $(Tf)(x)$ is determined by the values $f_k(x)$ where $k = m-1, m, m+1$. By Remark 11 those values are determined by the values $f(y)$ for $y \in \bigcup_{k=m-1}^{m+1} C_{x,k}$. Hence $|\bar{x} - \bar{y}| < 2^{-m+1} \leq 4\varrho_n(x)$, and

$$\tfrac{1}{4}A\varrho_n(x) \leq A2^{-m-2} < x_n - y_n < A2^{-m+1} \leq 4A\varrho_n(x).$$

Consequently $|x_n - \frac{9}{4}\varrho_n(x) - y_n| \leq 2A\varrho_n(x)$ and $y \in C_x$. Moreover, $\forall y \in C_x$

$$\varphi(\bar{y}) - y_n = \varphi(\bar{y}) - \varphi(\bar{x}) + \varphi(\bar{x}) - x_n + x_n - \tfrac{9}{4}A\varrho_n(x) - y_n + \tfrac{9}{4}A\varrho_n(x)$$

$$> (-4M - 1 + \tfrac{A}{4})\varrho_n(x) \geq 3\varrho_n(x) \tag{6.78}$$

because of (6.72). Therefore $\overline{C_x} \subset \Omega$. Note also that similarly

$$\varphi(\bar{y}) - y_n < 10A\varrho_n(x). \quad \square \tag{6.79}$$

Lemma 20 *Let $f \in L_1^{loc}(\Omega)$. Then $Tf \in C^\infty(G)$ and $\forall \alpha \in \mathbb{N}_0^n$*

$$D^\alpha(Tf)(x) = \sum_{0 \leq \beta \leq \alpha} \frac{\alpha!}{\beta!\,(\alpha - \beta)!} \sum_{k=-\infty}^{\infty} (D^{\alpha-\beta}\psi_k)(x)(D^\beta f_k)(x). \tag{6.80}$$

Idea of the proof. Apply Remark 11 and Lemma 18. $\square$

Proof. By Remark 11 $\forall k \in \mathbb{Z}$ and $\forall x \in \operatorname{supp} \psi_k$ we have $\overline{C_{x,k}} \subset \Omega$. Consequently, by the properties of mollifiers (see Section 1.1) $f_k \in C^\infty(H)$. By Lemma 18 $\forall x \in G$ there exists a ball, centered at x, which is contained in no more than 3 sets $\operatorname{supp} \psi_k$. Hence the series (6.70) can be differentiated term by term any number of times, and by the Leibnitz formula equality (6.80) follows. $\square$

Lemma 21 *Let $l \in \mathbb{N}, 1 \leq p \leq \infty$ and for $k \in \mathbb{Z}$*

$$\widetilde{G}_k = G_{k-1} \cup G_k \cup G_{k+1} = \{x \in G : 2^{-k-2} < \varrho_n(x) \leq 2^{-k+1}\}$$

and

$$\widetilde{\Omega}_k = \{x \in \Omega : 2^{-k-2} < |\varrho_n(x)| \leq b2^{-k+1}\}$$

where $b = 10.4$.

Then $\forall \alpha \in \mathbb{N}_0^n$ satisfying $|\alpha| \le l$

$$\|D^\alpha f_k\|_{L_p(\widetilde{G}_k)} \le c_{15}\|D_w^\alpha f\|_{L_p(\widetilde{\Omega}_k)} \tag{6.81}$$

where $c_{15} > 0$ depends only on n, l and M.

Moreover, $\forall \alpha \in \mathbb{N}_0^n$ there exists a function [5] g_α, independent of k, such that

$$\|D^\alpha f_k - g_\alpha\|_{L_p(\widetilde{G}_k)} \le c_{16} 2^{k(|\alpha|-l)}\|f\|_{w_p^l(\widetilde{\Omega}_k)}, \tag{6.82}$$

where $c_{16} > 0$ depend only on n, l, M and α.

Remark 12 It is important for the sequel that g_α should be independent of k and the multiplicities $\varkappa_G$ and $\varkappa_\Omega$ of both coverings $\{\widetilde{G}_k\}_{k\in\mathbb{Z}}$ and $\{\widetilde{\Omega}_k\}_{k\in\mathbb{Z}}$ be finite and bounded from above by quantities, which depend only on M. This follows since these mulitlicities coincide with the multiplicities of the one-dimensional coverings $\{(2^{-k-2}, 2^{-k+1})\}_{k\in\mathbb{Z}}$, $\{(2^{-k-2}, b2^{-k+1})\}_{k\in\mathbb{Z}}$ respectively, and because the multiplicity of the covering $\{(\mu 2^{-k-2}, \nu 2^{-k+1})\}_{k\in\mathbb{Z}}$, where $0 < \mu < \nu$, does not exceed $\log_2 \frac{\nu}{\mu}$. For, the inclusion $x \in (\mu 2^{-k-2}, \nu 2^{-k+1})$ is equivalent to $-\log_2 x - \log_2 \mu < k < -\log_2 x - \log_2 \nu$. Hence the length $\log_2 \frac{\nu}{\mu}$ of this interval is greater than or equal to the number of those k, for which $x \in (\mu 2^{-k-2}, \nu 2^{-k+1})$. Thus $\varkappa_G \le 3$ and $\varkappa_\Omega \le \log_2(8b)$.

Idea of the proof. Observe that $\forall x \in \widetilde{G}_k$

$$C_{x,k} \subset C_x \subset \widetilde{\Omega}_k. \tag{6.83}$$

To prove (6.81) for $\alpha = 0$ apply Minkowski's inequality, the substitution $\bar{x} - 2^{-k}\bar{z} = \bar{y}, x_n - Az_n = y_n$ and (6.83). To prove (6.82), in addition, expand the function $f(\bar{x} - 2^{-k}\bar{z}, x_n - A2^{-k}z_n)$ under the integral sign, applying the integral representation (3.38). Taking into account Remark 12 of Section 3.4, replace in (3.38) the ball B by the ball $B_x \equiv B(x^*, 4\varrho_n(x)) \subset C_x$ and ω by

$$\omega_x(y) = (4\varrho_n(x))^{-n}\mu((4\varrho_n(x))^{-1}(x - y)), \tag{6.84}$$

where μ is any fixed kernel of mollification satisfying (1.1). Apply also an analogue of inequality (3.56). $\square$

Proof. 1.The first inclusion (6.83) follows since for $y \in C_{x,k}$ we have $|\bar{x} - \bar{y}| < 2^{-k} \le 4\varrho_n(x)$ and

$$\tfrac{1}{4} A \varrho_n(x) \le A 2^{-k+1} < x_n - y_n < A 2^{-k} \le 4 A \varrho_n(x).$$

---

[5] If $|\alpha| \ge l$, (6.82) holds for $g_\alpha = 0$.

Consequently, as in the proof of Lemma 20, $|x_n - \frac{9}{4} A \varrho_n(x) - y_n| \leq 2 A \varrho_n(x)$. The second inclusion (6.83) follows since inequalities (6.78) and (6.79) hold $\forall y \in C_x$.

2. First let $\alpha = 0$. By Minkowski's inequality

$$\|f_k\|_{L_p(\tilde{G}_k)} \leq \int\limits_{B(0,1)} \|f(\bar{x} - 2^{-k}\bar{z}, \, x_n - A2^{-k}z_n)\|_{L_{p,x}(\tilde{G}_k)} \, |\omega(z)| \, dz$$

$$\leq \int\limits_{\mathrm{supp}\,\omega} \|f\|_{L_p(\bigcup_{x \in \tilde{G}_k} C_{x,k})} \, |\omega(z)| \, dz$$

since by Remark 11 $(\bar{x} - 2^{-k}\bar{z}, \, x_n - A\,2^{-k}z_n) \in C_{x,k}$ for $z \in \mathrm{supp}\,\omega$. Hence, by (6.83)

$$\|f_k\|_{L_p(\tilde{G}_k)} \leq c_{15} \|f\|_{L_p(\tilde{\Omega}_k)},$$

where $c_{15} = \|\omega\|_{L_1(\mathbb{R}^n)}$ and we have established (6.81) for $\alpha = 0$.

3. Let $\xi \in \mathbb{R}^n$ and let us consider the polynomial in $\xi_1, ..., \xi_n$ of order less than or equal to $l - 1$

$$P(\xi, x) = \int\limits_{B_x} \left(\sum_{|\gamma| < l} \frac{(-1)^{|\gamma|}}{\gamma!} D_y^\gamma [(\xi - y)^\gamma \omega_x(y)] \right) f(y) \, dy,$$

which is closely related to the first summand in the integral representation (3.51), where B, ω are replaced by B_x, ω_x respectively. Writing $u(z)$ for $(\bar{x} - 2^{-k}\bar{z}, x_n - A2^{-k}z_n)$, by (3.51) we have

$$f(\bar{x} - 2^{-k}\bar{z}, \, x_n - A2^{-k}z_n) \equiv f(u(z)) = P(u(z), x)$$

$$+ \sum_{|\gamma|=l} \int\limits_{V_{u(z)}} \frac{(D_w^\gamma f)(y)}{|u(z) - y|^{n-l}} \, w_{\gamma,x}(u(z), y) \, dy \equiv P(u(z), x) + \sum_{|\gamma|=l} r_\gamma(u(z), x).$$

Note that by (6.83) $u(z) \in C_x$ and hence $V_{u(z)} \subset C_x$. Furthermore,

$$f_k(x) = \int\limits_{B(0,1)} P(u(z), x) \, \omega(z) \, dz + \sum_{|\gamma|=l} \int\limits_{B(0,1)} r_\gamma(u(z), x) \, \omega(z) \, dz$$

$$\equiv R_{0,k}(x) + \sum_{|\gamma|=l} R_{\gamma,k}(x). \tag{6.85}$$

4. The function $P(u(z), x)$ is a polynomial in the variables $z_1, \ldots, z_n$ of degree less than or equal to $l - 1$:

$$P(u(z), x) = P(x, x) + \sum_{0 < |\beta| < l} c_\beta(x)\, z^\beta,$$

where $c_\beta(x)$ are independent of z. Note that by (6.74) $R_{0,k}(x) = r_0(x, x)$ and set

$$g_0(x) = P(x, x). \tag{6.86}$$

5. Since ω_x is defined by (6.84), from inequality (3.57) we get that $\forall y \in V_{u(z)}$

$$|w_{\gamma, x}(u(z), y)| \le M_1 \left(\frac{D}{d}\right)^{n-1}, \quad |\gamma| = l,$$

where M_1 depends only on n and l, $d = \operatorname{diam} B_x = 8\, \varrho_n(x)$ and by (6.77) $D \le \operatorname{diam} C_x \le 10\, A\, \varrho_n(x)$. Hence, $\forall y \in V_{u(z)}$

$$|w_{\gamma, x}(u(z), y)| \le M_2,$$

where M_2 depends only on n, l and M. Consequently,

$$|\, r_\gamma(u(z), x)\,| \le M_2 \int_{C_x} |(D_w^\alpha f)(y)| \cdot |u(z) - y|^{l-n}\, dy.$$

Let $\chi_{\widetilde{\Omega}_k}$ be the characteristic function of $\widetilde{\Omega}_k$ and $\Phi_\gamma(y) = |(D_w^\gamma f)(y)|\, \chi_{\widetilde{\Omega}_k}(y)$, $y \in \mathbb{R}^n$. (We assume that $\Phi_\gamma(y) = 0$ for $y \notin \Omega$.) Then

$$|R_{\gamma, k}(x)| \le M_2 \int_{B(0,1)} \left(\int_{C_x} \Phi_\gamma(y)\, |u(z) - y|^{l-n}\, dy \right) |\omega(z)|\, dz.$$

We set $\eta = u(z) - y$. Since both $u(z), y \in C_x$ we have $|\eta| \le \operatorname{diam} C_x \le 10\, A\, \varrho_n(x) \le 20\, A\, 2^{-k}$. Hence,

$$|R_{\gamma, k}(x)|$$

$$\le M_2 \int_{B(0,1)} \left(\int_{B(0, 20\, A\, 2^{-k})} \Phi_\gamma(\bar{x} - 2^{-k}\bar{z} - \bar{\eta},\, x_n - A 2^{-k} z_n - \eta_n)\, |\eta|^{l-n} d\eta \right) |\omega(z)|\, dz.$$

$$\tag{6.87}$$

Since,

$$\|\Phi_\gamma(\cdot - h)\|_{L_p(\widetilde{G}_k)} \le \|\Phi_\gamma(\cdot - h)\|_{L_p(\mathbb{R}^n)} = \|\Phi_\gamma\|_{L_p(\mathbb{R}^n)} = \|D_w^\gamma f\|_{L_p(\widetilde{\Omega}_k)},$$

applying Minkowski's inequality, we get

$$\|R_{\gamma,k}\|_{L_p(\tilde{G}_k)} \le M_3 \|D_w^\gamma f\|_{L_p(\tilde{\Omega}_k)} \int\limits_{B(0,20\,A\,2^{-k})} |\eta|^{l-n}\, d\eta \le M_4\, 2^{-kl} \|D_w^\gamma f\|_{L_p(\tilde{\Omega}_k)},$$

$$(6.88)$$

where $M_3 = M_2 \|\omega\|_{L_1(\mathbb{R}^n)}$ and $M_4 > 0$ depends only on n, l and M.

Inequality (6.82) with $\alpha = 0$ follows from (6.85), (6.86) and (6.88).

6. Now let $|\alpha| > 0$. Taking into account Lemma 3 of Chapter 1, we differentiate (6.71) and get

$$(D^\alpha f_k)(x) = \int\limits_{B(0,1)} (D_w^\alpha f)(\bar{x} - 2^{-k}\bar{z},\, x_n - A2^{-k}z_n)\, \omega(z)\, dz$$

$$= A^{-\alpha_n} 2^{k|\alpha|} \int\limits_{B(0,1)} f(\bar{x} - 2^{-k}\bar{z},\, x_n - A2^{-k}z_n)(D^\alpha \omega)(z)\, dz.$$

Inequality (6.81) with $|\alpha| > 0$ follows from the first equality and inequality (6.81) with $\alpha = 0$.

Next let

$$P_\alpha(\xi, x) = \int\limits_{B_x} \left(\sum_{|\gamma| < l - |\alpha|} \frac{(-1)^{|\alpha|+|\gamma|}}{\gamma!} D_y^{\alpha+\gamma}[(\xi - y)^\gamma \omega_x(y)] \right) f(y)\, dy.$$

By (3.52), as in steps 3–4, it follows from the second equality for $D^\alpha f_k$ that

$$(D^\alpha f_k)(x) = g_\alpha(x) + \sum_{|\gamma|=l-|\alpha|} A^{-\alpha_n} 2^{k|\alpha|} R_{\gamma,k}^{(\alpha)}(x),$$

where

$$g_\alpha(x) = A^{-\alpha_n} 2^{k|\alpha|} \int\limits_{B(0,1)} P_\alpha(u(z), x)(D^\alpha \omega)(z)\, dz$$

and $R_{\gamma,k}^{(\alpha)}(x)$ is obtained from $R_{\gamma,k}(x)$ by replacing $D_w^\gamma f$, l by $D_w^{\alpha+\gamma} f$, $l - |\alpha|$ respectively. By (6.74) $\int_{B(0,1)} z^\beta (D^\alpha \omega)(z)\, dz = 0$ if $|\alpha| \ge l$, or $|\alpha| < l$ and $\beta \ne \alpha$.

If $|\alpha| < l$ and $\beta = \alpha$, then $\int_{B(0,1)} z^\alpha (D^\alpha \omega)(z)\, dz = (-1)^{|\alpha|}\alpha!$. Hence $g_\alpha = 0$ for $|\alpha| \ge l$ and

$$g_\alpha(x) = A^{-\alpha_n} 2^{k|\alpha|} D_z^\alpha (P_\alpha(u(z), x))\Big|_{z=0} = P_\alpha(x, x) \qquad (6.89)$$

for $|\alpha| < l$. With this choice of g_α inequality (6.82) with $|\alpha| > 0$ follows as in step 5. $\square$

Remark 13 In the above proof the functions g_α defined for $\alpha = 0$ by (6.86) and for $0 < |\alpha| \leq l - 1$ by (6.89) are the first summands in the integral representations (3.51), (3.52) respectively, where B, ω are replaced by B_x, ω_x respectively. Since $\forall x \in \widetilde{G}_k$ we have $B(x^*(k), M_1 2^{-k}) \subset B_x = B(x^*(k), 4\,\varrho_n(x)) \subset B(x^*(k), M_2\, 2^{-k})$, where $x^*(k) = (\bar{x}, x_n - \frac{9}{4} A\, 2^{-k})$ and $M_1, M_2 > 0$ depend only on n. These inclusions explain why one may expect estimate (6.82) to hold with appropriate $g_{\alpha,k}$. The choice of the ball $B(x^*, 4\varrho_n(x))$, independent of k and "compatible" with $B(x^*(k), M_2\, 2^{-k})$, allows us to construct a function g_α, for which inequality (6.82) holds and which is independent of k.

Remark 14 In the proof of Lemma 22 (Section 4) we have applied property (6.73) for $|\alpha| \leq l - 1$. The fact that it holds also for $|\alpha| = l$ allows us to prove the following local variant of (6.82) for $p = \infty : \forall x \in \widetilde{G}_k$ and $\forall \alpha \in \mathbb{N}_0^n$ satisfying $|\alpha| \leq l$

$$|(D^\alpha f_k)(x) - \tilde{g}_\alpha(x)| \leq c_{17}\, 2^{-k(l+1-|\alpha|)} \|f\|_{C^{l+1}(\widetilde{\Omega}_k \cap B(x, a\, 2^{-k}))}, \qquad (6.90)$$

where $c_{17} > 0$ and $a > 0$ depend only on n, l and M. Here $\tilde{g}_\alpha$ is independent of k and is defined by (6.89) with $l + 1$ replacing l.

Estimate (6.90) follows from (6.87), where l is replaced by $l + 1$ and $|\gamma| = l + 1$, if to observe that $\forall z \in \operatorname{supp} \omega$ and $\forall \eta \in \widetilde{C}_k$ the point $(\bar{x} - 2^{-k}\bar{z} - \bar{\eta},\, x_n - A\, 2^{-k} z_n - \eta_n) \in B(x, a2^{-k})$ where $a = 22\, A$.

Lemma 22 *Let* $l \in \mathbb{N}, \alpha \in \mathbb{N}_0^n, |\alpha| \leq l$ *and* [6] $f \in C^\infty(\overline{\Omega})$. *Then the derivatives* $D^\alpha(Tf)$ *exist and are continuous on* $\mathbb{R}^n$.

Idea of the proof. By Lemma 7 $Tf \in C^\infty(\mathbb{R}^n \setminus \partial\Omega)$. Let $x \in \partial\Omega$. First show, by applying (6.90), that

$$\lim_{y \to x, y \in G} D^\alpha(Tf)(y) = (D^\alpha f_1)(x), \quad |\alpha| \leq l. \qquad (6.91)$$

Applying (6.91) and the definition of a derivative prove that $(D^\alpha(Tf))(x) = (D^\alpha f_1)(x)$ first for $|\alpha| = 1$ and then, by induction, for all $\alpha \in \mathbb{N}_0^n$ satisfying $|\alpha| \leq l \in \mathbb{N}$. $\square$

Proof. 1. Let $I_{\alpha,\beta} = \sum_{k=-\infty}^{\infty} D^{\alpha-\beta} \psi_k\, D^\beta f_k$. Then by (6.80)

$$D^\alpha(Tf)(y) = \sum_{0 \leq \beta \leq \alpha} \frac{\alpha!}{\beta!(\alpha - \beta)!} I_{\alpha\beta}(y), \quad y \in G, \qquad (6.92)$$

---

[6] I.e., there exists a domain $\Omega_1 \supset \overline{\Omega}$ and a function $f_1 \in C^\infty(\Omega_1)$ such that $f_1 = f$ on Ω.

where $I_{\alpha,\beta} = \sum\limits_{k=-\infty}^{\infty} D^{\alpha-\beta}\psi_k D^{\beta} f_k$. Let $x \in \partial\Omega$, i.e., $x = (\bar{x}, \varphi(\bar{x}))$. First we study the difference

$$I_{\alpha\alpha}(y) - D^{\alpha} f_1(x)$$

$$= \sum_{k=m-1}^{m+1} \psi_k(y) \int\limits_{B(0,1)} [(D^{\alpha} f_1)(\bar{y} - 2^{-k}\bar{z}, y_n - A\,2^{-k} z_n) - (D^{\alpha} f_1)(\bar{x}, \varphi(x))]\,\omega(z)\,dz,$$

where m is such that $y \in G_m$ (m is defined uniquely). Let $u = (\bar{y} - 2^{-k}\bar{z}, y_n - A\,2^{-k} z_n)$. then $|u - x| \le |x - y| + 2^{-k} + A\,2^{-k} \le |x - y| + (A+1)\,2^{-m+1} \le |x - y| + 4\,(A+1)\,\varrho_n(y)$. Since $\varrho_n(y) = y_n - \varphi(\bar{x}) = y_n - x_n + \varphi(\bar{x}) - \varphi(\bar{y}) \le (M+1)\,|x - y|$ we have $|u - x| \le M_1\,|y - x|$, where M_1 depends only on M. Consequently,

$$|I_{\alpha\alpha}(y) - (D^{\alpha} f_1)(x)| \le M_2 \sup_{|u-x| \le M_1\,|x-y|} |(D^{\alpha} f_1)(u) - (D^{\alpha} f_1)(x)| \to 0$$

as $y \to x$, $y \in G$. (Here M_2 depends only on n and M.)

Furthermore, when $\beta \ne \alpha$ we have $\sum\limits_{k=-\infty}^{\infty} (D^{\alpha-\beta}\psi_k)(y) = 0$ and

$$I_{\alpha\beta}(y) = \sum_{k=-\infty}^{\infty} (D^{\alpha-\beta}\psi_k)(y)((D^{\beta} f_k)(y) - g_{\beta}(y))$$

$$= \sum_{k=m-1}^{m+1} (D^{\alpha-\beta}\psi_k)(y)((D^{\beta} f_k)(y) - g_{\beta}(y)). \tag{6.93}$$

where g_{β} is the function constructed in Lemma 21 (see (6.89)). Applying (6.90) we get

$$|I_{\alpha\beta}(y)| \le M_3\,2^{-m(l+1-|\alpha|)}\|f\|_{C^{l+1}(\Omega \cap B(x, a2^{-m+1}))}$$

$$\le M_4\,|x - y|^{l+1-|\alpha|}\|f\|_{C^{l+1}(\Omega \cap B(x, M_5|x-y|))}$$

where M_3, M_4, M_5 depend only on n, l and M.

Therefore $I_{\alpha\beta}(y) \to 0$ as $y \to x, y \in G$, and this proves (6.91).

2. It follows from what has been proved in step 1 that the function Tf is continuous in $\mathbb{R}^n$ and $(Tf)(x) = f_1(x), x \in \partial\Omega$. Now we shall prove that $\frac{\partial(Tf)}{\partial x_1}(x) = \frac{\partial f_1}{\partial x_1}(x)$ for $x \in \partial\Omega$.

Consider the one-dimensional set $e_x = \Omega \cap l_x^{(1)}$, where $l_x^{(1)}$ is a straight line passing through the point x and parallel to the axis Ox_1. Let $x_2, \ldots, x_n$ be fixed and $\psi(x_1) = (Tf)(x_1, \ldots, x_n)$, $x_1 \in \mathbb{R}$, $\psi_1(x_1) = f_1(x_1, x_2, \ldots, x_n)$, $x_1 \in \overline{e_x}$. Consider

$$\frac{\partial(Tf)}{\partial x_1}(x) = \lim_{y_1 \to x_1} \frac{\psi(y_1) - \psi(x_1)}{y_1 - x_1} = \lim_{y_1 \to x_1} \frac{\psi(y_1) - \psi_1(x_1)}{y_1 - x_1}.$$

Note that

$$\lim_{y_1 \to x_1, y_1 \in \overline{e_x}} \frac{\psi(y_1) - \psi_1(x_1)}{y_1 - x_1} = \lim_{y_1 \to x_1} \frac{\psi_1(y_1) - \psi_1(x_1)}{y_1 - x_1} = \frac{\partial f_1}{\partial x_1}(x).$$

Let $y_1 \notin \overline{e_x}$. Denote by y_1^* the point in $\overline{e_x}$ lying between x_1 and y_1, which is closest to y_1. We obtain [7]

$$\frac{\psi(y_1) - \psi_1(x_1)}{y_1 - x_1} - \frac{\partial f_1}{\partial x_1}(x) = \frac{\psi(y_1) - \psi(y_1^*) + \psi_1(y_1^*) - \psi_1(x_1)}{y_1 - x_1} - \frac{\partial f_1}{\partial x_1}(x)$$

$$= \psi'(\xi_1)\frac{y_1 - y_1^*}{y_1 - x_1} + \frac{\psi_1(y_1^*) - \psi_1(x_1)}{y_1^* - x_1} \cdot \frac{y_1^* - x_1}{y_1 - x_1} - \frac{\partial f_1}{\partial x_1}(x)$$

$$= \left(\frac{\partial(Tf)}{\partial x_1}(\xi_1, x_2, \ldots, x_n) - \frac{\partial f_1}{\partial x_1}(x_1, x_2, \ldots, x_n)\right)\frac{y_1 - y_1^*}{y_1 - x_1}$$

$$+ \left(\frac{\psi_1(y_1^*) - \psi_1(x_1)}{y_1^* - x_1} - \frac{\partial f_1}{\partial x_1}(x)\right)\frac{y_1^* - x_1}{y_1 - x_1}.$$

If $y_1 \to x_1$ the first summand tends to zero because of (6.91) since $(\xi_1, x_2, \ldots, x_n) \in G$ and ξ_1 lies between x_1 and y_1, and the second summand tends to zero because $(y_1^*, x_2, \ldots, x_n) \in \overline{\Omega}$. This proves that $\frac{\partial(Tf)}{\partial x_1}(x) = \frac{\partial f_1}{\partial x_1}(x)$. The continuity of $\frac{\partial(Tf)}{\partial x_1}$ follows again from (6.91).

Similarly one can prove the existence and continuity of the derivatives $\frac{\partial(Tf)}{\partial x_i}$, $i = 2, \ldots, n$ (when $i = n$, the situation is simpler since $\Omega \cap l_x^{(n)}$ is a half-line), and, by induction, of the derivatives of higher orders. $\square$

Lemma 23 *Let $l \in \mathbb{N}, 1 \le p \le \infty, f \in C^\infty(\overline{\Omega})$. Then*

$$\|Tf\|_{L_p(\mathbb{R}^n)} \le c_{18} \|f\|_{L_p(\Omega)}, \tag{6.94}$$

$$\|Tf\|_{w_p^l(\mathbb{R}^n)} \le c_{19} \|f\|_{w_p^l(\Omega)}, \tag{6.95}$$

where $c_{18}, c_{19} > 0$ depend only on n, l and M, and

$$\|(x_n - \varphi(\bar{x}))^{|\alpha|-l} D^\alpha(Tf)\|_{L_p(^c\overline{\Omega})} \le c_{20} \|f\|_{w_p^l(\Omega)}, \quad |\alpha| > l, \tag{6.96}$$

where $c_{20} > 0$ depend only on n, l, M and α.

---

[7] If any neighbourhood of x contains infinitely many interval components of e_x, then $y_1^* \ne x$. Otherwise, for a point y_1, which is sufficiently close to x_1 we have $y_1^* = x_1$, and the argument becomes much simpler.

Idea of the proof. 1. To prove (6.94) first observe that, as in the proof of Lemma 13 of Chapter 2,

$$\|Tf\|_{L_p(\Omega)} \le 2\Big(\sum_{k=-\infty}^{\infty} \|f_k\|^p_{L_p(\widetilde{G}_k)} \Big)^{\frac{1}{p}}, \tag{6.97}$$

then apply inequality (6.81) and the fact that the multiplicity of the covering $\{\widetilde{\Omega}_k\}_{k\in\mathbb{Z}}$ is finite.

2. To prove (6.95) apply (6.92) and (6.93). Estimate $I_{\alpha\alpha}$ as in step 1. To estimate $I_{\alpha\beta}$ where $\beta \ne \alpha$ apply inequalities (6.69) and (6.82). In the case of inequality (6.96) use also the inequality $x_n - \varphi(\bar{x}) \le M_1 \, 2^{-k}$ on G_k where M_1 is independent of k. $\square$

Proof. 1. Since the sum (6.70) for each $x \in G$ contains at most two nonzero terms by Hölder's inequality

$$\|Tf\|^p_{L_p(G)} \le 2^{p-1} \int_G \Big(\sum_{k=-\infty}^{\infty} |\psi_k f_k|^p \Big)\, dx.$$

Furthermore,

$$\int_G \sum_{k=-\infty}^{\infty} = \sum_{k=-\infty}^{\infty} \int_{G_m} \sum_{k=m-1}^{m+1} = \sum_{m=-\infty}^{\infty} \sum_{k=m-1}^{m+1} \int_{G_m} = \sum_{k=-\infty}^{\infty} \sum_{m=k-1}^{k+1} \int_{G_m} = \sum_{k=-\infty}^{\infty} \int_{\widetilde{G}_k}$$

and inequality (6.97) follows since $0 \le \psi_k \le 1$. Consequently, by (6.81)

$$\|Tf\|_{L_p(G)} \le 2\,c_{15}\Big(\sum_{k=-\infty}^{\infty} \|f\|^p_{L_p(\widetilde{\Omega}_k)} \Big)^{\frac{1}{p}} \le 2\,c_{15}\varkappa_\Omega^{\frac{1}{p}} \|f\|_{L_p(\Omega)},$$

where $\varkappa_\Omega$ is the multiplicity of the covering $\{\widetilde{\Omega}_k\}_{k\in\mathbb{Z}}$, which, by Remark 12, does not exceed $\log_2(8b)$.

2. Suppose that $\alpha \in N_0^n$ satisfies $|\alpha| = l$. Then we consider equality (6.92). As in step 1

$$\|I_{\alpha\alpha}\|_{L_p(G)} \le c_{21}\|D_w^\alpha f\|_{L_p(\Omega)}.$$

To estimate $\|I_{\alpha\beta}\|_{L_p(G)}$ where $\beta \ne \alpha$ we can apply (6.93). First of all

$$\|I_{\alpha\beta}\|_{L_p(G)} \le 2\Big(\sum_{k=-\infty}^{\infty} \|D^{\alpha-\beta}\psi_k(D^\beta f_k - g_\beta)\|^p_{L_p(\widetilde{G}_k)} \Big)^{\frac{1}{p}}.$$

Furthermore, it follows, by (6.69), (6.82) and Remark 13, that

$$\|I_{\alpha\beta}\|_{L_p(G)} \leq M_2 \Big(\sum_{k=-\infty}^{\infty} (2^{k|\alpha-\beta|} \, 2^{-k(l-|\beta|)} \, \|f\|_{w_p^l(\widetilde{\Omega}_k)})^p \Big)^{\frac{1}{p}}$$

$$= M_2 \Big(\sum_{k=-\infty}^{\infty} \|f\|_{w_p^l(\widetilde{\Omega}_k)}^p \Big)^{\frac{1}{p}}$$

$$\leq M_3 \sum_{|\alpha|=l} \Big(\sum_{k=-\infty}^{\infty} \|D_w^\alpha f\|_{L_p(\widetilde{\Omega}_k)}^p \Big)^{\frac{1}{p}} \leq M_4 \, \varkappa_\Omega^{\frac{1}{p}} \sum_{|\alpha|=l} \|D_w^\alpha f\|_{L_p(\Omega)} \leq M_5 \, \|f\|_{w_p^l(\Omega)},$$

where $M_2, ..., M_5 > 0$ depend only on n, l and M, and inequality (6.95) follows. The proof of inequality (6.96) is similar. Let $|\alpha| > l$. Since $g_\alpha = 0$, for all β satisfying $0 \leq \beta \leq \alpha$ we have

$$\|(x_n - \varphi(\bar{x}))^{|\alpha|-l} I_{\alpha\beta}\|_{L_p(G)}$$

$$\leq 2 \Big(\sum_{k=-\infty}^{\infty} \|(x_n - \varphi(\bar{x}))^{|\alpha|-l} D^{\alpha-\beta} \psi_k \, (D^\beta f_k - g_\beta)\|_{L_p(\widetilde{G}_k)}^p \Big)^{\frac{1}{p}}$$

$$\leq M_6 \Big(\sum_{k=-\infty}^{\infty} (2^{-k(|\alpha|-l)} 2^{k|\alpha-\beta|} \, 2^{-k(l-|\beta|)} \, \|f\|_{w_p^l(\widetilde{\Omega}_k)})^p \Big)^{\frac{1}{p}} \leq M_7 \, \|f\|_{w_p^l(\Omega)},$$

where M_6, M_7 depend only on n, l, M and α. $\square$

Lemma 24 *For each polynomial p_l of degree less than or equal to l, $Tp_l = p_l$.*

Idea of the proof. Expand the polynomial $p_l(\bar{x} - 2^{-k}\bar{z}, x_n - A2^{-k}z_n)$ in (6.71) and apply [8] (6.74) and (6.66). $\square$

Lemma 25 *Let $l \in \mathbb{N}, 1 \leq p \leq \infty, f \in W_p^l(\Omega)$. Then there exists a sequence of functions $f_k \in C^\infty(\overline{\Omega})$ such that*

$$f_k \to f \quad \text{in} \quad W_p^l(\Omega), \quad 1 \leq p < \infty \tag{6.98}$$

and

$$f_k \to f \quad \text{in} \quad W_\infty^{l-1}(\Omega), \quad \|f_k\|_{W_\infty^l(\Omega)} \to \|f\|_{W_\infty^l(\Omega)} \tag{6.99}$$

as $k \to \infty$.

---

[8] If in (6.74) $|\alpha| \leq m$, then Lemma 25 is valid for polynomials of degree less than or equal to m. This lemma is similar to Lemma 15 of Chapter 2.

Idea of the proof. By Lemma 2 and Remark 2 of Chapter 2 it is enough to assume that supp f is compact in $\mathbb{R}^n$. Set

$$f_k = A_{\delta_k}(f(\cdot + \tfrac{1}{k}\, e_n)),$$

where $e_n = (0, ..., 0, 1)$ and A_{δ_k} is a mollifier with a non-negative kernel defined in Section 1.1 with step δ_k, which is such that $\delta_k < \mathrm{dist}\,(\mathrm{supp}\, f.\, \partial\Omega + \tfrac{1}{k}\, e_n)$, and apply the properties of mollifiers (see Sections 1.1 and 1.2). $\square$

Theorem 3 *Let* $l \in \mathbb{N}, 1 \leq p \leq \infty$ *and let* $\Omega \subset \mathbb{R}^n$ *be an open set with a Lipschitz boundary. Then there exists a bounded linear extension operator*

$$T : W_p^l(\Omega) \to W_p^l(\mathbb{R}^n) \bigcap C^\infty({}^c\overline{\Omega}) \tag{6.100}$$

such that

$$\|\varrho^{|\alpha|-l}D^\alpha(Tf)\|_{L_p({}^c\overline{\Omega})} \leq c_{21}\,\|f\|_{W_p^l(\Omega)}, \quad |\alpha| > l, \tag{6.101}$$

where $\varrho(x) = \mathrm{dist}\,(x, \partial\Omega)$ *and* $c_{21} > 0$ *is independent of* f.

There exists an open set Ω *having a Lipschitz boundary such that in* (6.101) *the exponent* $|\alpha| - l$ *cannot be replaced by* $|\alpha| - l - \varepsilon$ *for any* $\varepsilon > 0$ *and for any extension operator* (6.100).

Idea of the proof. Apply Lemmas 17, 23, 25 and note that for a domain Ω defined by (6.62), (6.63) [9]

$$\frac{x_n - \varphi(\bar{x})}{1 + M} \leq \varrho(x) \leq x_n - \varphi(\bar{x}). \tag{6.102}$$

To prove the last statement consider $\Omega = \mathbb{R}^n_- = \{x \in \mathbb{R}^n : x_n < 0\}$ and argue as in Remark 12 of Chapter 5. $\square$

Proof. First let Ω be a domain defined by (6.62), (6.63) and $f \in W_p^l(\Omega)$. By Lemma 25 there exists a sequence of fuctions $f_k \in C^\infty(\overline{\Omega})$ satisfying (6.97), (6.98). Consequently, by Lemma 23

$$\|Tf_k\|_{W_p^l(\mathbb{R}^n)} \leq M_1\,\|f_k\|_{W_p^l(\Omega)},$$

where M_1 depends only on n, l and M. Passing to the limit as $k \to \infty$ we establish this inequality with f replacing f_k. Applying Lemma 17 we get (6.100).

---

[9] The second inequality is obvious. To prove the first one we note that $\varrho(x) \geq \varrho_K(x)$, where $\varrho_K = \mathrm{dist}\,(x, \partial K)$ and $K \subset G$ is the infinite cone defined by $y_n > \varphi(\bar{x}) + M|\bar{x} - \bar{y}|, y \in \mathbb{R}^n$. The desired inequality follows since $B(x, (1 + M)^{-1}(x_n - \varphi(\bar{x}))) \subset K$, which is clear because $\forall y \in B(x, (1+M)^{-1}(x_n - \varphi(\bar{x})))$ we have $y_n - \varphi(\bar{x}) - M|\bar{x} - \bar{y}| = y_n - x_n + x_n - \varphi(\bar{x}) - M|\bar{x} - \bar{y}| \geq -(1 + M)|x - y| + (x_n - \varphi(\bar{x}))) > 0$.

In the case of inequality (6.101) the argument is similar. One should only take into consideration (6.102) and (6.96) and note that the appropriate weighted analogue of Lemma 17 is also valid.

Finally, let $\Omega = \mathbb{R}^n_-$ and suppose that for some $\varepsilon > 0$ and for some extension operator (6.100) we have $\||x_n|^{|\alpha|-l-\varepsilon} D^\alpha(Tf)\|_{L_p(\mathbb{R}^n_+)} < \infty$ for all $f \in W^l_p(\mathbb{R}^n_-)$ and for all $\alpha \in \mathbb{N}^n_0$ satisfying $|\alpha| = m > l + \varepsilon$. First suppose that $l > \frac{1}{p}$. Let $g \in B^{l-\frac{1}{p}}_p(\mathbb{R}^{n-1}) \setminus B^{l+\varepsilon-\frac{1}{p}}_p(\mathbb{R}^{n-1})$. By Theorem 3 of Chapter 5 there exists a function $f \in W^l_p(\mathbb{R}^n_-)$ such that $f\big|_{\mathbb{R}^{n-1}} = g$. By Lemma 2 $Tf\big|_{\mathbb{R}^{n-1}} = f\big|_{\mathbb{R}^{n-1}} = g$. Since $Tf \in W^m_{p,\,x_n^{m-l-\varepsilon}}(\mathbb{R}^n_+)$, by the trace theorem (5.68), $g \in B^{l+\varepsilon-\frac{1}{p}}_p(\mathbb{R}^{n-1})$ and we have arrived at a contradiction. If $l = p = 1$, the argument is similar: one should consider $g \in L_1(\mathbb{R}^{n-1}) \setminus B^\varepsilon_1(\mathbb{R}^{n-1})$ and apply Theorem 5 of Chapter 5 instead of Theorem 3 of that chapter. $\square$

Remark 15 The extension operator constructed in the proof of Theorem 3 satisfies (6.100) and (6.101). So, by the last statement of that theorem, it is the best possible extension operator in the sense that the derivatives of higher orders of Tf on $^c\overline{\Omega}$ have the minimal possible growth on approaching $\partial\Omega$.

Remark 16 The extension operator constructed in the proof of Theorem 3 is such that for all $m \in \mathbb{N}^n_0$ satisfying $m \leq l$ we have $T : W^m_p(\Omega) \to W^m_p(\mathbb{R}^n)$.

Remark 17 Now we describe an alternative way of proving of the first statement of Theorem 3. Let Ω be defined by (6.62) and (6.63). It is possible to get an extension operator (6.100) by "improving" the extension operator (6.61) . To do this we replace $x_n - \varphi(\bar{x})$, which in general is only a Lipschitz function, by the infinitely differentiable function $\Delta(x) = 2(1 + M)\varrho_{\frac{1}{2}}(x)$, where $\varrho_{\frac{1}{2}}$ is the regularized distance constructed in Theorem 10 of Section 2.6. By (6.102) we have

$$x_n - \varphi(\bar{x}) \leq \Delta(x) \leq 2\,(1 + M)(x_n - \varphi(\bar{x})) \tag{6.103}$$

and

$$|D^\alpha\Delta(x)| \leq c_\alpha\,(x_n - \varphi(\bar{x}))^{1-|\alpha|}, \quad \alpha \in \mathbb{N}^n_0. \tag{6.104}$$

So we set

$$(Tf)(x) = \sum_{k=1}^{l+1} \alpha_k f(\bar{x}, x_n - (1 + \beta_k)\Delta(x)), \tag{6.105}$$

where $\beta_k > 0$ and $\sum_{k=1}^{l+1} \alpha_k(-\beta_k)^s = 1$, $s = 0, ..., l$. (Hence $\sum_{k=1}^{l+1} \alpha_k(1 + \beta_k)^s = 0, s = 1, ..., l$.) By using formula (4.127), expanding for $f \in C^\infty(\overline{\Omega})$ the derivative $D^\beta f(\bar{x}, x_n - (1 + \beta_k)\Delta(x))$ by Taylor's formula with respect to the point

$(\bar{x}, \varphi(\bar{x})) \in \partial\Omega$ and applying (6.103), one can prove that Lemmas 22 and inequalities (6.94) and (6.95) are valid for this extension operator as well. The rest is the same as in the proof of Theorem 3.

The extension operator (6.105) cannot be "the best possible" because, in general, $Tf \notin C^\infty({}^c\overline{\Omega})$. On the other hand in (6.105) it is possible to replace the sum $\sum_{k=1}^{l+1}$ by the sum $\sum_{k=1}^{\infty}$ and [10] choose $\beta_k > 0$ and α_k in such a way that $\sum_{k=1}^{\infty} |\alpha_k|\,|\beta_k|^s < \infty$ and $\sum_{k=1}^{\infty} \alpha_k(-\beta_k)^s = 1$ for all $s \in \mathbb{N}_0$. This gives an operator (independent of l) such that (6.100) is satisfied for all $l \in \mathbb{N}$.

Remark 18 The extension operator (6.29) – (6.30) in contrast to the extension operator described in Remark 17, is also applicable to the spaces $W_p^{l,\ldots,l}(\Omega)$ defined in Remark 26 of Chapter 4, i.e., for $l \in \mathbb{N}$, $1 \le p \le \infty$ and an open set $\Omega \subset \mathbb{R}^n$ with a Lipschitz boundary

$$T : W_p^{l,\ldots,l}(\Omega) \to W_p^{l,\ldots,l}(\mathbb{R}^n). \tag{6.106}$$

To prove this for Ω defined by (6.62) – (6.63), following the same scheme, one needs to prove an analogue of (6.95) for $w_p^{l,\ldots,l}(\Omega)$. This can be established with the help of an integral representation, which involve only unmixed derivatives $\left(\frac{\partial^l f}{\partial x_j^l}\right)_w, j = 1, ..., n$.

Remark 19 The supposition "Ω has a Lipschitz boundary" in Theorem 3 is sharp in the following sense: for each $0 < \gamma < 1$ there exists an open set Ω with a boundary of the class [11] Lipγ, which is such that the extension operator (6.100) does not exist, as the following example shows.

Example 1 Let $n > 1$, $l \in \mathbb{N}$, $1 \le p < \infty$ and $\Omega_\gamma = \{x \in \mathbb{R}^n : |\bar{x}| < 1, |\bar{x}|^\gamma < x_n < 1\}$ where $0 < \gamma < 1$. Then $\partial\Omega \in \text{Lip}\,\gamma$, but $\partial\Omega \notin \text{Lip}1$. Suppose that there exists an extension operator (6.100), even nonlinear or unbounded. Then $\forall f \in W_p^l(\Omega_\gamma)$ we have $Tf \in W_p^l(\mathbb{R}^n)$. It follows, by the embedding theorems for $W_p^l(\mathbb{R}^n)$, that $Tf \in L_q(\mathbb{R}^n)$, hence $f \in L_q(\Omega_\gamma)$, where $q = \frac{np}{np-l}$ if $l < \frac{n}{p}$,

---

[10] Or by an appropriate integral. In that case $(Tf)(x) = \int_1^\infty f(\bar{x}, x_n - \lambda\Delta(x))\psi(\lambda)\,d\lambda$, where $\psi \in C^\infty([1,\infty))$ satisfies $\int_1^\infty |\psi(\lambda)|\lambda^s\,d\lambda < \infty$ for all $s \in \mathbb{N}_0$, $\int_1^\infty \psi(\lambda)\,d\lambda = 1$ and $\int_1^\infty \psi(\lambda)\lambda^s\,d\lambda = 0$ for all $s \in \mathbb{N}$.

[11] To obtain the definition of such sets one should replace in (4.89) $|\bar{x} - \bar{y}|$ by $|\bar{x} - \bar{y}|^\gamma$.

$q \in [1, \infty)$ is arbitrary if $l = \frac{n}{p}, p > 1$ and $q = \infty$ if $l > \frac{n}{p}, p > 1$ or $l \geq n, p = 1$. Consider the function $f_\delta(x) = x_n^\delta$ where $\delta \in \mathbb{R} \setminus \mathbb{N}_0$. Then $f_\delta \in W_p^l(\Omega_\gamma)$ if, and only if, $\delta > l - \frac{n}{p} + \frac{n-1}{p}(1 - \frac{1}{\gamma})$ because

$$\|f_\delta\|_{W_p^l(\Omega_\gamma)} < \infty \iff \int_0^1 \left(\int_{|\bar{x}| < x_n^\gamma} x_n^{(\delta-l)p} \, d\bar{x} \right) dx_n = v_{n-1} \int_0^1 x_n^{(\delta-l)p + \frac{n-1}{\gamma}} \, dx_n < \infty.$$

(This is also true for $l = 0$, i.e., for $L_p(\Omega_\gamma)$.) Let $l < \frac{n}{p}$, the cases $l = \frac{n}{p}$ and $l > \frac{n}{p}$ being similar. If $-\frac{n}{q} + \frac{n-1}{p}(1 - \frac{1}{\gamma}) = l - \frac{n}{p} + \frac{n-1}{p}(1 - \frac{1}{\gamma}) < \delta \leq -\frac{n}{q} + \frac{n-1}{q}\left(1 - \frac{1}{\gamma}\right)$, then $f_\delta \in W_p^l(\Omega_\gamma)$ but $f_\delta \notin L_q(\Omega_\gamma)$, and we have arrived at a contradiction.

Remark 20 If Ω has a boundary of the class Lip γ, where $0 < \gamma < 1$, then it is possible to construct an extension operator

$$T : W_p^l(\Omega) \to W_p^{\gamma l}(R^n), \tag{6.107}$$

where for noninteger γl $W_p^{\gamma l}(R^n) \equiv B_p^{\gamma l}(\mathbb{R}^n)$. The exponent γl is sharp. So the extension (6.107) is an extension with the minimal possible deterioration of smoothness. Moreover, if a bounded open set $\Omega \subset \mathbb{R}^n$ has a continuous boundary, then there exists an extension operator, which preserve some smoothness, i.e., for some $\lambda(\cdot)$

$$T : W_p^l(\Omega) \to B_{p,\infty}^{\lambda(\cdot)}(R^n). \tag{6.108}$$

Here $B_{p,\infty}^{\lambda(\cdot)}(R^n)$ is the space with the generalized smoothness, defined with the help of a function $\lambda(\cdot)$, which is positive, continuous, nondecreasing on $(0, \infty)$ and can tend to 0 arbitrarily slowly. To obtain the definition of the spaces $B_{p,\infty}^{\lambda(\cdot)}(\mathbb{R}^n)$ one should replace $|h|^l$ by $\lambda(|h|)$ in (5.8) – (5.9) with $\theta = \infty$ and suppose that $\lim_{t \to 0+} \lambda(t)t^{-\sigma} = \infty$.

Chapter 7

Comments

The first exposition of the theory of Sobolev spaces was given by S.L. Sobolev himself in his book [134] and later in his other book [135].

There are several books dedicated directly to different aspects of the theory of Sobolev spaces: R.A. Adams [2], V.G. Maz'ya [97], A. Kufner [85], S.V. Uspenskiĭ, G.V. Demidenko & V.G. Perepelkin [150]. V.G. Maz'ya & S.V. Poborchiĭ [100]. In some other books the theory of Sobolev spaces is included into a more general framework of the theory of function spaces: S.M. Nikol'skiĭ [114], O.V. Besov, V.P. Il'in & S.M. Nikol'skiĭ [16], A. Kufner, O. John & S. Fučik [86], E.M. Stein [138], H. Trielel [144], [145]. Moreover, in many other books, especially on the theory of partial differential equations, there are chapters containing exposition of different topics of the theory of Sobolev spaces, adjusted to the aims of those books. We name some of them: L.V. Kantorovich & G.P. Akilov [76], V.I. Smirnov [128], M. Nagumo [107], O.A. Ladyzhenskaya & N.N. Ural'tseva [88], C.B. Morrey [105], J. Nečas [108], J.-L. Lions & E. Magenes [92], V.M. Goldshtein & Yu.G. Reshetnyak [64], D.E. Edmunds & W.D. Evans [56], V.N. Maslennikova [96], E.H. Lieb & M. Loss [91]. Throughout the years several survey papers were published, containing exposition of the results on the theory of Sobolev spaces: S.L. Sobolev & S.M. Nikol'skiĭ [136], S.M. Nikol'skiĭ [113], V.I. Burenkov [20], O.V. Besov, V.P. Il'in, L.D. Kudryavtsev, P.I. Lizorkin & S.M. Nikol'skiĭ [15], S.K. Vodop'yanov, V.M. Gol'dshtein & Yu.G. Reshetnyak [152], L.D. Kudryavtsev & S.M. Nikol'skiĭ [84], V.G. Maz'ya [98]. We especially recommend the last two surveys containing updated information on Sobolev spaces.

We do not aim here to give a detailed survey of results on the theory of Sobolev spaces and their numerous generalizations, and we shall give only brief comments tightly connected with the material of Chapters 1–6.

Chapter 1

Section 1.1 The proofs of the properties of mollifiers A_δ can be found in the books S.L. Sobolev [134], S.M. Nikol'skiĭ [112] and E.M. Stein [138].

Section 1.2 The notion of the weak derivative plays a very important role in analysis. It ensures that function spaces of Sobolev type constructed on its base are complete. Many mathematicians arrived at this concept, friequently independently from their predessors. One can find it in investigations of B. Levi [89] at the beginning of the century. See also L. Tonelli [142], G.C. Evans [55], O.M. Nikodym [109].

S.L. Sobolev [131], [132] came to the definition of the weak derivative from the point of view of the concept of generalized function (distribution) introduced by him in [129], [130] and of the generalized solution of a differential equation. An approach to this notion, based on absolute continuity, was developed by J.W. Calkin [52], C.B. Morrey [104] and S.M. Nikol'skiĭ [112]. See the book S.M. Nikol'skiĭ [114] (Section 4.1) for details.

Lemma 3 is taken from [24]. Lemma 4 is due to S.L. Sobolev [134].

Section 1.3 S.L. Sobolev has introduced the spaces $W_p^l(\Omega)$ in [131], [132] and studied their different properties in those and later papers. (Some facts concerning these spaces, for particular values of parameters, were known earlier. See, for example, the papers B.Levi [89] and O.M. Nikodym [109].) In his book [134] S.L. Sobolev has pointed out that these spaces are essentially important for applications to various problems in mathematical physics. This book has given start to an intensive study of these and similar spaces, and to a wide usage of them in the theory of partial differential equations. Nowadays Sobolev spaces have become a standard tool in many topics of partial differential equations and analysis. S.L. Sobolev himself worked out deep applications of the spaces $W_p^l(\Omega)$ and their discrete analogues to numerical analysis. (See his book [135] on the theory of cubatures.)

Chapter 2

Section 2.2 Nonlinear mollifiers with variable step were first considered by H. Whitney [153] (their form is different from the mollifiers considered in Chapter 2), and later by J. Deny & J.-P. L. Lions [53] (the mollifiers $B_{\bar\delta}$) and N. Meyers & J. Serrin [102] (the mollifiers $C_{\bar\delta}$).

For a general lemma on partitions of unity, including Lemmas 3 – 5 see V.I. Burenkov [28]. That lemma is proved in the way which differs from the proofs of Lemmas 3–5 in Chapter 2. The idea of constructing the functions ψ_k by equality (2.10) has its own advantages: it is essentially used in the construction of the partition of unity in the proof of Theorem 5 of Chapter 5 satisfying inequality (5.71).

Section 2.3 For the spaces $\overline{C}^l(\Omega)$ Theorem 1 was proved by H. Whitney [153], for the spaces $W_p^l(\Omega)$ where $1 \leq p < \infty$ — by J. Deny & J.-P. L. Lions [53] and N. Meyers & J. Serrin [102]. The case of the spaces $W_\infty^l(\Omega)$ is new. Theorem 2 was proved by the author [24]. The statement mentioned in Remark 12 is proved in the same paper.

Section 2.4 For the spaces $\overline{C}^l(\Omega)$ Theorem 3 was proved in [153]. Theorem 3 (for $1 \leq p < \infty$) and Theorem 4 were proved by the author [24], [30].

Section 2.5 The linear mollifiers E_δ were introduced by the author [22]. In the case $\Omega = \mathbb{R}^n \setminus \mathbb{R}^m$ the linear mollifiers H_δ with variable step (see Remark 26) for some special kernels ω were considered and applied to the problem of extension of functions from $\mathbb{R}^m$ by A.A. Dezin [54] and L.D. Kudryavtsev [82], [83]. V.V. Shan'kov [126], [127] considered the linear mollifiers $\widetilde{H}_\delta$ with variable step and applied them to investigation of the trace theorems for weighted Sobolev spaces.

Theorems 5–9 are proved by the author [22], [30].

E.M. Popova [118] has proved that inequality (2.87) in Theorem 8 is sharp in a stronger sense, namely, the factor $\varrho^{|\alpha|-l}$ cannot be replaced by $\varrho^{|\alpha|-l}\nu(\varrho)$, where ν is an arbitrary positive continuous nonincreasing function, satisfying some regularity conditions, such that $\lim\limits_{u \to 0+} \nu(u) = \infty$.

Theorem 8 was generalized in different directions by the author [24], [30], V.V. Shan'kov [126], [127], E.M. Popova [118]. See survey [35] for details.

For a fixed ε Theorem 10 was proved by A.P. Calderón & A. Zygmund [51] (see detailed exposition in the book E.M. Stein [138]). For an arbitrary $\varepsilon \in (0,1)$ a direct proof of Theorem 10, without application of Theorem 9, was given by the author [21]. Later L.E. Fraenkel [59] gave another proof and considered the question of the sharpness of inequality (2.96). For the domain Ω defined by (6.62) and (6.63) Yu.V. Kuznetsov [87] (see also O.V. Besov [11]) constructed a regularized distance ϱ_δ, satisfying (2.93), (2.96) and, in addition, the inequality $(\frac{\partial \varrho_\delta}{\partial x_n})(x) \leq -b$, $x \in \Omega$, where b is a positive constant.

Chapter 3

Section 3.1 The idea of choosing the function ω in the integral representation (3.17) in an optimal way, which has been discussed in the simplest case in Remark 4, was used by the author in [29], [33], [34]. It gave possibility to establish a number of inequalities with sharp constants: for the norms of intermediate derivatives on a finite interval in [29], [33] and for the norms of polynomials in [34].

Section 3.2 In the case of bounded Ω Lemma 4 was proved by V.P. Glushko [63].

Section 3.4 Theorem 4 is due to S.L. Sobolev [131]–[133]. However, in those papers the first summand in (3.38) has the form of some polynomial in $x_1, ..., x_n$ of order less than or equal to $l-1$. The explicit form of that polynomial was found, and the tight connection of Sobolev's intergal representation to the multidimensional Taylor's formula was pointed out in O.V. Besov [9], [10], Yu.G. Reshetnyak [121] and V.I. Burenkov [23]. The proof in the text follows that of [23].

With the help of the integral representation (3.38) where $l = 1$ M.E. Bogovskiĭ [17], [18] constructed an explicit formula for the solution $v \in \mathring{W}_p^1(\Omega), 1 < p < \infty$, of the Cauchy problem: $\operatorname{div} v = f$, where $f \in L_p(\Omega), \int_\Omega f \, dx = 0$, for bounded domains star-shaped with respect to a ball.

The proof of the integral representation (3.67) on the base of (3.69) is given, for example, in the books M. Nagumo [107] and E.M. Stein [138].

For an arbitrary open set Ω an integral representation for functions $f \in w_1^l(\Omega) \cap \mathring{W}_1^k(\Omega)$, where $2k \geq l$ has been established by V.G. Maz'ya [98].

Finally, we note that in many cases it is important to have an integral representation, which involve only unmixed derivatives (see, for example Remark 17 of Chapter 6). A representation of such type was first obtained by V.P. Il'in [73]. In other cases it is desirable to get an integral representation via differences. Integral representations of both types may be deduced, in the simplest case, starting from the elementary identity $(A_\varepsilon f)(x) = (A_\delta f)(x) - \int_\varepsilon^\delta (\frac{\partial}{\partial t}(A_t f)(x)) \, dt$, where A_δ is a mollifier considered in Section 1.1. Detailed exposition of this topic can be found in the book O.V. Besov, V.P. Il'in & S.M. Nikol'skiĭ [16] (Sections 7–8).

Chapter 4

Section 4.1 Lemma 1 is a variant of Theorem 2 of Section 7.6 in the book S.M. Nikol'skiĭ [114]. We discuss in more detail the case of semi-Banach spaces (see Lemmas 2–3).

Section 4.2 Inequality (4.49) for $p = \infty$ is due to A.N. Kolmogorov [77]. E.M. Stein [138] proved that $c_{l,m,1} = c_{l,m,\infty}$ and $c_{l,m,p} \leq c_{l,m,\infty}$ for $p \in (1, \infty)$.

Theorem 4 and Corollaries 10, 11 contain all the cases, known to the author, in which the constants are sharp. If $(b - a) > (p' + 1)^{\frac{1}{p'}}$, in (4.57) the sharp value of the constant multiplying $\|f'_w\|_{L_p(a,b)}$ is not known.

Section 4.4 For open sets with quasi-continuous boundaries inequalities (4.105) and (4.107) in Theorem 6 are proved in the book J. Nečas [108]. The first proof and application of a theorem similar to Theorem 8 was given by R. Rellich [120].

In V.I. Burenkov & A.L. Gorbunov [43] it is proved that in inequality (4.112) $c_{31} \leq M^l l^{|\beta|}$, where M depends only on n.

Formula (4.127) for weak derivatives is proved, for example, in the book S.M. Nikol'skiĭ [114] (Section 4.4.9).

One can find the detailed proof of the Marcinkiewicz multiplicator theorem, formulated in footnote 21, in [114] (Sections $1.5.3 - 1.5.5$).

Section 4.5 Theorem 10 was proved by G.H. Hardy & J.E. Littlewood [66] for $n = 1$ and S.L. Sobolev [131], [132] for $n > 1$. The proof discussed in Section 4.5 is taken from L.I. Hedberg [68]. One can find proofs of the properties of the maximal functions, formulated in footnote 22, in the books E.M. Stein [138] and E.M. Stein & G. Weiss [140]. The proof of the Theorem 11 in the case $\beta < \frac{1}{v_n}$ is a modification of the proof given by L.I. Hedberg [68]. In the case $\beta = \frac{1}{v_n}$ Theorem 11 was proved by D.R. Adams [1]. Counter-example in the case $\beta > \frac{1}{v_n}$ was constructed in J.A. Hempel, G.I. Morris & N.S. Trudinger [69].

Section 4.6 Theorem 12 is due to S.L. Sobolev [131], [132], [133]. The statement of Remark 33 was established by V.I. Burenkov & V.A. Gusakov [44].

Section 4.7 Theorem 13 for $p > 1$ was proved by S.L. Sobolev [131], [132], for $p = 1$ — by E. Gagliardo [61]. The case in which $p = 1$ and in (4.149) q_* is replaced by $q < q_*$ was also considered in [131], [132], [133] (see Remark 36). The second proof of Theorem 13 is a modification of the proof given in [61]. For further modifications of this proof see V.I. Burenkov & N.B. Victorova [49].

The statement of Remark 38 was proved by V.G. Maz'ya [97] and H. Federer & W.H. Fleming [58] for $p = 1$, and by E. Rodemich [123], T. Aubin [4] and G. Talenti [141] for $p > 1$. (For detailed exposition see [141].) The statement of Remark 39 was proved by V.I. Burenkov & V.A. Gusakov [45], [46].

The compactness of embedding (4.16), under assumptions (4.169), was proved by V.I. Kondrashov [78].

Theorem 15 was independently proved by V.I. Yudovič [154], S.I. Pokhozhaev [117] and N.S. Trudinger [146]. The sharp value of c_{54} in (4.170) for the case of the space $\overset{\circ}{W}{}^1_n(\Omega)$, was computed by J. Moser [106].

In Theorems $12 - 13$ sufficient conditions on Ω weaker that the cone condition, and in some cases necessary and sufficient conditions on Ω, in terms of capacity were obtained by V.G. Maz'ya [97], [98], [99]. The case of degenerated open sets Ω is investigated in detail in V.G. Maz'ya & S.V. Poborchiĭ [100].

Chapter 5

Section 5.1 Definition 1 is close to the definition of a trace given in the book S.M. Nikol'skiĭ [114]. Theorem 1 is similar to Lemma 6.10.1 of that book and to Theorem 10.10 of the book O.V. Besov, V.P. Il'in and S.M. Nikol'skiĭ [16].

Section 5.2 Theorem 2 is an updated version of the theorem proved by S.L. Sobolev [133], [134].

Section 5.3 The spaces $B_{p,\infty}^l(\mathbb{R}^n) \equiv H_p^l(\mathbb{R}^n)$ were introduced and studied by S.M. Nikol'skiĭ [110], the spaces $B_{p,\theta}^l(\mathbb{R}^n)$, where $1 \le \theta < \infty$, – by O.V. Besov [7], [8]. Of possible equivalent norms we have chosen, as the main norm, the norm (5.8), which contains only differences. This definition appeared to be convenient in the approach which is used in the proofs of the direct and inverse trace theorems in this book. In this section we prove only those properties of the spaces $B_{p,\theta}^l(\mathbb{R}^n)$, which are necessary in order to prove the trace theorems for Sobolev spaces. Detailed exposition of the theory of the spaces $B_{p,\theta}^l(\mathbb{R}^n)$ can be found in the books S.M. Nikol'skiĭ [114] (including the case $l \le 0$), O.V. Besov, V.P. Il'in & S.M. Nikol'skiĭ [16] and H. Triebel [143], [144] (for $-\infty < l < \infty, 0 < p, \theta \le \infty$).

The usefulness of the simple identity (5.12) was pointed out by A. Marchoud [95]. The proof of Lemma1 is a modification of known proofs. We note that it works for all $1 \le p, \theta \le \infty$ and does not use the density of $C_0^\infty(\mathbb{R}^n)$ in $B_{p,\theta}^l(\mathbb{R}^n)$ for $1 \le p, \theta < \infty$.

In the one-dimensional case the proof of the inequality (5.19), based on an integral representation via differences, is given in [16] (Section 16.1). The identity (5.16) and the proof of (5.19) are taken from [36].

One can find the proofs of the facts stated in Remarks $5-8$ in [114] and [16].

Section 5.4 Lemma 10 may be considered as one of possible generalizations of Hardy's inequalities (5.13), (5.14). The proof of the direct trace theorem for Sobolev spaces (the first part of Theorem 3) is based on the identities for differences (5.31), (5.43) and (5.36) and Lemma 10. In the case $l = 1$, $m = n-1$ it is due to E. Gagliardo [60]. In the rest of the cases it seems to be new.

Theorem 3 was proved by the efforts of many mathematicians: N. Aronszajn [3], V.M. Babič & L.N. Slobodetskiĭ [6], E. Gagliardo [60], O.V.Besov [7], [8], P.I. Lizorkin [93], S.V. Uspenskiĭ [147], [148], V.A. Solonnikov [137]. The final step was done by O.V. Besov [7], [8]. Theorem 3 was preceeded by a similar theorem for the spaces $B_{p,\infty}^{l}(R^n)$ established by S.M. Nikol'skiĭ [110].

The trace theorem (5.68) was proved by S.V. Uspenskiĭ [149].

Theorem 5 is due to E. Gagliardo [60]. Nonexistence of a bounded linear extension operator was proved by J. Peétre [116]. Existence of a bounded linear extension operator $T : L_1(R^m) \to B_{1,\theta}^{n-m}(\mathbb{R}^n)$, where $\theta > 1$, was established in V.I. Burenkov & M.L. Gol'dman [41].

The extension operators constructed in the proofs of Theorems 4, 6 and Remark 15 in the case of Sobolev spaces $W_p^l(\mathbb{R}^n)$ are the best possible (see Remark 16). In the case of Nikol'skiĭ Besov spaces $B_{p,q}^l(\mathbb{R}^n)$ the best possible extension operators were constructed by L.D. Kudryavtsev [83], Ya.S. Bugrov [19] and S.V. Uspenskiĭ [149].

Section 5.5 Detailed exposition of the trace theorem in the case of smooth m-dimentional manifolds, where $m < n - 1$, is given in the book O.V. Besov, V.P. Il'in & S.M. Nikol'skiĭ [16] (Chapter 5). The trace theorem in the case of Lipschitz $(n-1)$-dimensional manifolds was proved by O.V. Besov [11], [12] (see also [16], Section 20). In more general case of the so-called d-sets, $0 < d \le n$ the trace is studied in the book A. Jonsson & H. Wallin [75].

Chapter 6

Section 6.1 The idea of defining an extension operator by (6.6) is due to M.P. Hestenes [70]. Estimate (6.4) can be found in V.I. Burenkov & A.L. Gorbunov [43]. Lemmas 5–6 are proved by V.I. Burenkov & G.A. Kalyabin [47]. Inequality (6.25) is taken from V.I. Burenkov A.L. Gorbunov [42], [43]. For $b - a = 1$ Theorem 1 is formulated in V.I. Burenkov [31], in the general case it is proved in V.I. Burenkov & A.L. Gorbunov [43].

Section 6.3 Theorem 2 is proved independently by V.M. Babič [5] and S.M. Nikol'skiĭ [111].

Section 6.4 If $1 < p < \infty$, then the existence of an extension operator (6.100) for Lipschitz boundaries was proved by A.P. Calderón [52]. His extension operator makes use of an integral representation of functions. (In the simplest case this possibility was discussed in Remark 2 of Chapter 3.) To prove (6.100) L_p-estimates of singular integrals are used, which is possible only if $1 < p < \infty$.

For $1 \le p \le \infty$ the existence of an extension operator (6.100) is proved by E.M. Stein [138]. The idea of his method is discussed in Remark 17. The construction used in [138], which is independent of the soomthness exponent l, is given in footnote 10. Another construction of an extension operator of such type is given by V.S. Rychkov [124]. In the case of the halfspace the existence of an extension operator T, independent of l and satisfying (4.100) for every $l \in \mathbb{N}_0$, follows from earlier papers by B.S. Mityagin [103] and R.T. Seeley [125].

The best possible extension operator, satisfying inequality (6.101), is constructed by the author [25], [26]. It satisfies also (6.106). Further generalizations of the methods and results of Section 6.4 for anisotropic Sobolev spaces are given in V.I. Burenkov & B.L. Fain [39], [40].

There is an alternative way of constructing the best possible extension operator. One may start from an arbitrary extension operator T (6.100) and improve it by applying the linear mollifier E_δ with variable step of Chapter 2, constructed for $^c\overline{\Omega}$, i.e., by considering the extension operator defined by $E_\delta T$ on $^c\overline{\Omega}$. See V.I. Burenkov & E.M. Popova [48] and E.M. Popova [119].

For open sets Ω with a Lipschitz boundary the multidimensional analogue of Theorem 1 is proved in V.I. Burenkov & A.L. Gorbunov [42], [43].

The problem of extension with preservation of Sobolev semi-norm $\| \cdot \|_{w_p^l(\Omega)}$ is considered in [27], [28].

The condition $\partial\Omega \in \mathrm{Lip}\, 1$ in Theorem 3 is essential, as Example 1 shows, but it is not necessary. For a wider class of open sets satisfying the so-called $\varepsilon - \delta$ condition the existence of an extension operator (6.100) was proved for $l = 1, n = 2$ by V.M. Gol'dshtein [65] and in the general case by P.W. Jones [74].

We emphasize that the important problem of finding necessary and sufficient conditions on Ω for the existence of an extension operator (6.100) is still open. Answers are known only in some particular cases. If Ω is a siply connected domain, then for $l = 1, n = 2, p = 2$ in S.K. Vodop'yanov, V.M. Gol'dshtein & T.G. Latfullin [151] and V.M. Gol'dshtein & S.K. Vodop'yanov [65] it is proved that the $\varepsilon - \delta$ condition is necessary and sufficient. In the case $l \in \mathbb{N}, n = 2$ and $p = \infty$ necessary and sufficient conditions for simply connected domains are found by V.N. Konovalov [80], [81].

The existence of an extension operator (6.107) is proved by the author [25], [26]. The fact that for bounded open sets Ω with continuous boundaries the extension by zero from Ω to $\mathbb{R}^n$ satisfies (6.108) for some $\lambda(\cdot)$ is proved in V.I. Burenkov [32].

Another types of extensions with deterioration of the class in the case $\partial\Omega \in \mathrm{Lip}\,\gamma$ into $W_q^l(\mathbb{R}^n)$ where $q < p$ and into a weighted space $W_{p,\phi}^l(\mathbb{R}^n))$ were obtained by B.L. Fain [57], V.G. Maz'ya & S.V. Poborchiĭ [100].

Finally, we note that the idea of constructing extension operators with the help of appropriate partitions of unity, which is used in [25], [26], [39], [40], [74] and in Section 6.4, goes back to H. Whitney [153].

Bibliography

[1] D.R. ADAMS, *A sharp inequality of J. Moser for higher order derivatives*, Ann. Math. **79** (1978), 385-398.

[2] R.A. ADAMS, *Sobolev spaces*, Academic Press, 1975.

[3] N. ARONSZAJN, *Boundary values of functions with finite Dirichlet integral*, Conference on Partial Differential Equations, Studies in Eigenvalue Problems, University of Kansas (1956).

[4] T. AUBIN, *Problémes isopérimètrique et espaces de Sobolev*, C. R. Acad. Sci. Sér. A. **280** (1975), 279-281.

[5] V.M. BABIČ *On the extension of functions*, Uspekhi Mat. Nauk **8** (1953), 111-113 (Russian).

[6] V.M. BABIČ & L.N. SLOBODETSKIĬ, *On boundness of the Dirichlet integral*, Dokl. Akad. Nauk SSSR **106** (1956), 604-607 (Russian).

[7] O.V. BESOV, *On a certain family of function spaces. Embedding and extension theorems*, Dokl. Akad. Nauk SSSR **126** (1959), 1163-1165 (Russian).

[8] O.V. BESOV, *Investigation of a certain family of function spaces in connection with embedding and extension theorems*, Trudy Mat. Inst. Steklov **60 (1961)**, 42-81 (Russian); English transl. in Amer. Math. Soc. Transl. (2) **40** (1964).

[9] O.V. BESOV, *Extension of functions in L_p^l and W_p^l*, Trudy Mat. Inst. Steklov **89 (1967)**, 5-17 (Russian); English transl. in Proc. Steklov Inst. Math. **89** (1967).

[10] O.V. BESOV, *The behaviour of differentiable functions at infinity and the density of functions with compact support*, Trudy Mat. Inst. Steklov **105** (1969), 3-14 (Russian); English transl. in Proc. Steklov Inst. Math. **105** (1969).

[11] O.V. BESOV, *Behaviour of differentiable functions on a nonsmooth surface*, Trudy Mat. Inst. Steklov **117** (1972), 3-10 (Russian); English transl. in Proc. Steklov Inst. Math. **117** (1972).

[12] O.V. BESOV, *On traces on a nonsmooth serface of classes of differentiable functions*, Trudy Mat. Inst. Steklov **117** (1972), 11-21 (Russian); English transl. in Proc. Steklov Inst. Math. **117** (1972).

[13] O.V. BESOV, *On the density of functions with compact support in the weighted space of S.L. Sobolev*, Trudy Mat. Inst. Steklov **161** (1983), 29-47 (Russian); English transl. in Proc. Steklov Inst. Math. **161** (1983).

[14] O.V. BESOV, *Integral representations of functions and embedding theorems for a domain with the flexible horn condition*, Trudy Mat. Inst. Steklov **170** (1984), 12-30 (Russian); English transl. in Proc. Steklov Inst. Math. **170** (1984).

[15] O.V. BESOV, V.P. IL'IN, L.D. KUDRYAVTSEV, P.I. LIZORKIN & S.M. NIKOL'SKIĬ, *Embedding theory for classes of differentiable functions of several variables*, Proc. Sympos. in honour of the 60th birthday of academician S.L.Sobolev, Inst. Mat. Sibirsk. Otdel. Akad. Nauk. SSSR, "Nauka", Moscow, 1970, 38-63 (Russian).

[16] O.V. BESOV, V.P. IL'IN & S.M. NIKOL'SKIĬ, *Integral representation of functions and embedding theorems*, 1-st ed. – "Nauka", Moscow, 1975 (Russian); 2-nd ed. –"Nauka", Moscow, 1996 (Russian); English transl. of 1-st ed., Vols. 1, 2, Wiley, 1979.

[17] M.E. BOGOVSKIĬ, *Solution of the first boundary value problem for the equation of continuity of an incompressible medium*, Dokl. Akad. Nauk SSSR **248** (1979), 1037-1040 (Russian); English transl. in Soviet Math. Dokl. **20** (1979).

[18] M.E. BOGOVSKIĬ, *Solution of some vector analysis problems connected with operators div and grad*, Trudy Sem. S.L.Sobolev, Novosibirsk **1 (80)** (1980), 5-40 (Russian).

[19] YA.S. BUGROV, *Differential properties of solutions of a certain class of differenital equations of higher order*, Mat. Sb. **63** (1963), 69-121.

[20] V.I. BURENKOV, *Embedding and extension theorems for classes of differentiable functions of several variables defined on the whole space*, Itogi Nauki i Tekhniki: Mat. Anal. **1965**, VINITI, Moscow, 1966, 71-155 (Russian); English transl. in Progress in Math. **2**, Plenum Press. 1968.

[21] V.I. BURENKOV, *On regularized distance*, Trudy MIREA. Issue **67**. Mathematics (1973), 113-117 (Russian).

[22] V.I. BURENKOV, *On the density of infinitely differentiable functions in Sobolev spaces for an arbitrary open set*, Trudy Mat. Inst. Steklov **131** (1974), 39-50 (Russian); English transl. in Proc. Steklov Inst. Math. **131** (1974).

[23] V.I. BURENKOV, *The Sobolev integral representation and the Taylor formula*, Trudy Mat. Inst. Steklov **131** (1974), 33-38 (Russian); English transl. in Proc. Steklov Inst. Math. **131** (1975).

[24] V.I. BURENKOV, *On the density of infinitely differentiable functions in spaces of functions defined on an arbitrary open set*, In: *"Theory of cubatures and applications of functional analysis to problems in mathematical physics"*, Novosibirsk, 1975, 9-22 (Russian).

[25] V.I. BURENKOV, *On the extension of functions with preservation and with deterioration of the differential properties*, Dokl. Akad. Nauk SSSR **224** (1975), 269-272 (Russian); English transl. in Soviet Math. Dokl. **16** (1975).

[26] V.I. BURENKOV, *On a certain method for extending differentiable functions*, Trudy Mat. Inst. Steklov **140** (1976), 27-67 (Russian); English transl. in Proc. Steklov Inst. Math. **140** (1976).

[27] V.I. BURENKOV, *On the extension of functions with preservation of semi-norm*, Dokl. Akad. Nauk SSSR **228** (1976), 779-782 (Russian); English transl. in Soviet Math. Dokl. **17** (1976).

[28] V.I. BURENKOV, *On partitions of unity*, Trudy Mat. Inst. Steklov **150** (1979), 24-30 (Russian); English transl. in Proc. Steklov Inst. Math. **150** (1979).

[29] V.I. BURENKOV, *On sharp constants in the inequalities for the norms of intermediate deriavtives on a finite inteval*, Trudy Mat. Inst. Steklov **156** (1980), 23-29 (Russian); English transl. in Proc. Steklov Inst. Math. **156** (1980).

[30] V.I. BURENKOV, *Mollifying operators with variable step and their applications to approximation by infintely differentiable functions*, In: *"Nonlinear analysis, function spaces and applications"*, Teubner-Texte zur Mathematik, Leipzig, **49** (1982), 5-37.

[31] V.I. BURENKOV, *On estimates of the norms of extension operators*, 9-th All-Union school on the theory of operators in function spaces. Ternopol. Abstracts (1984), 19-20 (Russian).

[32] V.I. BURENKOV, *Extension of functions with preservation of Sobolev semi-norm*, Trudy Mat. Inst. Steklov **172** (1985), 81-95 (Russian); English transl. in Proc. Steklov Inst. Math. **172** (1985).

[33] V.I. BURENKOV, *On sharp constants in the inequalities for the norms of intermediate derivatives on a finite inteval.* **II**, Trudy Mat. Inst. Steklov **173** (1986), 38-49 (Russian); English transl. in Proc. Steklov Inst. Math. **173** (1986).

[34] V.I. BURENKOV, *On sharp constants in the inequalities of different metrics for polynomials*, In: *"Theory of approximation of functions."* Proc. of international conference on approximation of functions. "Nauka", Moscow (1987), 79 (Russian).

[35] V.I. BURENKOV, *Approximation of functions, defined on an arbitrary open set in $\mathbb{R}^n$, by infintely differentiable functions with preservation of boundary values*, In: *"Methods of functional analysis in mathematical physics"*. Peoples' Friendship University of Russia, Moscow (1987), 20-43 (Russian).

[36] V.I. BURENKOV, *On an equality for differences and on an estimate of the norm of the difference via its average value*, 13-th All-Union school on the theory of operators in function spaces. Kuibyshev. Abstracts (1988), 37 (Russian).

[37] V.I. BURENKOV, *Compactness of embeddings for Sobolev and more general spaces and extensions with preservation of some smoothness* (to appear).

[38] V.I. BURENKOV & P.M. BADJRACHARIA, *Approximation by infinitely differentiable vector-functions with preservation of image for different function spaces*, Trudy Mat. Inst. Steklov **201** (1992), 152-164 (Russian); English transl. in Proc. Steklov Inst. Math. **201** (1992).

[39] V.I. BURENKOV & B.L. FAIN, *On the extension of functions in anisotropic spaces with preservation of class*, Dokl. Akad. Nauk SSSR **228** (1976), 525-528 (Russian); English transl. in Soviet Math. Dokl. **17** (1976).

[40] V.I. BURENKOV & B.L. FAIN, *On the extension of functions in anisotropic classes with preservation of the class*, Trudy Mat. Inst. Steklov **150** (1979), 52-66 (Russian); English transl. in Proc. Steklov Inst. Math. **150** (1979).

[41] V.I. BURENKOV & M.L. GOL'DMAN, On the extension of functions in L_p, Trudy Mat. Inst. Steklov **150** (1979), 31-51 (Russian); English transl. in Proc. Steklov Inst. Math. **150** (1979).

[42] V.I. BURENKOV & A.L. GORBUNOV, *Sharp estimates for the minimal norm of an extension operator for Sobolev spaces*, Dokl. Akad. Nauk SSSR **330** (1993), 680-682 (Russian); English transl. in Soviet Math. Dokl. **47** (1993).

[43] V.I. BURENKOV & A.L. GORBUNOV, *Sharp estimates for the minimal norm of an extension operator for Sobolev spaces*, Izvestiya Ross. Akad. Nauk **61** (1997), 1-44 (Russian).

[44] V.I. BURENKOV & V.A. GUSAKOV, *On sharp constants in Sobolev embedding theorems*, Dokl. Akad. Nauk SSSR **294** (1987), 1293-1297 (Russian); English transl. in Soviet Math. Dokl. **35** (1987).

[45] V.I. BURENKOV & V.A. GUSAKOV, *On sharp constants in Sobolev embedding theorems*. **II**, Dokl. Akad. Nauk SSSR **324** (1992), 505-510 (Russian); English transl. in Soviet Math. Dokl. **45** (1992).

[46] V.I. BURENKOV & V.A. GUSAKOV, *On sharp constants in Sobolev embedding theorems*. **III**, Trudy Mat. Inst. Steklov **204** (1993), 68-80 (Russian); English transl. in Proc. Steklov Inst. Math. **204** (1993).

[47] V.I. BURENKOV & G.A. KALYABIN, *Lower estimates of the norms of extension operators for Sobolev spaces on the halfline* (to appear).

[48] V.I. BURENKOV & E.M. POPOVA, *On improving extension operators with the help of the operators of approximation with preservation of the boundary values*, Trudy Mat. Inst. Steklov **173** (1986), 50-54 (Russian); English transl. in Proc. Steklov Inst. Math. **173** (1986).

[49] V.I. BURENKOV & N.B. VICTOROVA, *On an embedding theorem for Sobolev spaces with mixed norm for limiting exponents*, Mat. Zametki **59** (1996), 62-72 (Russian).

[50] A.P. CALDERÓN, *Lebesgue spaces of differentiable functions and distributions*, Proc. Symp. Pure Math. **IV** (1961), 33-49.

[51] A.P. CALDERÓN, A. ZYGMUND, *Local properties of solutions of elliptic partial differential equations*, Studia Math. **20** (1961), 171-225.

[52] J.W. CALKIN, *Functions of several variables and absolute continuity. I*, Duke Math. J. **6** (1940), 176-186.

[53] J. DENY & J.L. LIONS, *Les espaces du type de Beppo Levi*, Ann. Inst. Fourier, Grenoble **5** (1953-54), 305-370.

[54] A.A. DEZIN, *On embedding theorems and the extension problem*, Dokl. Akad. Nauk SSSR **88** (1953), 741-743 (Russian).

[55] G.C. EVANS, *Complements of potential theory. II*, Amer. J. Math **55** (1933), 29-49.

[56] D.E. EDMUNDS & W.D. EVANS, *Spectral theory and differential operators*, Clarendon Press, Oxford, 1987.

[57] B.L. FAIN, *On extension of functions in Sobolev spaces for irregular domains with preservation of the smoothness exponent*, Dokl. Akad. Nauk SSSR **285** (1985), 296-301 (Russian); English transl. in Soviet Math. Dokl. **30** (1985).

[58] H. FEDERER & W.H. FLEMING, *Normal and integral currents*, Ann. Math. **72** (1960), 458-520.

[59] L.E. FRAENKEL, *On regularized distance and related functions*, Proc. Royal Soc. Edinburg **83a** (1979), 115-122.

[60] E. GAGLIARDO, *Caratterizzazione delle tracce sulla frontiera relative ad alcune classi di funzioni in n variabili*, Rend. Sem. Mat. Univ. Padova **27** (1957), 284-305.

[61] E. GAGLIARDO, *Proprietà di alcune classi di funzioni in più variabili*, Ricerche Mat. **7** (1958), 102-137.

[62] E. GAGLIARDO, *Ulteriori proprietà di alcune classi di funzioni in più variabili*, Ricerche. Mat. **8** (1959), 24-51.

[63] V.P. GLUSHKO, *On domains starlike with respect to a ball*, Dokl. Akad. Nauk SSSR **144** (1962), 1215-1216 (Russian); English transl. in Soviet Math. Dokl. **3** (1962).

[64] V.M. GOL'DSHTEIN & YU.G. RESHETNYAK, *Foundations of the theory of functions with generalized derivatives and quasiconformal mappings*, "Nauka", Moscow, 1983 (Russian); English transl. , Reidel, Dordrecht, 1989.

[65] V.M. GOL'DSHTEIN & S.K. VODOP'YANOV, *Prolongement des fonctions de classe L_p^1 et applications quasiconformes*, C. R. Acad sci. Sér. A **290** (1983), 453-456.

[66] G.H. HARDY & J.L. LITTLEWOOD, *Some properties of fractional integrals. I*, Math. Zeit. **27** (1928), 565-606.

[67] G.H. HARDY & J.L. LITTLEWOOD & G. POLYA, *Inequalities*, Cambridge, 1934.

[68] L. HEDBERG, *On certain convolution inequalities*, Proc. Amer. Math. Soc. **36** (1972), 505-510.

[69] J.A. HEMPEL, G.P. MORRIS & N.S. TRUDINGER, *On the sharpness of a limiting case of the Sobolev imbedding theorem*, Bull. Austral. Math. Soc. **3** (1970), 369-373.

[70] M.R. HESTENES, *Extension of the range of a differentiable function*, Duke Math. J. **8** (1941), 183-192.

[71] V.P. IL'IN, *On an embedding theorem for a limiting exponent*, Dokl. Akad. Nauk SSSR **96** (1954), 905-908 (Russian).

[72] V.P. IL'IN, *On a theorem due to G.H. Hardy and J.L. Littlewood*, Trudy Mat. Inst. Steklov **53** (1959), 128-144 (Russian).

[73] V.P. IL'IN, *Properties of some classes of diffrentiable functions of several variables defined in n-dimensional regions*, Trudy Mat. Inst. Steklov **66** (1962), 227-363 (Russian).

[74] P.W. JONES, *Quasiconformal mappings and extendability of functions in Sobolev spaces*, Acta Math. **147** (1981), 71-88.

[75] A. JONSSON & H. WALLIN, *Function spaces on subsets of $\mathbb{R}^n$*, Mathematical Reports. Harwood Acad. Publ. New York. V. **2**, 1984.

[76] L.V. KANTOROVICH & G.P. AKILOV, *Functional analysis in normed spaces*, 2-nd ed., "Nauka", Moscow, 1977 (Russian); 3-rd ed., "Nauka", Moscow, 1985 (Russian); English transl. of 2-nd ed., Pergamon Press, Oxford, 1982.

[77] A.N. KOLMOGOROV, *On inequalities for upper bounds of successive derivatives of an arbitrary function on an infinite interval*, Uchenye Zapiski MGU. Mathematics **30** (1939), Vol. 3, 3-16 (Russian).

[78] V.I. KONDRASHOV, *On certain estimates of families of functions satisfying integral inequalities*, Dokl. Akad. Nauk SSSR **48** (1938), 235-239 (Russian).

[79] V.I. KONDRASHOV, *Behaviour of functions in L_p^ν on the manifolds of different dimensions*, Dokl. Akad. Nauk SSSR **72** (1950), 1009-1012 (Russian).

[80] V.N. KONOVALOV, *Description of traces for some classes of functions of several variables*, Preprint 84.21. Math. Inst. Ac. sci. Ukraine. Kiev, 1984 (Russian).

[81] V.N. KONOVALOV, *A criterion for extension of Sobolev spaces $W_\infty^{(r)}$ on bounded planar domains*, Dokl. Akad. Nauk SSSR **289** (1986), 36-39 (Russian); English transl. in Soviet Math. Dokl. **33** (1986).

[82] L.D. KUDRYAVTSEV, *On extention and embedding of certain classes of functions*, Dokl. Akad. Nauk SSSR **107** (1956), 501-505 (Russian).

[83] L.D. KUDRYAVTSEV, *Direct and inverse embedding theorems. Applications to the solution of elliptic equations by the variational method*, Trudy Mat. Inst. Steklov **48** (1959), 1-181 (Russian); English transl. , Amer. Math. Soc., Providence, R.I., 1974.

[84] L.D. KUDRYAVTSEV & S.M. NIKOL'SKIĬ, *Spaces of differentiable functions of several variables and the embedding theorems*, Contemprorary problems in mathematics. Fundamental directions. V. **26** (1988). Analysis - 3. VINITI, Moscow, 5-157 (Russian).

[85] A. KUFNER, *Weighted Sobolev spaces*, John Wiley and Sons, 1983.

[86] A. KUFNER, O. JOHN & S. FUČIK, *Function spaces*, Academia, Prague, 1977.

[87] YU.V. KUZNETSOV, *On regularized distance for one class of domains*, VINITI, Moscow, No 2582-74 Dep (1974), 1-12 (Russian).

[88] O.A. LADYZHENSKAYA & N.N. URAL'TSEVA, *Linear and quasilinear elliptic equations*, "Nauka", Moscow, 1964 (Russian); English transl. , Academic Press, 1968.

[89] B. LEVI, *Sul principio di Dirichlet*, Rend. Ciec. Mat. Palermo **22** (1906), 293-359.

[90] E.H. LIEB, *Sharp constants in the Hardy-Littlewood-Sobolev and related inequalities*, Ann. of Math., **118** (1963), 349-374.

[91] E.H. LIEB & M. LOSS, *Analysis*, Amer. Math. Soc., 1997.

[92] J.-L. LIONS & E. MAGENES, *Problèmes aux limites non homogènes et applications*, vol. **I**, Dunod, Paris (1968).

[93] P.I. LIZORKIN, *Boundary properties of functions in weighted classes*, Dokl. Akad. Nauk SSSR **132** (1960), 514-517 (Russian); English transl. in Soviet Math. Dokl. **1** (1960).

[94] P.I. LIZORKIN, *Generalized Liouville differentiation and the function spaces $L_p^r(E_n)$. Embedding theorems*, Mat. Sb. **60 (102)** (1963), 325-353 (Russian).

[95] A. MARCHOUD, *Sur les dérivées et sur les différences des fonctions de variables réelles*, J. Math. pures et appl. **6** (1927), 337-425.

[96] *V.N. MASLENNIKOVA, Partial differential equations*, Peoples' Friendship University of Russia, Moscow, 1996 (Russian).

[97] V.G. MAZ'YA, *Classes of domains and embedding theorems for function spaces*, Dokl. Akad. Nauk SSSR **133** (1960), 527-530 (Russian); English transl. in Soviet Math. Dokl. **1** (1960).

[98] V.G. MAZ'YA, *Sobolev spaces*, LGU, Leningrad, 1984 (Russian); English transl. , Springer-Verlag, Springer Series in Soviet Mathematics, 1985.

[99] V.G. MAZ'YA, *Classes of domains, measures and capacities in the theory of spaces of differentiable functions*, Contemprorary problems in mathematics. Fundamental directions. V. **26.** (1988) Analysis - 3. VINITI, Moscow, 158-228 (Russian).

[100] V.G. MAZ'YA & S.V. POBORCHIĬ, *On extension of functions in Sobolev spaces to the exterior of a domain with a peak vertex on the boundary*, Dokl. Akad. Nauk SSSR **275** (1984), 1066-1069 (Russian); English transl. in Soviet Math. Dokl. **29** (1984).

[101] V.G. MAZ'YA & S.V. POBORCHIĬ, *Differentiable functions on bad domains*, World Scientific Publishing, 1997.

[102] N. MEYERS & J. SERRIN, $H = W$, Proc. Nat. Acad. Sci. USA, **51** (1964), 1055-1056.

[103] B.S. MITYAGIN, *Approximative dimension and bases in nuclear spaces*, Uspekhi Mat. Nauk, **164** (1961), 63-132 (Russian).

[104] S.B. MORREY, JR., *Functions of several variables and absolute continuity.* **II**, Duke Math. J. **6** (1940), 187-215.

[105] S.B. MORREY, JR., *Multiple integrals in the calculus of variations*, Springer-Verlag, 1966.

[106] J. MOSER, *A sharp form of an inequality by N. Trudinger*, Ind. Univ. Math. J., **20** (1971), 1077-1092.

[107] M. NAGUMO, *Lectures on contemprorary theory of partial differential equations*, Kyoritsu Suppan, Tokyo (1957) (Japanese).

[108] J. NEČAS, *Les méthodes directes en théorie des équations élliptiques*, Academia, Prague, 1967.

[109] O.M. NIKODYM, *Sur une classe de fonctions considérée dans le problème de Dirichlet*, Fund. Math. **21** (1933), 129-150.

[110] S.M. NIKOL'SKIĬ, *Inequalities for entire functions of exponential type and their applications in the theory of differentiable functions of several variables,* Trudy Mat. Inst. Steklov **38** (1951), 244-278 (Russian); English transl. in Amer. Math. Soc. Transl. (2) **80** (1969).

[111] S.M. NIKOL'SKIĬ, *On the solutions of the polyharmonic equation by a variational method,* Dokl. Akad. Nauk SSSR **88** (1953), 409-411 (Russian).

[112] S.M. NIKOL'SKIĬ, *Properties of some classes of functions of several variables on differentiable manifolds,* Mat. Sb. **33 (75)** (1953), 261-326 (Russian).

[113] S.M. NIKOL'SKIĬ, *On embedding, extension and approximation theorems for differentiable functions of several variables,* Uspekhi Mat. Nauk **16** (1961), 63-114 (Russian); English transl. in Russian Math. Surveys **16** (1961).

[114] S.M. NIKOL'SKIĬ, *Approximation of functions of several variables and embedding theorems,* 1-st ed., "Nauka", Moscow, 1969 (Russian); 2-nd ed., "Nauka", Moscow, 1977 (Russian); English transl. of 1-st ed., Springer-Verlag, 1975.

[115] J. PEÉTRE, *The trace of Besov spaces - a limiting case.* Technical Report. Lund (1975), 1-32.

[116] J. PEÉTRE, *A counter-example connected with Gagliardo's trace theorem,* Comment Math. Prace Mat. **2** (1979), 277-282.

[117] S.I. POKHOZHAEV, *On the Sobolev embedding theorem in the case $pl = n$,* In: *"Reports of scientific-technical conference of MEI"*, Moscow, 1965, 158-170 (Russian).

[118] E.M. POPOVA, *Supplements to theorem of V.I. Burenkov on approximation of functions in Sobolev spaces with preservation of boundary values,* In: *"Differential equations and functional analysis"*. Peoples' Friendship University of Russia, Moscow (1985), 86-98 (Russian).

[119] E.M. POPOVA, *On improving extension operators with the help of the operators of approximation with preservation of the boundary values,* In: *"Function spaces and applications to differential equations"*, Peoples' Friendship University of Russia, Moscow (1992), 154-165 (Russian).

[120] R. RELLICH, *Ein Satz über mittlere Konvergenz*, Nachr. Akad. Wiss. Göttingen, Math.-Phys. K1., (1930), 30-35.

[121] YU.G. RESHETNYAK, *Estimates for certain differential operators with finite-dimensional kernel*, Sibirsk. Mat. Zh. **11** (1970), 414-428 (Russian); English transl. in Siberian Math. J. **11** (1970).

[122] YU.G. RESHETNYAK, *Certain integral representations of differentiable functions*, Sibirsk. Mat. Zh. **12** (1971), 420-432 (Russian); English transl. in Siberian Math. J. **12** (1971).

[123] E. RODEMICH, *The Sobolev inequalities with best possible constants*, In:*"Analysis Seminar at California Institute of Technology"* (1966).

[124] V.S. RYCHKOV, *On restrictions and extensions of Besov and Triebel-Lizorkin spaces with respect to Lipschitz domains* (to appear).

[125] R.T. SEELEY, *Extention of C^∞-functions defined in halfspace*, Proc. Amer. Math. Soc. **15** (1964), 625-626.

[126] V.V. SHAN'KOV, *An averaging operator with variable radius and the inverse trace theorem*, Dokl. Akad. Nauk SSSR **273** (1983), 307-310 (Russian); English transl. in Soviet Math. Dokl. **27** (1983).

[127] V.V. SHAN'KOV, *An averaging operator with variable radius and the inverse trace theorem*, Sibirsk. Mat. Zh. **6** (1985) (Russian); English transl. in Siberian Math. J. **6** (1985).

[128] V.I. SMIRNOV, *Course in higher mathematics, Vol.* **5** rev. ed., Fizmatgiz, Moscow, 1959 (Russian); English transl. , Pergamon Press, Oxford and Eddison-Wesley, Reading, Mass., 1964.

[129] S.L. SOBOLEV, *Le problème de Cauchy dans l'espace des fonctionelles*, Dokl. Akad. Nauk SSSR, **3** (1935), 291-294.

[130] S.L. SOBOLEV, *Métode nouvelle à résourdre le problème de Cauchy pour les équations lineéaires hyperboliques normales*, Mat. Sb., **1 (43)** (1936), 39-72.

[131] S.L. SOBOLEV, *On certain estimates relating to families of functions having derivatives that are square integrable*, Dokl. Akad. Nauk SSSR, **1** (1936), 267-270. (Russian).

[132] S.L. SOBOLEV, *On a theorem of functional analysis*, Dokl. Akad. Nauk SSSR **20** (1938), 5-10, (Russian).

[133] S.L. SOBOLEV, *On a theorem of functional analysis,*. Mat. Sb. **46** (1938), 471-497, (Russian). English translation: Amer. Math. Soc. Translations, 2(**34**) (1963), 39-68.

[134] S.L. SOBOLEV, *Applications of functional analysis in mathematical physics*, 1-st ed.: LGU, Leningrad, 1950 (Russian). 2-nd ed.: NGU, Novosibirsk, 1963 (Russian). 3-rd ed.: "Nauka", Moscow, 1988 (Russian). English translation of 3-rd ed.: Translations of Mathematical Monographs, Amer. Math. Soc., **90**, 1991.

[135] S.L. SOBOLEV, *Introduction to the theory of cubature formulae*, "Nauka", Moscow, 1974 (Russian).

[136] S.L. SOBOLEV & S.M. NIKOL'SKIĬ, *Embedding theorems*, Vol. **1**, Proc. Fourth All-Union Math. Congress, 1961, "Nauka", Leningrad, 1963, 227-242 (Russian); English transl. in Amer. Math. Soc. Transl. **(2) 87** (1970).

[137] V.A. SOLONNIKOV, *On some properties of the spaces M_p^l of fractional order*, Dokl. Akad. Nauk SSSR **134** (1960), 282-285 (Russian); English transl. in Soviet Math. Dokl. **1** (1960).

[138] E.M. STEIN, *Functions of exponential type*, Ann. Math **65** (1957), 582-592.

[139] E.M. STEIN, *Singular integrals and differentiability properties of functions*, Princeton Univ. Press, 1970.

[140] E.M. STEIN & G. WEISS, *Introduction to Fourier analysis on Euclidean spaces*, Princeton Univ. Press, 1971.

[141] G. TALENTI, *Best constant in Sobolev inequality*, Ann. Mat. Pura Appl., **110** (1976), 353-372.

[142] L. TONELLI, *Sulla quadrature della superficie*, Rend. R. Accad. Lincei **6** (1926), 633-638.

[143] H. TRIEBEL, *Interpolation theory, function spaces, differential operators*, Berlin, VEB Deutscher Verlag der Wissenschaften 1978, Amsterdam, North-Holland, 1978.

[144] H. TRIEBEL, *Theory of function spaces*, Birkhäuser–Verlag, Basel, Boston, Stuttgart, 1983; and Akad. Verlagsges. Geest & Portig, Leipzig, 1983.

[145] H. TRIEBEL, *Theory of function spaces. II*, Birkhäuser–Verlag, Basel, Boston, Stuttgart, 1992.

[146] N.S. TRUDINGER, *On imbeddings into Orlicz spaces and some applications*, J. Math. Mech. **17** (1967), 473-483.

[147] S.V. USPENSKIĬ, *An embedding theorem for the fractional order classes of S.L. Sobolev*, Dokl. Akad. Nauk SSSR **130** (1960), 992-993 (Russian); English transl. in Soviet Math. Dokl. **1** (1960).

[148] S.V. USPENSKIĬ, *Properties of the classes $W_p^{(r)}$ with fractional derivatives on differentiable manifods*, Dokl. Akad. Nauk SSSR **132** (1960), 60-62 (Russian); English transl. in Soviet Math. Dokl. **1** (1960).

[149] S.V. USPENSKIĬ, *On embedding theorems for weighted classes*, Trudy Mat. Inst. Steklov **50** (1961), 282-303 (Russian).

[150] S.V. USPENSKIĬ, G.V. DEMIDENKO & V.G. PEREPELKIN, *Embedding theorems and their applications to differential equations*, "Nauka", Novosibirsk, 1984 (Russian).

[151] S.K. VODOP'YANOV, V.M. GOL'DSHTEIN & T.G. LATFULLIN, *A criterion for extension of functions in class L_2^1 from unbounded planar domains*, Sibirsk. Mat. Zh. **34** (1979), 416-419 (Russian); English transl. in Siberian Math. J. **34** (1979).

[152] S.K. VODOP'YANOV, V.M. GOL'DSHTEIN & YU.G. RESHETNYAK, *On geometric properties of functions with first generalized derivatives*, Uspekhi Mat. Nauk **34** (1979), no. **1 (205)**, 3-74 (Russian); English transl. in Russian Math. Surveys **20** (1985).

[153] H. WHITNEY, *Analyticic extension of differentiable functions defined in closed sets*, Trans. Amer. Math. Soc. **36** (1934), 63-89.

[154] V.I. YUDOVIČ, *On certain estimates related to integral operators and solutions of elliptic equations*, Dokl. Akad. Nauk SSSR **138** (1961), 805-808 (Russian); English transl. in Soviet Math. Dokl. **2** (1961).

Index